Heat Conduction

Latif M. Jiji

Heat Conduction

Third Edition

 Springer

Professor Latif M. Jiji
Department of Mechanical Engineering
Grove School of Engineering
The City College of
The City University of New York
New York, New York 10031
USA
E-Mail: jiji@ccny.cuny.edu

"Additional material to this book can be downloaded from http://extra.springer.com"

ISBN 978-3-642-42488-5   ISBN 978-3-642-01267-9 (eBook)

DOI 10.1007/978-3-642-01267-9

Typesetting by the Author.
Production: Scientific Publishing Services Pvt. Ltd., Chennai, India.
Cover Design: WMX Design GmbH, Heidelberg.

Printed in acid-free paper

30/3100/as 5 4 3 2 1 0

springer.com

This book is dedicated to my wife Vera for opening many possibilities and providing balance in my life.

PREFACE

This book is designed to:

- Provide students with the tools to model, analyze and solve a wide range of engineering applications involving conduction heat transfer.

- Introduce students to three topics not commonly covered in conduction heat transfer textbooks: perturbation methods, heat transfer in living tissue, and microscale conduction.

- Take advantage of the mathematical simplicity of one-dimensional conduction to present and explore a variety of physical situations that are of practical interest.

- Present textbook material in an efficient and concise manner to be covered in its entirety in a one semester graduate course.

- Drill students in a systematic problem solving methodology with emphasis on thought process, logic, reasoning and verification.

To accomplish these objectives requires judgment and balance in the selection of topics and the level of details. Mathematical techniques are presented in simplified fashion to be used as tools in obtaining solutions. Examples are carefully selected to illustrate the application of principles and the construction of solutions. Solutions follow an orderly approach which is used in all examples. To provide consistency in solutions logic, I have prepared solutions to all problems included in the first ten chapters myself. Instructors are urged to make them available electronically rather than posting them or presenting them in class in an abridged form.

This edition adds a new chapter, "Microscale Conduction." This is a new and emerging area in heat transfer. Very little is available on this subject as textbook material at an introductory level. Indeed the preparation of such a chapter is a challenging task. I am fortunate that Professor Chris Dames of the University of California, Riverside, agreed to take on this responsibility and prepared all the material for chapter 11.

Now for the originality of the material in this book. Much that is here was inspired by publications on conduction. I would like to especially credit *Conduction Heat Transfer* by my friend Vedat S Arpaci. His book contains a wealth of interesting problems and applications. My original notes on conduction contained many examples and problems taken from the literature. Not having been careful in my early years about recording references, I tried to eliminate those that I knew were not my own. Nevertheless, a few may have been inadvertently included.

ACKNOWLEDGMENTS

First I would like to acknowledge the many teachers who directly or indirectly inspired and shaped my career. Among them I wish to single out Professors Ascher H. Shapiro of the Massachusetts Institute of Technology, Milton Van Dyke of Stanford University, D.W. Ver Planck of Carnegie Institute of Technology and Gordon J. Van Wylen and John A. Clark of the University of Michigan. I only wish that I had recognized their lasting contributions to my education decades earlier.

Chapter 11 was carefully reviewed by Professor Gang Chen of the Massachusetts Institute of Technology. The chapter author Chris Dames and I are grateful for his technical comments which strengthened the chapter.

My wife Vera read the entire manuscript and made constructive observations. I would like to thank her for being a supportive, patient and understanding partner throughout this project.

Latif M. Jiji
New York, New York
March, 2009

CONTENTS

1

BASIC CONCEPTS

1.1 Examples of Conduction Problems

Conduction heat transfer problems are encountered in many engineering applications. The following examples illustrate the broad range of conduction problems.

(1) *Design*. A small electronic package is to be cooled by free convection. A heat sink consisting of fins is recommended to maintain the electronic components below a specified temperature. Determine the required number of fins, configuration, size and material.

(2) *Nuclear Reactor Core*. In the event of coolant pump failure, the temperature of a nuclear element begins to rise. How long does it take before meltdown occurs?

(3) *Glaciology*. As a glacier advances slowly due to gravity, its front recedes due to melting. Estimate the location of the front in the year 2050.

(4) *Re-entry Shield*. A heat shield is used to protect a space vehicle during re-entry. The shield ablates as it passes through the atmosphere. Specify the required shield thickness and material to protect a space vehicle during re-entry.

(5) *Cryosurgery*. Cryoprobes are used in the treatment of certain skin cancers by freezing malignant tissue. However, prolonged contact of a cryoprobe with the skin can damage healthy tissue. Determine tissue temperature history in the vicinity of a cold probe subsequent to contact with the skin.

(6) *Rocket Nozzle*. One method for protecting the throat of a supersonic rocket nozzle involves inserting a porous ring at the throat. Injection of helium through the ring lowers the temperature and protects the nozzle. Determine the amount of helium needed to protect a rocket nozzle during a specified trajectory.

(7) *Casting*. Heat conduction in casting is accompanied by phase change. Determine the transient temperature distribution and the interface motion of a solid-liquid front for use in thermal stress analysis.

(8) *Food Processing*. In certain food processing operations conveyor belts are used to move food products through a refrigerated room. Use transient conduction analysis to determine the required conveyor speed.

1.2 Focal Point in Conduction Heat Transfer

What do all these examples have in common? If you guessed temperature, you are on the right track. However, what drives heat is *temperature difference* and not temperature. Refining this observation further, we state that conduction heat transfer problems involve the determination of *temperature distribution* in a region. Once temperature distribution is known, one can easily determine heat transfer rates. This poses two questions: (1) how is the rate of heat transfer related to temperature distribution? And (2) what governs temperature distribution in a region? *Fourier's law of conduction* provides the answer to the first question while the principle of *conservation of energy* gives the answer to the second.

1.3 Fourier's Law of Conduction

Our experience shows that if one end of a metal bar is heated, its temperature at the other end will eventually begin to rise. This transfer of energy is due to molecular activity. Molecules at the hot end exchange their kinetic and vibrational energies with neighboring layers through random motion and collisions. A temperature gradient, or slope, is established with energy continuously being transported in the direction of decreasing temperature. This mode of energy transfer is called *conduction*.

We now turn our attention to formulating a law that will help us determine the rate of heat transfer by conduction. Consider the wall shown in Fig.1.1. The temperature of one surface $(x = 0)$ is T_{si} and of the other surface $(x = L)$ is T_{so}. The wall thickness is L and its surface area is A. The remaining four surfaces are well insulated and thus heat is transferred in the x-direction only.

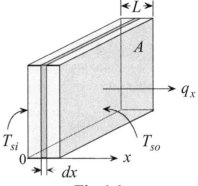

Fig. 1.1

Assume steady state and let q_x be the rate of heat transfer in the x-direction. Experiments have shown that q_x is directly proportional to A and $(T_{si} - T_{so})$ and inversely proportional to L. That is

$$q_x \propto \frac{A\,(T_{si} - T_{so})}{L}.$$

Introducing a proportionality constant k, we obtain

$$q_x = k\frac{A\,(T_{si} - T_{so})}{L}, \tag{1.1}$$

where k is a property of material called *thermal conductivity*. We must keep in mind that eq. (1.1) is valid for: (i) steady state, (ii) constant k and (iii) one-dimensional conduction. These limitations suggest that a re-formulation is in order. Applying eq. (1.1) to the element dx shown in Fig.1.1 and noting that T_{si} becomes $T(x)$, T_{so} becomes $T(x+dx)$, and L is replaced by dx, we obtain

$$q_x = k\,A\frac{T(x) - T(x+dx)}{dx} = -k\,A\frac{T(x+dx) - T(x)}{dx}.$$

Since $T(x+dx) - T(x) = dT$, the above gives

$$q_x = -k\,A\,\frac{dT}{dx}. \tag{1.2}$$

It is useful to introduce the term *heat flux* q_x'', which is defined as the heat flow rate per unit surface area normal to x. Thus,

$$q_x'' = \frac{q_x}{A}. \tag{1.3}$$

Therefore, in terms of heat flux, eq. (1.2) becomes

$$q_x'' = -k\,\frac{dT}{dx}. \tag{1.4}$$

Although eq. (1.4) is based on one-dimensional conduction, it can be generalized to three-dimensional and transient conditions by noting that heat flow is a vector quantity. Thus, the temperature derivative in eq. (1.4) is changed to a partial derivative and adjusted to reflect the direction of heat flow as follows:

$$q_x'' = -k\,\frac{\partial T}{\partial x}, \quad q_y'' = -k\,\frac{\partial T}{\partial y}, \quad q_z'' = -k\,\frac{\partial T}{\partial z}, \tag{1.5}$$

where x, y, and z are the rectangular coordinates. Equation (1.5) is known as *Fourier's law of conduction*. Three observations are worth making: (i) The negative sign indicates that when the gradient is negative, heat flow is in the positive direction, i.e., towards the direction of decreasing temperature, as dictated by the second law of thermodynamics. (ii) The conductivity k need not be uniform since eq. (1.5) applies at a point in the material and not to a finite region. In reality the thermal conductivity varies with temperature. However, eq. (1.5) is limited to *isotropic material*, i.e., k is invariant with direction. (iii) Returning to our previous observation that the focal point in conduction is the determination of temperature distribution, we now recognize that once $T(x,y,z,t)$ is known, the heat flux in any direction can be easily determined by simply differentiating the function T and using eq. (1.5).

1.4 Conservation of Energy: Differential Formulation of the Heat Conduction Equation in Rectangular Coordinates

What determines temperature distribution in a region? Is the process governed by a fundamental law and therefore predictable? The answer is that the temperature at each point is adjusted such that the principle of conservation of energy is satisfied everywhere. Thus the starting point in *differential formulation* must be based on an infinitesimal element. Fig.1.2 shows a region described by rectangular coordinates in which heat conduction is three-dimensional. To generalize the formulation, the material is assumed to be moving. In addition, energy is generated throughout the material at a rate q''' per unit volume. Examples of volumetric energy generation include heat conduction in nuclear elements, metabolic heat production in tissue, and electrical energy loss in transmission lines,

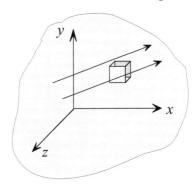

Fig. 1.2

We select an infinitesimal element *dxdydz* and apply the principle of conservation of energy (first law of thermodynamics)

Rate of energy added + Rate of energy generated – Rate of energy removed

= Rate of energy change within element

Denoting these terms by the symbols $\dot{E}_{in}$, $\dot{E}_{g}$, $\dot{E}_{out}$, and $\dot{E}$, respectively, we obtain

$$\dot{E}_{in} + \dot{E}_{g} - \dot{E}_{out} = \dot{E}. \qquad (1.6)$$

This form of conservation of energy is not helpful in solving conduction problems. Specifically, temperature, which is the focal point in conduction, does not appear explicitly in the equation. The next step is to express eq.(1.6) in terms of the dependent variable T. To simplify the formulation, the following assumptions are made: (1) uniform velocity, (2) constant pressure, (3) constant density and (4) negligible changes in potential energy.

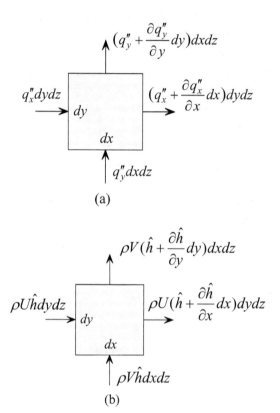

(a)

(b)

Fig. 1.3

Energy is exchanged with the element by conduction and mass motion. These two modes of energy transfer are shown in Fig. 1.3a and Fig. 1.3b, respectively. Not shown in these figures are the z-components, which can be formulated by analogy with the x and y-components. Energy enters the element by conduction at fluxes q_x'', q_y'' and q_z'', in the x, y and z directions, respectively. Since each flux represents energy per unit area per unit time, it must be multiplied by the area normal to it. Energy also enters the element through mass flow. The mass flow rate entering the element in the x-direction is $\rho U dy dz$, where ρ is density and U is the velocity component in the x-direction. The rate of energy carried by this mass is $\rho U \hat{h} dy dz$, where $\hat{h}$ is enthalpy per unit mass. The corresponding components in the y and z directions are $\rho V \hat{h} dx dz$ and $\rho W \hat{h} dx dy$, where V and W are the velocity components in the y and z directions, respectively. Thus, $\dot{E}_{in}$ is given by

$$\dot{E}_{in} = q''_x \, dy \, dz + q''_y \, dx \, dz + q''_z \, dx \, dy + \rho U \hat{h} \, dy \, dz + \rho V \hat{h} \, dx \, dz + \rho W \hat{h} \, dx \, dy. \tag{a}$$

Energy generation $\dot{E}_g$ is

$$\dot{E}_g = q''' dx dy dz . \tag{b}$$

Formulation of energy leaving the element is constructed using Taylor series expansion

$$\dot{E}_{out} = (q''_x + \frac{\partial q''_x}{\partial x} dx) dy dz + (q''_y + \frac{\partial q''_y}{\partial y} dy) dx dz + (q''_z + \frac{\partial q''_z}{\partial z} dz) dx dy + \rho U (\hat{h} + \frac{\partial \hat{h}}{\partial x} dx) dy dz + \rho V (\hat{h} + \frac{\partial \hat{h}}{\partial y} dy) dx dz + \rho W (\hat{h} + \frac{\partial \hat{h}}{\partial z} dz) dx dy. \tag{c}$$

Note that U, V and W are constant since material motion is assumed uniform. Energy change within the element $\dot{E}$ is expressed as

$$\dot{E} = \rho \frac{\partial \hat{u}}{\partial t} dx dy dz , \tag{d}$$

where $\hat{u}$ is internal energy per unit mass and t is time. Substituting (a)-(d) into eq. (1.6)

$$-\frac{\partial q''_x}{\partial x} - \frac{\partial q''_y}{\partial y} - \frac{\partial q''_z}{\partial z} - \rho U \frac{\partial \hat{h}}{\partial x} - \rho V \frac{\partial \hat{h}}{\partial y} - \rho W \frac{\partial \hat{h}}{\partial z} + q''' = \rho \frac{\partial \hat{u}}{\partial t}. \tag{e}$$

Enthalpy $\hat{h}$ is defined as

$$\hat{h} = \hat{u} + \frac{P}{\rho}, \tag{f}$$

here P is pressure, assumed constant. Substituting (f) into (e) and rearranging

$$-\frac{\partial q''_x}{\partial x} - \frac{\partial q''_y}{\partial y} - \frac{\partial q''_z}{\partial z} + q''' = \rho\left(\frac{\partial \hat{h}}{\partial t} + U\frac{\partial \hat{h}}{\partial x} + V\frac{\partial \hat{h}}{\partial y} + W\frac{\partial \hat{h}}{\partial z}\right). \tag{g}$$

The next step is to express the heat flux and enthalpy in terms of temperature. Fourier's law, eq.(1.5), relates heat flux to temperature gradient. Enthalpy change for constant pressure is given by

$$d\hat{h} = c_p dT, \tag{h}$$

where c_p is specific heat at constant pressure. Substituting eqs. (1.5) and (h) into (g)

$$\frac{\partial}{\partial x}\left(k\frac{\partial T}{\partial x}\right) + \frac{\partial}{\partial y}\left(k\frac{\partial T}{\partial y}\right) + \frac{\partial}{\partial z}\left(k\frac{\partial T}{\partial z}\right) + q''' =$$

$$\rho c_p\left(\frac{\partial T}{\partial t} + U\frac{\partial T}{\partial x} + V\frac{\partial T}{\partial y} + W\frac{\partial T}{\partial z}\right). \tag{1.7}$$

Although eq. (1.7) is based on uniform velocity, it can be shown that it is also applicable to incompressible flow with variable velocity, as long as dissipation, which is work done due to viscous forces, is negligible [1]. For constant conductivity and stationary material, eq. (1.7) simplifies to

$$\alpha\left(\frac{\partial^2 T}{\partial x^2} + \frac{\partial^2 T}{\partial y^2} + \frac{\partial^2 T}{\partial z^2}\right) + \frac{q'''}{\rho c_p} = \frac{\partial T}{\partial t}, \tag{1.8}$$

where α is a material property called *thermal diffusivity*, defined as

$$\alpha = \frac{k}{\rho c_p}. \tag{1.9}$$

1.5 The Heat Conduction Equation in Cylindrical and Spherical Coordinates

To analyze conduction problems in cylindrical and spherical geometries requires the formulation of the heat conduction equation in the cylindrical and spherical coordinates shown in Fig. 1.4. For constant properties and no dissipation, the heat equation takes the following forms:

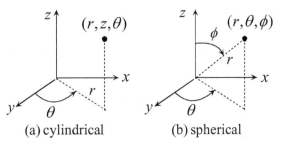

(a) cylindrical   (b) spherical

Fig. 1.4

Cylindrical Coordinates (r,θ,z):

$$\alpha\left[\frac{1}{r}\frac{\partial}{\partial r}\left(r\frac{\partial T}{\partial r}\right)+\frac{1}{r^2}\frac{\partial^2 T}{\partial \theta^2}+\frac{\partial^2 T}{\partial z^2}\right]+\frac{q'''}{\rho c_p}=\left[\frac{\partial T}{\partial t}+V_r\frac{\partial T}{\partial r}+\frac{V_\theta}{r}\frac{\partial T}{\partial \theta}+V_z\frac{\partial T}{\partial z}\right],$$

(1.10)

where V_r, V_θ and V_z are the velocity components in the r, θ and z directions, respectively. For a stationary material this equation simplifies to

$$\alpha\left[\frac{1}{r}\frac{\partial}{\partial r}\left(r\frac{\partial T}{\partial r}\right)+\frac{1}{r^2}\frac{\partial^2 T}{\partial \theta^2}+\frac{\partial^2 T}{\partial z^2}\right]+\frac{q'''}{\rho c_p}=\frac{\partial T}{\partial t}.$$

(1.11)

Spherical Coordinates (r,θ,ϕ):

$$\alpha\left[\frac{1}{r^2}\frac{\partial}{\partial r}\left(r^2\frac{\partial T}{\partial r}\right)+\frac{1}{r^2\sin^2\phi}\frac{\partial^2 T}{\partial \theta^2}+\frac{1}{r^2\sin\phi}\frac{\partial}{\partial \phi}\left(\sin\phi\frac{\partial T}{\partial \phi}\right)\right]+\frac{q'''}{\rho c_p}=$$

$$\frac{\partial T}{\partial t}+V_r\frac{\partial T}{\partial r}+\frac{V_\phi}{r}\frac{\partial T}{\partial \phi}+\frac{V_\theta}{r\sin\phi}\frac{\partial T}{\partial \theta},$$

(1.12)

where V_ϕ is the velocity component in the ϕ direction. For stationary material this equation reduces to

$$\alpha\left[\frac{1}{r^2}\frac{\partial}{\partial r}\left(r^2\frac{\partial T}{\partial r}\right)+\frac{1}{r^2\sin^2\phi}\frac{\partial^2 T}{\partial\theta^2}+\frac{1}{r^2\sin\phi}\frac{\partial}{\partial\phi}\left(\sin\phi\frac{\partial T}{\partial\phi}\right)\right]+\frac{q'''}{\rho c_p}=$$

$$\frac{\partial T}{\partial t}. \qquad (1.13)$$

1.6 Boundary Conditions

Although the heat conduction equation governs temperature behavior in a region, it does not give temperature distribution. To obtain temperature distribution it is necessary to solve the equation. However, to construct a complete solution, boundary and initial conditions must be specified. For example, eq.(1.8) requires six boundary conditions, two in each of the variables x, y and z, and one initial condition in time t. Boundary conditions are mathematical equations describing what takes place physically at a boundary. Similarly, an initial condition describes the temperature distribution at time $t = 0$.

Since boundary conditions involve thermal interaction with the surroundings, it is necessary to first describe two common modes of surface heat transfer: convection and radiation.

1.6.1 Surface Convection: Newton's Law of Cooling

In this mode of heat transfer, energy is exchanged between a surface and a fluid moving over it. Based on experimental observations, it is postulated that the flux in convection is directly proportional to the difference in temperature between the surface and the streaming fluid. That is

$$q''_s \propto (T_s - T_\infty),$$

where q''_s is surface flux, T_s is surface temperature and T_∞ is the fluid temperature far away from the surface. Introducing a proportionality constant to express this relationship as equality, we obtain

$$q''_s = h(T_s - T_\infty). \qquad (1.14)$$

This result is known as *Newton's law of cooling*. The constant of proportionality h is called *heat transfer coefficient*. This coefficient depends on geometry, fluid properties, motion, and in some cases temperature difference ($T_s - T_\infty$). Thus, unlike thermal conductivity, h is not a property of material.

1.6.2 Surface Radiation: Stefan-Boltzmann Law

While conduction and convection require a medium to transport energy, radiation does not. Furthermore, radiation energy is transmitted by electromagnetic waves, which travel best in a vacuum. The maximum possible radiation is described by the *Stefan-Boltzmann law*, which gives surface radiation flux for an ideal body called *blackbody* as

$$q_b'' = \sigma T_s^4, \tag{1.15}$$

where q_b'' is blackbody radiation flux, T_s is surface temperature, measured in absolute degrees, and σ is the *Stefan-Boltzmann constant* given by

$$\sigma = 5.67 \times 10^{-8} \text{ W/m}^2\text{-K}^4. \tag{1.16}$$

To determine the radiation flux q_r'' emitted from a real surface, a radiation property called *emissivity*, ε, is defined as

$$\varepsilon = \frac{q_r''}{q_b''}. \tag{1.17}$$

Combining (1.15) and (1.17)

$$q_r'' = \varepsilon \sigma T_s^4. \tag{1.18}$$

From the definition of ε, it follows that its maximum value is unity (*blackbody*).

Radiation energy exchange between two surfaces depends on the geometry, shape, area, orientation and emissivity of the two surfaces. In addition, it depends on the *absorptivity* α of each surface. Absorptivity is a surface property defined as the fraction of radiation energy incident on a surface which is absorbed by the surface. Although the determination of the net heat exchange by radiation between two surfaces, q_{12}, can be complex, the analysis is simplified for an ideal model for which $\varepsilon = \alpha$. Such an ideal surface is called a *gray* surface. For the special case of a gray surface which is completely enclosed by a much larger surface, q_{12} is given by

$$q_{12} = \varepsilon_1 \sigma A_1 (T_1^4 - T_2^4), \tag{1.19}$$

where ε_1 is the emissivity of the small surface, A_1 its area, T_1 its absolute temperature and T_2 is the absolute temperature of the surrounding surface. Note that for this special case neither the area A_2 of the large surface nor its emissivity ε_2 affects the result.

1.6.3 Examples of Boundary Conditions

There are several common physical conditions that can take place at boundaries. Fig. 1.5 shows four typical boundary conditions for two-dimensional conduction in a rectangular plate. Fig. 1.6 shows an interface of two materials. Two boundary conditions are associated with this case.

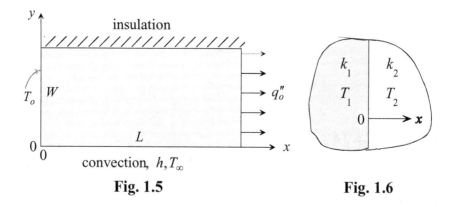

Fig. 1.5   **Fig. 1.6**

Before writing boundary conditions, an origin and coordinate axes must be selected. Boundary conditions for Fig. 1.5 and Fig. 1.6 are expressed mathematically as follows:

(1) Specified temperature. Along boundary $(0, y)$ the temperature is T_o. This temperature can be uniform or can vary along y as well as with time. Mathematically this condition is expressed as

$$T(0, y) = T_o .$$

(1.20)

(2) Specified flux. The heat flux along boundary (L, y) is q_o''. According to Fourier's law this condition is expressed as

$$q_o'' = -k \frac{\partial T(L, y)}{\partial x} .$$

(1.21)

(3) Convection. Neither the temperature nor the flux are known along boundary $(x,0)$. Instead, heat is exchanged by convection with the ambient fluid. The simplest way to formulate this condition mathematically is to pretend that heat flows in the positive coordinate direction. In this example the fluid, which is at temperature T_∞, adds energy by convection to the boundary $(x,0)$. Conservation of energy requires that the added energy be conducted to the interior in the positive y-direction. Therefore, equating Newton's law of cooling with Fourier's law of conduction gives

$$h[T_\infty - T(x,0)] = -k \frac{\partial T(x,0)}{\partial y} .$$

(1.22)

Note that surface temperature $T(x,0)$ is not equal to the ambient temperature T_∞.

(4) Insulated boundary. The boundary at (x,W) is thermally insulated. Thus, according to Fourier's law, this condition is expressed as

$$\frac{\partial T(x,W)}{\partial y} = 0 .$$

(1.23)

Note that this is a special case of eq. (1.21) with $q_o'' = 0$, or of eq. (1.22) with $h = 0$.

(5) Interface. Fig.1.6 shows a composite wall of two materials with thermal conductivities k_1 and k_2. For a perfect interface contact, the two temperatures must be the same at the interface. Thus

$$T_1(0, y) = T_2(0, y). \tag{1.24}$$

Conservation of energy at the interface requires that the two fluxes be identical. Application of Fourier's law gives

$$k_1 \frac{\partial T_1(0, y)}{\partial x} = k_2 \frac{\partial T_2(0, y)}{\partial x}. \tag{1.25}$$

(6) Interface with a heat source. One example of this case is an electrical heating element which is sandwiched between two non-electrically conducting materials as shown in Fig. 1.7. Another example is frictional heat, which is generated by relative motion between two surfaces. Energy dissipated at an interface can be conducted through both materials or through either one of the two. Boundary conditions at the interface are again based on the continuity of temperature and conservation of energy. Assuming perfect contact, continuity results in eq.(1.24). To apply conservation of energy at the interface, it is convenient to pretend that heat is conducted in the positive coordinate direction through both materials. Referring to Fig. 1.7, conservation of energy requires that heat flux added to the interface by conduction through material 1, plus heat flux generated at the interface, q_i'', be equal to the flux removed by conduction through material 2. Thus

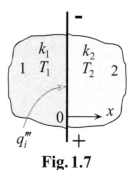

Fig. 1.7

$$-k_1 \frac{\partial T_1(0, y)}{\partial x} + q_i'' = -k_2 \frac{\partial T_2(0, y)}{\partial x}. \tag{1.26}$$

(7) Radiation. To illustrate how a radiation boundary condition is formulated, consider Fig. 1.5. Assume that the boundary $(x,0)$ exchanges heat by radiation in addition to convection. Again we pretend that net radiation energy is added to the surface in the positive y-direction. Conservation of energy at this boundary requires that energy added at the surface by convection and radiation be equal to energy conducted in the positive y-direction. Thus

$$q''_{conv} + q''_{rad} = q''_{cond}.$$

Using Fourier's law of conduction for q'''_{cond}, Newton's law of cooling for q''_{conv} and assuming that Stefan-Boltzmann radiation result for q''_{rad}, eq. (1.19), is applicable, the above gives

$$h\left[T_\infty - T(x,0)\right] + \sigma\varepsilon\left[T_{sur}^4 - T^4(x,0)\right] = -k\frac{\partial T(x,0)}{\partial y}, \qquad (1.27)$$

where T_{sur} is the surroundings temperature. To formulate the boundary condition for the case of energy exchange by radiation only, set $h = 0$ in eq. (1.27).

1.7 Problem Solving Format

Conduction problems lend themselves to a systematic solution procedure. The following basic format which builds on the work of Ver Plank and Teare [2] is used throughout the book.

(1) Observations. Read the problem statement very carefully and note essential facts and features such as geometry, temperature symmetry, heat flow direction, number of independent variables, etc. Identify characteristics that require special attention such as variable properties, composite domain, non-linearity, etc. Note conditions justifying simplified solutions. Show a schematic diagram.

(2) Origin and Coordinates. Select an origin and a coordinate system appropriate to the geometry under consideration.

(3) Formulation. In this stage the conduction problem is expressed in mathematical terms. Define all terms. Assign units to all symbols whenever numerical computations are required. Proceed according to the following steps:

(i) **Assumptions.** Model the problem by making simplifications and approximations. List all assumptions.

(ii) **Governing Equations.** Based on the coordinate system chosen and the assumptions made, select an appropriate form of the heat equation. Note that composite domains require a heat equation for each layer.

(iii) **Boundary Conditions.**   Examine the governing equations and decide on the number of boundary conditions needed based on the number of independent variables and the order of the highest derivative. Identify the physical nature of each boundary condition and express it mathematically. Use the examples of Section 1.6.3 as a guide in formulating boundary conditions.

(4) **Solution.** Examine the equation to be solved and select an appropriate method of solution. Apply boundary conditions to determine constants of integration.

(5) **Checking.** Check each step of the analysis as you proceed. Apply dimensional checks and examine limiting cases.

(6) **Computations.** Execute the necessary computations and calculations to generate the desired numerical results.

(7) **Comments.** Review your solution and comment on such things as the role of assumptions, the form of the solution, the number of governing parameters, etc.

1.8 Units

SI units are used throughout this textbook. The basic units in this system are

Length (L): meter (m)
Time (t): second (s)
Mass (m): kilogram (kg)
Temperature (T): kelvin (K)

Temperature on the *Celsius* scale is related to the *kelvin* scale by

$$T(^\circ C) = T(K) - 273.15. \qquad (1.28)$$

Note that temperature difference on the two scales is identical. Thus, a change of one kelvin is equal to a change of one Celsius. This means that quantities that are expressed per unit kelvin, such as thermal conductivity, heat transfer coefficient and specific heat, are numerically the same as per degree Celsius. That is, $W/m^2\text{-}K = W/m^2\text{-}°C$.

The basic units are used to derive units for other quantities. Force is measured in *newtons* (N). One newton is the force needed to accelerate a mass of one kilogram one meter per second per second:

$$Force = mass \times acceleration$$

$$N = kg.m/s^2.$$

Energy is measured in *joules* (J). One joule is the energy associated with a force of one newton moving a distance of one meter.

$$J = N.m = kg.m^2/s^2.$$

Power is measured in *watts* (W). One watt is energy rate of one joule per second.

$$W = J/s = N.m/s = kg.m^2/s^3.$$

REFERENCES

[1] Bejan, A., *Convection Heat Transfer*, 2nd Edition, Wiley, New York, 1995.

[2] Ver Planck, D.W., and Teare Jr., B.R., *Engineering Analysis: An Introduction to Professional Method*, Wiley, New York, 1952

PROBLEMS

1.1   Write the heat equation for each of the following cases:

[a]   A wall, steady state, stationary, one-dimensional, incompressible and no energy generation.

[b]   A wall, transient, stationary, one-dimensional, incompressible, constant k with energy generation.

[c]   A cylinder, steady state, stationary, two-dimensional (radial and axial), constant k, incompressible, with no energy generation.

[d]   A wire moving through a furnace with constant velocity, steady state, one-dimensional (axial), incompressible, constant k and no energy generation.

[e]   A sphere, transient, stationary, one-dimensional (radial), incompressible, constant k with energy generation.

1.2   A long electric wire of radius r_o generates heat at a rate q'''.   The surface is maintained at uniform temperature T_o. Write the heat equation and boundary conditions for steady state one-dimensional conduction.

1.3   You are interested in analyzing the rate at which a spherical ice ball melts. What heat equation should you use for the ice?   List all assumptions.

1.4   Consider axial flow of water in a cold tube. Write the heat conduction equation for the ice forming axisymmetrically on the inside surface of the tube.

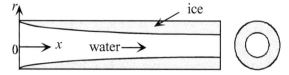

1.5   Consider two-dimensional conduction in the semi-circular cylinder shown. The cylinder is heated with uniform flux along its outer surface and is maintained at a variable temperature

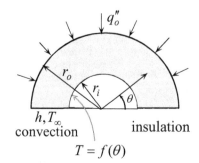

along its inner surface. One of the plane surfaces is insulated while the other exchanges heat by convection with an ambient fluid at T_∞. The heat transfer coefficient is h. Write the heat equation and boundary conditions for steady state conduction.

1.6   A long hollow cylinder exchanges heat by radiation and convection along its outside surface with an ambient fluid at T_∞. The heat transfer coefficient is h. The surroundings is at T_{sur} and surface emissivity is ε. Heat is removed from the inside surface at a uniform flux q_i''. Use a simplified radiation model and write the heat equation and boundary conditions for one-dimensional steady state conduction.

1.7   Write the heat equation and boundary conditions for steady state two-dimensional conduction in the rectangular plate shown.

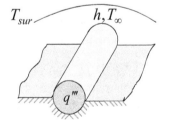

1.8   Heat is generated at a rate q''' in a spherical shell with inner radius r_i and outer radius r_o. Heat is added at the outer surface at a uniform flux q_o''. The inside surface is maintained at uniform temperature T_i. Write the heat equation and boundary conditions for steady state conduction.

1.9   A long electric cable generates heat at a rate q'''. Half the cable is buried underground while the other half exchanges heat by radiation and convection. The ambient and surroundings temperatures are T_∞ and T_{sur}, respectively. The heat transfer coefficient is h. Neglecting the thickness of the electrical insulation layer and heat loss to the ground, select a model to analyze the temperature distribution in the cable and write the heat equation and boundary conditions.

1.10   A shaft of radius r_o rotates inside a sleeve of thickness δ. Frictional heat is generated at the interface at a rate q_i''. The outside surface of the sleeve is cooled by convection. The ambient temperature is T_∞ and the heat transfer coefficient is h. Consider steady state one-dimensional conduction in the radial direction. Write the heat equations and boundary conditions for the temperature distribution in the shaft and sleeve.

1.11   A rectangular plate of length L and height H slides down an inclined surface with a velocity U. Sliding friction results in surface heat flux q''_o. The front and top sides of the plate exchange heat by convection. The heat transfer coefficient is h and

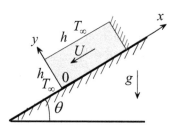

the ambient temperature is T_∞. Neglect heat loss from the back side and assume that no frictional heat is conducted through the inclined surface. Write the two-dimensional steady state heat equation and boundary conditions.

1.12   Consider a semi-circular section of a tube with inside radius R_i and outside radius R_o. Heat is exchanged by convection along the inside and outside cylindrical surfaces. The inside temperature and heat transfer coefficient are T_i and h_i. The outside temperature and heat transfer coefficient are T_o and h_o. The two plane surfaces are maintained at uniform temperature T_b. Write the two-dimensional steady state heat equation and boundary conditions.

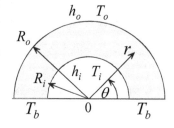

1.13   Two large plates of thicknesses L_1 and L_2 are initially at temperatures T_1 and T_2. Their respective conductivities are k_1 and k_2. The two plates are pressed together and insulated along their exposed surfaces. Write the heat equation and boundary conditions.

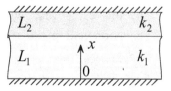

1.14 Heat is generated at a volumetric rate q''' in a rod of radius r_o and length L. Half the cylindrical surface is insulated while the other half is heated at 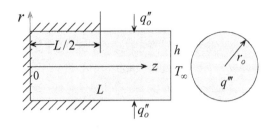 a flux q_o''. One end is insulated and the other exchanges heat by convection. Write the heat equation and boundary conditions for steady state two-dimensional conduction.

1.15 A cable of radius r_o, conductivity k and specific heat c_p moves through a furnace with velocity U and leaves at temperature T_o. Electric energy is dissipated in the cable resulting in a 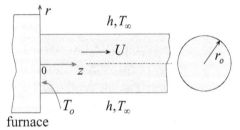 volumetric energy generation rate of q'''. Outside the furnace the cable is cooled by convection and radiation. The heat transfer coefficient is h, emissivity ε, ambient temperature T_∞ and surroundings temperature is T_{sur}. Use a simplified radiation model and assume that the cable is infinitely long. Write the steady state two-dimensional heat equation and boundary conditions.

1.16 Radiation is suddenly directed at the surface of a semi-infinite plate of conductivity k and thermal diffusivity α. Due to the thermal absorption characteristics of the material the radiation results in a variable energy generation rate given by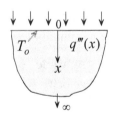

$$q'''(x) = q_o''' e^{-bx},$$

where q_o''' and b are constants and x is distance along the plate. Surface temperature at $x = 0$ is T_o. Initially the plate is at uniform temperature T_i. Write the heat equation and boundary conditions.

1.17 A section of a long rotating shaft of radius r_o is
buried in a material of very low thermal
conductivity. The length of the buried section is
L. Frictional heat generated in the buried
section can be modeled as surface heat flux.
Along the buried surface the radial heat
flux, q_r'', is assumed uniform. However, at the
flat end the axial heat flux, q_z'', varies with
radius according to

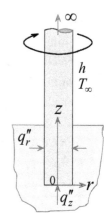

$$q_z'' = \beta r,$$

where β is constant. The exposed surface exchanges heat by
convection with the ambient. The heat transfer coefficient is h and
the ambient temperature is T_∞. Treat the shaft as semi-infinite and
assume that all frictional heat is conducted through the shaft. Write
the steady state heat equation and boundary conditions.

1.18 A plate of thickness $2L$ moves through a furnace with velocity U and
leaves at temperature T_o. Outside the furnace it is cooled by
convection and radiation. The heat transfer coefficient is h,
emissivity ε, ambient temperature T_∞
and surroundings temperatures is T_{sur}.
Write the two-dimensional steady
state heat equation and boundary
conditions. Use a simplified radiation
model and assume that the plate is
infinitely long.

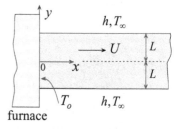

furnace

1.19 A cable of radius r_o moves with
velocity U through a furnace of length
L where it is heated at uniform flux
q_o''. Far away from the inlet of the
furnace the cable is at temperature T_i.
Assume that no heat is exchanged
with the cable before it enters and
after it leaves the furnace. Write the

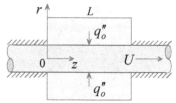

two-dimensional steady state heat equation and boundary conditions.

1.20   A hollow cylinder of inner radius R_i and outer radius R_o is heated with uniform flux q_i'' at its inner surface. The lower half of the cylinder is insulated and the upper half exchanges heat by convection and radiation. The heat transfer coefficient is h, ambient temperature T_∞, surroundings temperature T_{sur} and surface emissivity is ε. Neglecting axial conduction and using a simplified radiation model, write the steady state heat equations and boundary conditions.

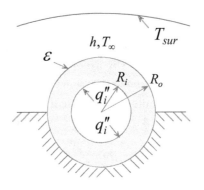

ONE-DIMENSIONAL STEADY STATE CONDUCTION

One-dimensional conduction, with its simple mathematical level, is used in this chapter to present the essential steps in the analysis of conduction heat transfer problems. The objective is to learn to model problems, set up the applicable governing equations and boundary conditions and construct solutions. The simplicity of the mathematical treatment of one-dimensional conduction enables us to explore a variety of practical applications.

2.1 Examples of One-dimensional Conduction

Example 2.1: Plate with Energy Generation and Variable Conductivity

Consider a plate with internal energy generation q''' and a variable thermal conductivity k given by

$$k = k_o (1 - \gamma T),$$

where k_o and γ are constant and T is temperature. Both surfaces in Fig. 2.1 are maintained at $0\,^\circ C$. Determine the temperature distribution in the plate.

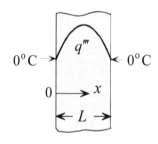

Fig. 2.1

(1) Observations. (i) Thermal conductivity is variable. (ii) Temperature distribution is symmetrical about the center plane. (iii) The geometry can be described by a rectangular coordinate system. (iv) The temperature is specified at both surfaces.

(2) Origin and Coordinates. A rectangular coordinate system is used with the origin at one of the two surfaces. The coordinate x is oriented as shown in Fig. 2.1.

(3) Formulation.

(i) **Assumptions.** (1) One-dimensional, (2) steady state, (3) stationary and (4) uniform energy generation.

(ii) **Governing Equations.** Introducing the above assumptions into eq. (1.7) gives

$$\frac{d}{dx}(k\frac{dT}{dx})+q''' = 0 , \tag{2.1}$$

where

k = thermal conductivity
q''' = rate of energy generation per unit volume
T = temperature
x = independent variable

The thermal conductivity k varies with temperature according to

$$k = k_o(1-\gamma T), \tag{a}$$

where k_o and γ are constant. Substituting (a) into eq. (2.1) gives

$$\frac{d}{dx}[(1-\gamma T)\frac{dT}{dx}]+\frac{q'''}{k_o} = 0 . \tag{b}$$

(iii) **Boundary Conditions.** Since (b) is second order with a single independent variable x, two boundary conditions are needed. The temperature is specified at both boundaries, $x = 0$ and $x = L$. Thus

$$T(0) = 0 , \tag{c}$$

and

$$T(L) = 0 . \tag{d}$$

(4) Solution. Integrating (b) twice

$$T - \frac{\gamma}{2}T^2 = -\frac{q'''}{2k_o}x^2 + C_1 x + C_2 , \tag{e}$$

where C_1 and C_2 are constants of integration. Application of boundary conditions (c) and (d) gives C_1 and C_2

$$C_1 = \frac{q'''L}{2k_o}, \quad C_2 = 0. \tag{f}$$

Substituting (f) into (e)

$$T^2 - \frac{2}{\gamma}T + \frac{q'''Lx}{\gamma k_o}\left(1 - \frac{x}{L}\right) = 0. \tag{g}$$

Recognizing that (g) is a quadratic equation, its solution is

$$T = \frac{1}{\gamma} \pm \sqrt{\frac{1}{\gamma^2} - \frac{q'''Lx}{\gamma k_o}\left(1 - \frac{x}{L}\right)}. \tag{h}$$

To satisfy boundary conditions (c) and (d), the minus sign in (h) must be considered. The solution becomes

$$T = \frac{1}{\gamma} - \sqrt{\frac{1}{\gamma^2} - \frac{q'''Lx}{\gamma k_o}\left(1 - \frac{x}{L}\right)}. \tag{i}$$

(5) Checking. *Dimensional check*: Assigning units to the quantities in (i) shows that each term has units of temperature.

Boundary conditions check: Setting $x = 0$ or $x = L$ in (i) gives $T = 0$. Thus, boundary conditions (c) and (d) are satisfied.

Limiting check: For the special case of $q''' = 0$, the temperature distribution should be uniform given by $T = 0$. Setting $q''' = 0$ in (i) gives

$$T = \frac{1}{\gamma} - \sqrt{\frac{1}{\gamma^2}} = 0.$$

Symmetry check: Symmetry requires that the temperature gradient be zero at the center, $x = L/2$. Differentiating (i) with respect to x

$$\frac{dT}{dx} = \frac{1}{2}\left[\frac{1}{\gamma^2} - \frac{q'''Lx}{\gamma k_o}\left(1 - \frac{x}{L}\right)\right]^{-\frac{1}{2}}\left(\frac{q'''L}{\gamma k_o}\right)\left(\frac{2x}{L} - 1\right). \tag{j}$$

Setting $x = \dfrac{L}{2}$ in (j) gives $\dfrac{dT}{dx} = 0$.

Quantitative check: Conservation of energy and symmetry require that half the total energy generated in the plate leaves in the negative x-direction at $x = 0$ and half in the positive x-direction at $x = L$. That is, for a wall of surface area A the heat transfer rate q is

$$q(0) = -\frac{q'''AL}{2},$$ (k)

and

$$q(L) = \frac{q'''AL}{2}.$$ (l)

The heat transfer rate at $x = 0$ and $x = L$ is determined by applying Fourier's law

$$q(0) = -A k_o [1 - \gamma T(0)]\frac{dT(0)}{dx},$$ (m)

and

$$q(L) = -A k_o [1 - \gamma T(L)]\frac{dT(L)}{dx}.$$ (n)

Evaluating the gradient (j) at $x = 0$, noting that $T(0) = 0$ and substituting into (m) gives the same result as (k). Similarly, results in (n) agree with (l).

(6) Comments. The solution to the special case of constant thermal conductivity corresponds to $\gamma = 0$. Since γ can not be set equal to zero in (i), the solution to this case must be obtained from (g) by first multiplying through by γ and then setting $\gamma = 0$.

Example 2.2: Radial Conduction in a Composite Cylinder with Interface Friction

A shaft of radius R_s rotates inside a sleeve of inner radius R_s and outer radius R_o. Frictional heat is generated at the interface at a flux q_i''. The outside surface of the sleeve is cooled by convection with an ambient fluid at T_∞. The heat transfer coefficient is h. Consider one-dimensional steady state conduction in the radial direction, determine the temperature distribution in the shaft and sleeve.

(1) Observations. (i) The shaft and sleeve form a composite cylindrical wall. (ii) The geometry can be described by a cylindrical coordinate system. (iii) Heat conduction is assumed to be in the radial direction only. (iv) Steady state requires that all energy generated at the interface be conducted through the sleeve.   Thus, no heat is conducted through the shaft. It follows that the shaft must be at a uniform temperature. (v) The heat flux is specified at the inner radius of the sleeve and convection takes place at the outer radius.

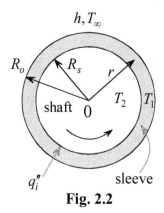

Fig. 2.2

(2) Origin and Coordinates. A cylindrical coordinate system is used with the origin at the center of the shaft as shown in Fig. 2.2.

(3) Formulation.

(i) Assumptions.   (1) One-dimensional radial conduction, (2) steady state, (3) constant thermal conductivities, (4) no energy generation, (5) uniform frictional energy flux at the interface and (6) stationary material (no motion in the radial direction).

(ii) Governing Equations.   Physical consideration requires that the shaft temperature be uniform. Thus, although this is a composite domain problem, only one equation is needed to determine the temperature distribution. Introducing the above assumptions into eq.(1.11) gives

$$\frac{d}{dr}\left(r\frac{dT_1}{dr}\right) = 0,\tag{2.2}$$

where

r = radial coordinate

T_1 = temperature distribution in the sleeve

(iii) Boundary Conditions. Equation (2.2) is second order with a single independent variable. Thus two boundary conditions are needed. Since all frictional energy is conducted through the sleeve, it follows that the heat flux added to the sleeve is specified at the inside radius R_s.

Conservation of energy at the interface and Fourier's law of conduction give

$$q_i'' = -k_1 \frac{dT_1(R_s)}{dr}, \qquad \text{(a)}$$

where

k_1 = thermal conductivity of the sleeve

q_i'' = interface frictional heat flux

Convection at the outer surface gives the second boundary condition

$$-k_1 \frac{dT_1(R_o)}{dr} = h[T_1(R_o) - T_\infty]. \qquad \text{(b)}$$

(4) Solution. Equation (2.2) is solved by integrating it twice with respect to r to obtain

$$T_1 = C_1 \ln r + C_2, \qquad \text{(c)}$$

where C_1 and C_2 are constants of integration. Application of boundary conditions (a) and (b) gives C_1 and C_2

$$T_1 = -\frac{q_i'' R_s}{k_1} \ln r + C_2$$

$$C_1 = -\frac{q_i'' R_s}{k_1}, \qquad \text{(d)}$$

and

$$C_2 = T_\infty + \frac{q_i'' R_s}{k_1}\left(\ln R_o + \frac{k_1}{hR_o}\right). \qquad \text{(e)}$$

Substituting (d) and (e) into (c) and rearranging gives the dimensionless temperature distribution in the sleeve

$$\frac{T_1(r) - T_\infty}{q_i'' R_s / k_1} = \ln \frac{R_o}{r} + \frac{k_1}{hR_o}. \qquad \text{(f)}$$

The dimensionless ratio, hR_o / k_1, appearing in the solution is called the *Biot number.* It is associated with convection boundary conditions.

The temperature of the shaft, T_2, is obtained from the interface boundary condition

$$T_2(r) = T_2(R_s) = T_1(R_s). \qquad \text{(g)}$$

Evaluating (f) at $r = R_s$ and substituting into (g) gives the dimensionless shaft temperature

$$\frac{T_2(r) - T_\infty}{q_i'' R_s / k_1} = \ln \frac{R_o}{R_s} + \frac{k_1}{hR_o} . \qquad (h)$$

(5) Checking. *Dimensional check*: Each term in solution (f) is dimensionless

Boundary conditions check: Substituting (f) into boundary conditions (a) and (b) shows that they are satisfied.

Limiting check: If frictional heat vanishes, both shaft and sleeve will be at the ambient temperature. Substituting $q_i'' = 0$ into (f) and (h) gives $T_1(r) = T_2(r) = T_\infty$.

(6) Comments. (i) Shaft conductivity plays no role in the solution. (ii) The solution is characterized by two dimensionless parameters: the Biot number hR_o / k_1 and the geometric ratio R_o / R_s. (iii) This problem can also be treated formally as a composite cylindrical wall. This requires writing two heat equations and four boundary conditions. The heat equation for the sleeve is given by eq. (2.2). A second similar equation applies to the shaft

$$\frac{d}{dr}\left(r \frac{dT_2}{dr}\right) = 0 . \qquad (i)$$

Boundary condition (b) is still applicable. The remaining three conditions are: Symmetry of temperature at the center

$$\frac{dT_2(0)}{dr} = 0 . \qquad (j)$$

Conservation of energy and equality of temperature at the interface give two conditions

$$-k_2 \frac{dT_2(R_s)}{dr} + q_i'' = -k_1 \frac{dT_1(R_s)}{dr} , \qquad (k)$$

and

$$T_1(R_s) = T_2(R_s) . \qquad (l)$$

where k_2 is the thermal conductivity of the shaft. This approach yields the solutions to the temperature distribution in the sleeve and shaft obtained above.

Example 2.3: Composite Wall with Energy Generation

A plate of thickness L_1 and conductivity k_1 generates heat at a volumetric rate of q'''. The plate is sandwiched between two plates of conductivity k_2 and thickness L_2 each. The exposed surfaces of the two plates are maintained at constant temperature T_o. Determine the temperature distribution in the three plates.

(1) Observations. (i) The three plates form a composite wall. (ii) The geometry can be described by a rectangular coordinate system. (iii) Due to symmetry, no heat is conducted through the center plane of the heat generating plate. (iv) Heat conduction is assumed to

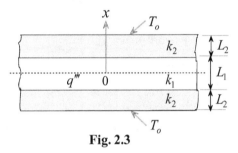

Fig. 2.3

be in the direction normal to the three plates. (v) Steady state and symmetry require that the total energy generated be conducted equally through each of the outer plates.

(2) Origin and Coordinates. To take advantage of symmetry, the origin is selected at the center of the heat generating plate. The composite wall is shown in Fig. 2.3.

(3) Formulation.

(i) Assumptions. (1) One-dimensional, (2) steady state, (3) constant thermal conductivities and (4) perfect interface contact.

(ii) Governing Equations.

Since temperature is symmetrical about the origin, only half the composite system is considered (see Fig. 2.4). Thus two heat equations are required. Let subscripts 1 and 2 refer to the center plate and outer plate,

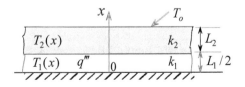

Fig. 2.4

respectively. Based on the above assumptions, eq.(1.8) gives

$$\frac{d^2 T_1}{dx^2} + \frac{q'''}{k_1} = 0,$$ (a)

and

$$\frac{d^2 T_2}{dx^2} = 0.$$ (b)

where

k_1 = thermal conductivity of the heat generating plate
q''' = rate of energy generation per unit volume
T = temperature
x = coordinate

(iii) Boundary Conditions. Since equations (a) and (b) are second order with a single independent variable x, four boundary conditions are needed. They are:

Symmetry of temperature at the center

$$\frac{dT_1(0)}{dx} = 0 .$$ (c)

Conservation of energy and equality of temperature at the interface

$$k_1 \frac{dT_1(L_1/2)}{dx} = k_2 \frac{dT_2(L_1/2)}{dx},$$ (d)

and

$$T_1(L_1/2) = T_2(L_1/2),$$ (e)

where

L_1 = thickness of the center plate
L_2 = thickness of the outer plate

Specified temperature at the outer surface

$$T_2(L_1/2 + L_2) = T_o .$$ (f)

(4) Solution. Integrating (a) twice gives

$$T_1(x) = -\frac{q'''}{2k_1} x^2 + Ax + B .$$ (g)

Integration of (b) gives

$$T_2(x) = Cx + D,$$ (h)

where A, B, C and D are constants of integration. Application of the four boundary conditions gives the constants of integration. Solutions (g) and (h), expressed in dimensionless form, become

$$\frac{T_1(x) - T_o}{q'''L_1^2 / k_1} = \frac{1}{8} + \frac{1}{2}\frac{k_1 L_2}{k_2 L_1} - \frac{1}{2}\frac{x^2}{L_1^2},$$ (i)

and

$$\frac{T_2(x) - T_o}{q'''L_1^2 / k_1} = \frac{1}{4}\frac{k_1}{k_2} + \frac{1}{2}\frac{k_1}{k_2}\frac{L_2}{L_1} - \frac{1}{2}\frac{k_1}{k_2}\frac{x}{L_1}.$$ (j)

(5) Checking. *Dimensional check*: Each term in solutions (i) and (j) is dimensionless.

Boundary conditions check: Substitution of solutions (i) and (j) into (c), (d), (e) and (f) shows that they satisfy the four boundary conditions.

Quantitative check: Conservation of energy and symmetry require that half the energy generated in the center plate must leave at the interface $x = L_1 / 2$

$$\frac{L_1}{2}q''' = -k_1 \frac{dT_1(L_1/2)}{dx} .$$ (k)

Substituting (i) into (k) shows that this condition is satisfied. Similarly, half the heat generated in the center plate must be conducted through the outer plate

$$\frac{L_1}{2}q''' = -k_2 \frac{dT_2(L_1/2 + L_2)}{dx} .$$ (l)

Substituting (j) into (l) shows that this condition is also satisfied.

Limiting check: (i) If energy generation vanishes, the composite wall should be at a uniform temperature T_o. Setting $q''' = 0$ in (i) and (j) gives $T_1(x) = T_2(x) = T_o$.

(ii) If the thickness of the center plate vanishes, no heat will be generated and consequently the outer plate will be at uniform temperature T_o. Setting $L_1 = 0$ in eq. (j) gives $T_2(x) = T_o$.

(6) Comments. (i) The solution is characterized by two dimensionless parameters: a geometric parameter L_2 / L_1 and a conductivity parameter k_1 / k_2. (ii) An alternate approach to solving this problem is to consider the outer plate first and note that the heat flux at the interface is equal to half the energy generated in the center plate. Thus the outer plate has two known boundary conditions: a specified flux at $x = L_1 / 2$ and a specified temperature at $x = (L_1 / 2) + L_2$. The solution gives the temperature distribution in the outer plate. Equality of temperatures at the interface gives the temperature of the center plate at $x = L_1 / 2$. Thus the center plate has two known boundary conditions: an insulated condition at $x = 0$ and a specified temperature at $x = L_1 / 2$.

2.2 Extended Surfaces: Fins

2.2.1 The Function of Fins

We begin with Newton's law of cooling for surface heat transfer by convection

$$q_s = hA_s (T_s - T_\infty) . \tag{2.3}$$

Eq. (2.3) provides an insight as to the options available for increasing surface heat transfer rate q_s. One option is to increase the heat transfer coefficient h by changing the fluid and/or manipulating its motion. A second option is to lower the ambient temperature T_∞. A third option is to increase surface area A_s. This option is exercised in many engineering applications in which the heat transfer surface is "extended" by adding fins. Inspect the back side of your refrigerator where the condenser is usually placed and note the many thin rods attached to the condenser's tube. The rods are added to increase the rate of heat transfer from the tube to the surrounding air and thus avoid using a fan. Other examples include the honeycomb surface of a car radiator, the corrugated surface of a motorcycle engine, and the disks attached to a baseboard radiator.

2.2.2 Types of Fins

Various geometries and configurations are used to construct fins. Examples are shown in Fig. 2.5. Each fin is shown attached to a wall or surface. The end of the fin which is in contact with the surface is called the *base* while the free end is called the *tip*. The term *straight* is used to indicate that the base extends along the wall in a straight fashion as shown in (a) and (b). If the cross-sectional area of the fin changes as one moves from the base towards the tip, the fin is characterized as having a *variable cross-sectional area*. Examples are the fins shown in (b), (c) and (d). A *spine* or a *pin* fin is distinguished by a circular cross section as in (c). A variation of the pin fin is a bar with a square or other cross-sectional geometry. An *annular* or *cylindrical* fin is a disk which is mounted on a tube as shown in (d). Such a disk can be either of uniform or variable thickness.

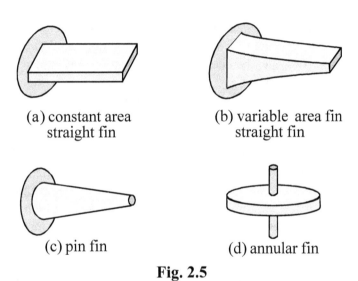

(a) constant area   (b) variable area fin
    straight fin   straight fin

(c) pin fin   (d) annular fin

Fig. 2.5

2.2.3 Heat Transfer and Temperature Distribution in Fins

In the pin fin shown in Fig. 2.6, heat is removed from the wall at the base and is carried through the fin by conduction in both the axial and radial directions. At the fin surface, heat is exchanged with the surrounding fluid by convection. Thus the direction of heat flow is two-dimensional. Examining the

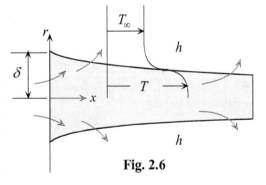

Fig. 2.6

temperature profile at any axial location x we note that temperature variation in the lateral or radial direction is barely noticeable near the center of the fin. However, it becomes more pronounced near its surface. This profile changes as one proceeds towards the tip. Thus temperature distribution is also two-dimensional.

2.2.4 The Fin Approximation

An important simplification made in the analysis of fins is based on the assumption that temperature variation in the lateral direction is negligible. That is, the temperature at any cross section is uniform. This assumption vastly simplifies the mathematical treatment of fins since it not only transforms the governing equation for steady state from partial to ordinary, but also makes it possible to analytically treat fins having irregular cross-sectional areas. The question is, under what conditions can this approximation be made? Let us try to develop a criterion for justifying this assumption. First, the higher the thermal conductivity is the more uniform the temperature will be at a given cross section. Second, a low heat transfer coefficient tends to act as an insulation layer and thus forcing a more uniform temperature in the interior of the cross section. Third, the smaller the half thickness δ is the smaller the temperature drop will be through the cross section. Assembling these three factors together gives a dimensionless ratio, $h\delta/k$, which is called the *Biot number*. Therefore, based on the above reasoning, the criterion for assuming uniform temperature at a given cross section is a Biot number which is small compared to unity. That is

$$\text{Biot number} = Bi = \frac{h\delta}{k} \ll 1. \qquad (2.4)$$

Comparisons between exact and approximate solutions have shown that this simplification is justified when the Biot number is less than 0.1. Note that δ/k represents the internal conduction resistance and $1/h$ is the external convection resistance. Rewriting the Biot number in eq.(2.4) as $Bi = (\delta/k)/(1/h)$ shows that it represents the ratio of the internal and external resistances.

2.2.5   The Fin Heat Equation: Convection at Surface

To determine the rate of heat transfer from fins it is first necessary to obtain the temperature distribution. As with other conduction problems, temperature distribution is determined by solving an appropriate heat equation based on the principle of conservation of energy.   Since conduction in fins is two-dimensional which is modeled mathematically as one-dimensional, it is necessary to formulate the principle of conservation of energy specifically for fins.

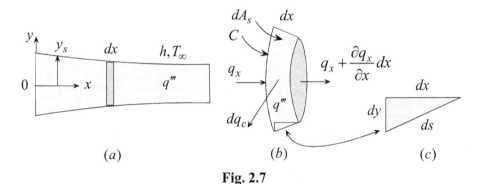

Fig. 2.7

We consider a general case of a variable area fin with volumetric energy generation q''' under transient conditions. The fin exchanges heat with an ambient fluid by convection. The heat transfer coefficient is h and the ambient temperature is T_∞. We select an origin at the base and a coordinate axis x as shown in Fig.2.7. We limit the analysis to stationary material and assume that the Biot number is small compared to unity and thus invoke the fin approximation that the temperature does not vary within a cross section. Since the temperature depends on a single spatial variable x, our starting point should be the selection of an appropriate element. We select a fin slice at location x of infinitesimal thickness dx which encompasses the entire cross section.   This element is enlarged in Fig.2.7b. Conservation of energy for the element requires that

$$\dot{E}_{in} + \dot{E}_g - \dot{E}_{out} = \dot{E},$$
(a)

where

   $\dot{E}$ = rate of energy change within the element

   $\dot{E}_{in}$ = rate of energy added to the element

   $\dot{E}_g$ = rate of energy generation per unit volume

$\dot{E}_{out}$ = rate of energy removed from the element

Energy enters the element by conduction at a rate q_x and leaves at a rate $q_x + (dq_x / dx)dx$. Energy also leaves by convection at a rate dq_c. Note that this rate is infinitesimal because the surface area of the element is infinitesimal. Neglecting radiation, we write

$$\dot{E}_{in} = q_x,\tag{b}$$

and

$$\dot{E}_{out} = q_x + \frac{\partial q_x}{\partial x}dx + dq_c,\tag{c}$$

Substituting (b) and (c) into (a)

$$\dot{E}_g - \frac{\partial q_x}{\partial x}dx - dq_c = \dot{E}.\tag{d}$$

We introduce Fourier's law and Newton's law to eliminate q_x and dq_c, respectively. Thus,

$$q_x = -kA_c\frac{\partial T}{\partial x},\tag{e}$$

and

$$dq_c = h(T - T_\infty)dA_s,\tag{f}$$

where $A_c = A_c(x)$ is the cross-sectional area through which heat is conducted. We should keep in mind that this area may change with x and that it is normal to x. The infinitesimal area dA_s is the surface area of the element where heat is exchanged by convection with the ambient. The volumetric energy generation is given by

$$\dot{E}_g = q'''A_c(x)dx.\tag{g}$$

The rate of energy change within the element is given by (see Section 1.4)

$$\dot{E} = \rho c_p A_c(x)\frac{\partial T}{\partial t}dx.\tag{h}$$

Substituting (e)-(h) into (d)

$$\frac{\partial}{\partial x}\left(kA_c(x)\frac{\partial T}{\partial x}\right)dx - h(T-T_\infty)dA_s + q'''A_c(x)\,dx = \rho c_p A_c(x)\frac{\partial T}{\partial t}\,dx, \quad (2.5a)$$

or, for constant k

$$\frac{\partial^2 T}{\partial x^2} + \frac{1}{A_c(x)}\frac{dA_c}{dx}\frac{\partial T}{\partial x} - \frac{h}{kA_c(x)}(T-T_\infty)\frac{dA_s}{dx} + \frac{q'''}{k} = \frac{1}{\alpha}\frac{\partial T}{\partial t}, \quad (2.5b)$$

where α is thermal diffusivity. The quantities $A_c(x)$, dA_c/dx, and dA_s/dx are determined from the geometry of the fin. Equation (2.5b) is based on the following assumptions: (1) stationary material, (2) constant k, (3) no surface radiation and (4) $Bi \ll 1$.

2.2.6 Determination of $\dfrac{dA_s}{dx}$

Given the geometry of a fin, the coefficient dA_s/dx in eq. (2.5b) can be easily determined. Referring to Fig. 2.7b, the surface area of the element, dA_s, is given by

$$dA_s = C(x)\,ds, \qquad (a)$$

where $C(x)$ is the circumference of the element in contact with the ambient fluid and ds is the slanted length of the element, shown in Fig. 2.7b and (c). Equation (a) assumes that ds is constant along $C(x)$. For a right angle triangle

$$ds = [dx^2 + dy_s^2]^{1/2}, \qquad (b)$$

where y_s is the coordinate normal to x which describes the fin profile. Substituting (b) into (a) and rearranging gives

$$\frac{dA_s}{dx} = C(x)\left[1+\left(\frac{dy_s}{dx}\right)^2\right]^{1/2}, \qquad (2.6a)$$

In most applications, the slope dy_s/dx is small compared to unity and may be neglected. Equation (2.6a) simplifies to

$$\frac{dA_s}{dx} = C(x). \qquad (2.6b)$$

2.2.7 Boundary Conditions

Fin equation (2.5) is second order and therefore it requires two boundary conditions which must be described mathematically at known values of the independent variable x. In the majority of cases, one condition is specified at the base and the other at the tip. Typical conditions may be among those described in Section 1.6.3. Note that convection along the fin surface is accounted for in the differential equation itself and is not a boundary condition.

2.2.8 Determination of Fin Heat Transfer Rate q_f

Once a solution is obtained for the temperature distribution in a fin, the heat transfer rate can be easily determined.   For steady state, no energy generation and no radiation, conservation of energy requires that energy added by conduction at the base be equal to the energy removed at the surface. Referring to Fig. 2.8

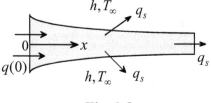

Fig. 2.8

$$q_f = q(0) = q_s ,$$ (a)

where

q_f = fin heat transfer rate

$q(0)$ = heat conducted at base

q_s = heat convected at surface

Therefore, q_f can be determined by one of the following two methods:

(1) *Conduction at the fin base.* Applying Fourier's law at $x = 0$ gives $q(0)$

$$q_f = q(0) = -kA_c(0)\frac{\partial T(0)}{\partial x} .$$ (2.7)

(2) *Convection at the fin surface.* Newton's law is applied at the fin surface to determine the total heat transfer by convection to the ambient fluid. However, since surface temperature varies along the length of the fin, it is necessary to carry out an integration procedure. Thus,

$$q_f = q_s = \int_{A_s} h[T(x) - T_\infty] dA_s .$$ (2.8)

The use of eq.(2.7) is recommended since it is easier to apply than eq.(2.8). However, care should be exercised in using these equations in certain applications. (i) If a fin is attached at both ends, eq.(2.7) must be modified accordingly. (ii) If a fin exchanges heat by convection at the tip, the integral in eq.(2.8) must include this component of heat loss or gain. (iii) If a fin exchanges heat with the surroundings by convection and radiation, eq.(2.7) is still applicable if the solution $T(x)$ takes into account radiation. However, eq.(2.8) must be modified to include heat exchange by radiation.

2.2.9 Steady State Applications: Constant Area Fins with Surface Convection

Mathematically, the simplest fin problems are those with constant cross-sectional area A_C. Note that the area need not be circular. It can be triangular, rectangular, square or any irregular shape as long as it does not vary along its length. An example of such a fin is shown in Fig. 2.9.

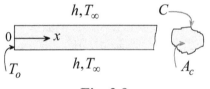

Fig. 2.9

A. Governing Equation.   The heat equation for this class of fins for constant conductivity is given by eq.(2.5b). Since the area A_c is constant it follows that

$$dA_c / dx = 0 .$$ (a)

Furthermore, surface coordinate y_s is constant, that is $dy_s / dx = 0$, eq.(2.6a) gives

$$dA_s / dx = C ,$$ (b)

where the circumference C is constant. Substituting (a) and (b) into eq.(2.5b) and noting that for steady state $\partial T / \partial t = 0$, gives

$$\frac{d^2 T}{dx^2} - \frac{hC}{kA_c}(T - T_\infty) = 0 .$$ (2.9)

Equation (2.9) is written in an alternate form by introducing the following definitions of θ and m

$$\theta = T - T_\infty,$$
(c)

and

$$m^2 = \frac{hC}{kA_c}.$$
(d)

Assuming that T_∞ is constant, substituting (c) and (d) into (2.9), gives

$$\frac{d^2\theta}{dx^2} - m^2\theta = 0.$$
(2.10)

Eq. (2.10) is valid under the following conditions: (1) steady state, (2) no energy generation, (3) constant k, (4) no radiation, (5) $Bi << 1$, (6) constant fin area, (7) stationary fin and (8) constant ambient temperature T_∞.

B. Solution. The simplest solution to eq. (2.10) is obtained if we further assume that the heat transfer coefficient h is constant. This is a reasonable approximation which is usually made. Thus m is constant and eq.(2.10) becomes a second order differential equation with constant coefficients whose solution is (Appendix A)

$$\theta(x) = A_1\exp(mx) + A_2\exp(-mx),$$
(2.11a)

or

$$\theta(x) = B_1\sinh mx + B_2\cosh mx.$$
(2.11b)

where A_1, A_2, B_1 and B_2 are constants of integration. These constants depend on the location of the origin, direction of coordinate axis x and on the boundary conditions.

C. Special Case (i): Finite length fin with specified temperature at the base and convection at the tip. Fig.(2.10) shows a constant area fin of length L. At the tip the fin exchanges heat with the ambient fluid through a heat transfer coefficient h_t which is different from that along the surface. The two boundary conditions for this case are

$$T(0) = T_o,$$
(e)

and

$$-k\frac{dT(L)}{dx} = h_t[T(L) - T_\infty].$$
(f)

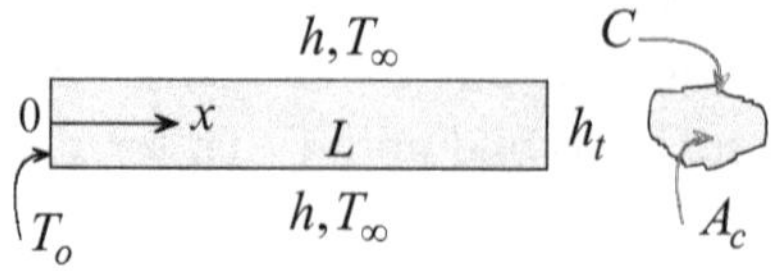

Fig. 2.10

Expressing these boundary conditions in terms of θ, we obtain

$$\theta(0) = \theta_o, \tag{g}$$

and

$$-k\frac{d\theta(L)}{dx} = h_t\theta(L). \tag{h}$$

These two boundary conditions give the constants B_1 and B_2. Equation (2.11b) becomes

$$\frac{\theta(x)}{\theta_o} = \frac{T(x) - T_\infty}{T_o - T_\infty} = \frac{\cosh m(L-x) + (h_t/mk)\sinh m(L-x)}{\cosh mL + (h_t/mk)\sinh mL}. \tag{2.12}$$

Using this solution for the temperature distribution, eq.(2.7) gives the fin heat transfer rate

$$q_f = [hCkA_c]^{1/2}\frac{(T_o - T_\infty)[\sinh mL + (h_t/mk)\cosh mL]}{\cosh mL + (h_t/mk)\sinh mL}. \tag{2.13}$$

D. Special Case (ii): Finite length fin with specified temperature at the base and insulated tip. This is the same as the previous case except that there is no convection at the tip (insulated). Therefore, boundary condition (h) is replaced by

$$\frac{d\theta(L)}{dx} = 0. \tag{i}$$

Note that this boundary condition is a special case of the more general condition where heat is exchanged by convection as described by (h). If we set $h_t = 0$ in (h), we obtain the condition for an insulated surface represented by (i).

To proceed formally we can apply eq.(2.11b) to boundary conditions (g) and (i) and solve for the constants B_1 and B_2. However, a much simpler approach is to set $h_t = 0$ in eq.(2.12) and obtain the solution to the insulated boundary. The result is

$$\frac{\theta(x)}{\theta_o} = \frac{T(x) - T_\infty}{T_o - T_\infty} = \frac{\cosh m(L-x)}{\cosh mL}. \tag{2.14}$$

Similarly, setting $h_t = 0$ in eq.(2.13) gives q_f

$$q_f = [hCkA_c]^{1/2}(T_o - T_\infty)\tanh mL.$$ (2.15)

It is worth reviewing the conditions leading to this simple fin solution. The assumptions are: (1) steady state, (2) constant k, (3) no energy generation, (4) no radiation, (5) $Bi \ll 1$, (6) constant fin area, (7) stationary fin, (8) constant ambient temperature T_∞, (9) constant h, (10) specified base temperature and (11) insulated tip.

2.2.10 Corrected Length L_c

Fins with insulated tips have simpler temperature and heat transfer solutions than fins with convection at the tip. This is evident when we compare eq.(2.12) with eq.(2.14) and eq.(2.13) with eq.(2.15). It is therefore desirable to use the simpler solutions if the error is negligible. Because tip area is small compared with fin surface area, and because the temperature difference $(T - T_\infty)$ is lowest at the tip, heat transfer by convection from the tip is small compared to heat transfer from the surface of the fin. Therefore little error is introduced by assuming that the tip is insulated. In practice this assumption is often made. Nevertheless, to compensate for ignoring heat loss from the tip by using the insulated model, the fin length is increased by a small increment ΔL_c. Thus, the *corrected length* L_c is

$$L_c = L + \Delta L_c.$$ (2.16)

The *correction increment* ΔL_c depends on the geometry of the fin. It is determined by requiring that the increase in surface area due to ΔL_c is equal to the tip area. Thus for a circular fin of radius r_o

$$\pi r_o^2 = 2\pi r_o \Delta L_c.$$

Therefore

$$\Delta L_c = r_o / 2.$$

Similarly, for a square bar of side t

$$\Delta L_c = t / 4.$$

2.2.11 Fin Efficiency η_f

Fin *efficiency* η_f is defined as the ratio of heat transfer from the fin to the maximum heat that can be transferred from the fin, q_{max}. That is

$$\eta_f = \frac{q_f}{q_{max}}.$$ (2.17)

The maximum heat, q_{max}, is an ideal state in which the entire surface of the fin is at a uniform temperature equal to the base temperature. Thus

$$q_{max} = hA_s(T_o - T_\infty),$$

where A_s is the total surface area of the fin. Introducing the above into eq.(2.17)

$$\eta_f = \frac{q_f}{hA_s(T_o - T_\infty)}.$$ (2.18)

2.2.12 Moving Fins

There are applications where a material exchanges heat with the surroundings while moving through a furnace or a channel. Examples include the extrusion of plastics, drawing of wires and sheets and the flow of liquids. Such problems can be modeled as moving fins as long as the criterion for fin approximation is satisfied. Fig. 2.11a shows a sheet being drawn with velocity U through rollers. The sheet exchanges heat with the surroundings by radiation. It also exchanges heat with an ambient fluid by convection. Thus its temperature varies with distance from the rollers.

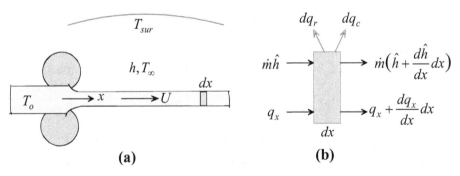

Fig. 2.11

Formulation of the heat equation for moving fins must take into consideration energy transport due to material motion. Consider a constant area fin moving with constant velocity U. For steady state, conservation of energy for the element dx of Fig.2.11b gives

$$q_x + \dot{m}\hat{h} = q_x + \frac{dq_x}{dx}dx + \dot{m}\hat{h} + \dot{m}\frac{d\hat{h}}{dx}dx + dq_c + dq_r, \qquad \text{(a)}$$

where $\hat{h}$ is enthalpy per unit mass, $\dot{m}$ is mass flow rate and q_x, q_c and q_r are heat transfer rates by conduction, convection and radiation, respectively. Mass flow rate is given by

$$\dot{m} = \rho U A_c, \qquad \text{(b)}$$

where A_c is the cross-sectional area and ρ is density. For constant pressure process the change in enthalpy is

$$d\hat{h} = c_p dT, \qquad \text{(c)}$$

where c_p is specific heat. Using Fourier's and Newton's laws, we obtain

$$q_x = -kA_c \frac{dT}{dx}, \qquad \text{(d)}$$

and

$$dq_c = h(T - T_\infty)dA_s, \qquad \text{(e)}$$

where dA_s is the surface area of the element and T_∞ is the ambient temperature. Surface area is expressed in terms of fin circumference C as

$$dA_s = C\,dx. \qquad \text{(f)}$$

Considering a simplified radiation model of a gray body, which is completely enclosed by a much larger surface, radiation heat transfer is given by

$$dq_r = \varepsilon \sigma (T^4 - T_{sur}^4)\,dA_s, \qquad \text{(g)}$$

where ε is emissivity, σ is Stefan-Boltzmann constant and T_{sur} is surroundings temperature. Substituting (b)-(g) into (a) and assuming constant k

$$\frac{d^2 T}{dx^2} - \frac{\rho c_p U}{k}\frac{dT}{dx} - \frac{hC}{kA_c}(T - T_\infty) - \frac{\varepsilon \sigma C}{kA_c}(T^4 - T_{sur}^4) = 0. \quad (2.19)$$

The assumptions leading to eq.(2.19) are: (1) steady state, (2) no energy generation, (3) constant k, (4) constant velocity, (5) constant pressure, (6) gray body, (7) small surface completely enclosed by a much larger surface and (8) $Bi \ll 1$.

2.2.13 Application of Moving Fins

Example 2.4: Moving Fin with Surface Convection

A thin plastic sheet of thickness t and width W is heated in a furnace to temperature T_o. The sheet moves on a conveyor belt traveling with velocity U. It is cooled by convection outside the furnace by an ambient fluid at T_∞. The heat transfer coefficient is h. Assume steady state, Bi<0.1, negligible radiation and no heat transfer from the sheet to the conveyor belt. Determine the temperature distribution in the sheet.

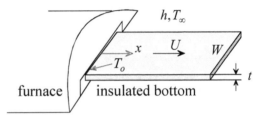

Fig. 2.12

(1) Observations. (i) This is a constant area moving fin problem. Temperature distribution can be assumed one-dimensional. (ii) Heat is exchanged with the surroundings by convection. (iii) The temperature is specified at the outlet of the furnace. (iv) The fin is semi-infinite.

(2) Origin and Coordinates. A rectangular coordinate system is used with the origin at the exit of furnace, as shown in Fig. 2.12.

(3) Formulation.

  (i) Assumptions. (1) One-dimensional, (2) steady state, (3) constant pressure, (4) constant k, (5) constant conveyor speed, (6) constant h, (7) no radiation and (8) $Bi \ll 1$.

  (ii) Governing Equation. Introducing assumption (7) into eq.(2.19)

$$\frac{d^2 T}{dx^2} - \frac{\rho c_p U}{k}\frac{dT}{dx} - \frac{hC}{kA_c}(T - T_\infty) = 0. \qquad (2.20)$$

The cross-sectional area and circumference are given by

$$\frac{d^2 T}{dx^2} - \frac{\rho c_p U}{k}\frac{dT}{dx} - \frac{h(w+2t)}{kWt}(T - T_\infty) = 0$$

$$\frac{d^2 T}{dx^2} - \frac{\rho c_p U}{k}\frac{dT}{dx} - \frac{Th(W+2t)}{kWt} + \frac{T_\infty h(W+2t)}{kWt}$$

$$A_c = Wt,$$ (a)

and

$$C = W + 2t.$$ (b)

Substituting (a) and (b) into eq. (2.20)

$$\frac{d^2T}{dx^2} + 2b\frac{dT}{dx} + m^2T = c,$$ (c)

where the constants b, c and m^2 are defined as

$$b = -\frac{\rho c_p U}{2k}, \quad c = -\frac{h(W + 2t)}{kWt}T_\infty, \quad m^2 = -\frac{h(W + 2t)}{kWt}.$$ (d)

(iii) Boundary Conditions. Since equation (c) is second order with a single independent variable x, two boundary conditions are needed. The temperature is specified at $x = 0$. Thus

$$T(0) = T_o.$$ (e)

Far away from the furnace the temperature is T_∞. Thus, the second boundary condition is

$$T(\infty) = \text{finite}.$$ (f)

(4) Solution. Equation (c) is a linear, second order differential equation with constant coefficients. Its solution is presented in Appendix A. Noting that $b^2 > m^2$, the solution is given by equation (A-6c)

$$T = C_1 \exp(-bx + \sqrt{b^2 - m^2}\,x) + C_2 \exp(-bx - \sqrt{b^2 - m^2}\,x) + \frac{c}{m^2}, \quad (2.21)$$

where C_1 and C_2 are constants of integration. Since $m^2 < 0$, equation (f) requires that

$$C_1 = 0.$$ (g)

Equation (e) gives C_2

$$C_2 = T_o - \frac{c}{m^2}.$$ (h)

Substituting (d), (g) and (h) into (2.21) gives the temperature distribution in the sheet

$$\frac{T(x)-T_\infty}{T_o-T_\infty} = \exp\left[\frac{\rho c_p U}{2k} - \sqrt{(\frac{\rho c_p U}{2k})^2 + \frac{h(W+2t)}{kWt}}\,x\right]. \quad (2.22)$$

(5) Checking. *Dimensional check*: Each term of the exponent in eq.(2.22) must be dimensionless:

$$\frac{\rho(\text{kg/m}^3)c_p(\text{J/kg}-°\text{C})U(\text{m/s})x(\text{m})}{k(\text{W/m}-°\text{C})} = \text{dimensionless}$$

and

$$\left[\frac{h(\text{W/m}^2-°\text{C})(W+2t)(\text{m})}{k(\text{W/m}-°\text{C})W(\text{m})t(\text{m})}\right]^{1/2}x(\text{m}) = \text{dimensionless}$$

Boundary conditions check: Substitution of eq.(2.22) into (e) and (f) shows that the two boundary conditions are satisfied.

Limiting checks: (i) If $h = 0$, no energy can be removed from the sheet after leaving the furnace and therefore its temperature must remain constant. Setting $h = 0$ in eq.(2.22) gives $T(x) = T_o$. (ii) If the velocity U is infinite, the time it takes the fin to reach a distance x vanishes and therefore the energy removed also vanishes. Thus at $U = \infty$ the temperature remains constant. Setting $U = \infty$ in eq.(2.22) gives $T(x) = T_o$.

(6) Comments. (i) The temperature decays exponentially with distance x. (ii) Sheet motion has the effect of slowing down the decay. In the limit, $U = \infty$, no decay takes place and the temperature remains constant.

2.2.14 Variable Area Fins

An example of a variable area fin is the cylindrical or annular fin shown in Fig.2.5d and Fig. 2.13. The conduction area in this fin increases as r is increased. In applications where weight is a factor, tapering a fin reduces its weight.

Mathematically, fins with variable cross-sectional areas are usually governed by differential equations having variable coefficients. In this section we will present the derivation of the fin equation for two specific variable area fins. Solutions to certain ordinary differential equations with variable coefficients will be outlined in Sections 2.3 and 2.4.

Case (i): The annular fin. We wish to formulate the governing equation for temperature distribution for the fin shown in Fig. 2.13. For a constant thermal conductivity fin the starting point is eq.(2.5b). Since we are dealing with a cylindrical geometry, the independent variable x in eq.(2.5b) is replaced by the radial coordinate r. Assuming no energy generation, eq.(2.5b) becomes

$$\frac{d^2T}{dr^2} + \frac{1}{A_c(r)}\frac{dA_c}{dr}\frac{dT}{dr} - \frac{h}{kA_c(r)}(T - T_\infty)\frac{dA_s}{dr} = 0. \qquad (2.23)$$

Heat is conducted in the radial direction and convected from the upper and lower surfaces of the fin. The heat transfer coefficient is h and the ambient temperature is T_∞. To determine $A_c(r)$ we select a circular element (ring) of radius r, thickness dr and height t. The area through which heat is conducted is normal to r, given by

$$A_c(r) = 2\pi rt . \qquad (a)$$

Differentiating (a)

$$\frac{dA_c}{dr} = 2\pi t . \qquad (b)$$

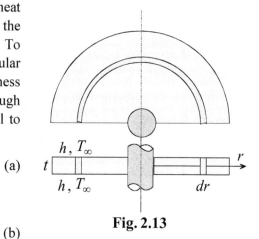

h, T_∞

h, T_∞

t

r

dr

Fig. 2.13

The term dA_s/dr is given by eq.(2.6a)

$$\frac{dA_s}{dr} = C(r)\left[1 + \left(\frac{dy_s}{dr}\right)^2\right]^{1/2}, \qquad (2.6a)$$

where y_s is the coordinate describing the surface. For a disk of uniform thickness, y_s is constant and $dy_s/dr = 0$. The circumference $C(r)$ of the element which is in contact with the fluid is given by

$$C(r) = 2(2\pi r) . \qquad (c)$$

Thus

$$\frac{dA_s}{dr} = 2(2\pi r) . \qquad (d)$$

The factor 2 in (c) is introduced to account for heat convected along the top and bottom surfaces. If the disk is insulated along one surface, this factor will be unity. Substituting (a), (b) and (d) into eq.(2.23) and rearranging, we obtain

$$\frac{d^2T}{dr^2} + \frac{1}{r}\frac{dT}{dr} - (2h/kt)(T - T_\infty) = 0. \tag{2.24}$$

The coefficient of the second term, $1/r$, is what makes eq. (2.24) a differential equation with variable coefficients.

Case (ii): The straight triangular fin. This fin is shown in Fig. 2.14. For convenience, the origin is placed at the tip. The length of the fin is L and its thickness at the base is t. The fin extends a distance W in the direction normal to the paper. The triangular end surfaces are assumed insulated. The heat transfer coefficient is h and the ambient temperature is T_∞. We select an element dx as shown. The conduction area is

$$A_c = 2W y_s(x), \qquad \text{(e)}$$

where $y_s(x)$ is half the element height given by

$$y_s = (x/L)(t/2). \qquad \text{(f)}$$

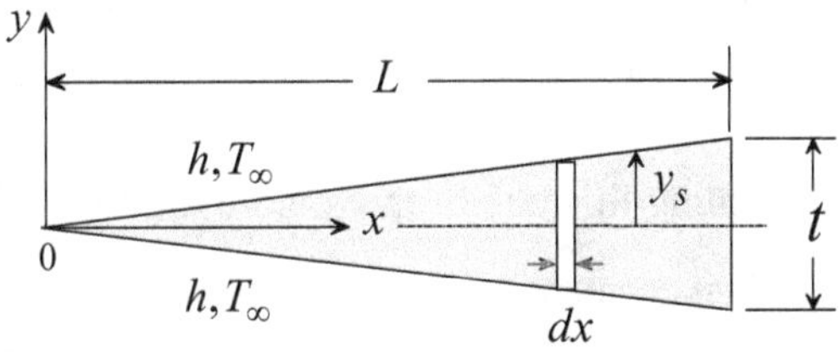

Therefore A_c becomes

$$A_c = (Wt/L)x, \qquad \text{(g)}$$

It follows that

$$\frac{dA_c}{dx} = \frac{Wt}{L}. \tag{h}$$

Fig. 2.14

The geometric factor dA_s/dx is given by eq. (2.6a)

$$\frac{dA_s}{dx} = 2W[1 + (dy_s/dx)^2]^{1/2} = 2W[1 + (t/2L)^2]^{1/2}. \tag{i}$$

Substituting (g), (h) and (i) into eq. (2.5b), we obtain

$$\frac{d^2T}{dx^2} + \frac{1}{x}\frac{dT}{dx} - (2hL/kt)[(1+(t/2L)^2]^{1/2}\frac{1}{x}(T-T_\infty) = 0. \quad (2.25)$$

Note that the second and third terms in eq. (2.25) have variable coefficients.

2.3 Bessel Differential Equations and Bessel Functions

2.3.1 General Form of Bessel Equations

A class of linear ordinary differential equations with variable coefficients is known as *Bessel differential equations*. Such equations are encountered in certain variable area fins and in multi-dimensional conduction problems. Equations (2.24) and (2.25) are typical examples. A general form of Bessel differential equations is given by [1]

$$x^2\frac{d^2y}{dx^2} + \left[(1-2A)x - 2Bx^2\right]\frac{dy}{dx} +$$

$$\left[c^2D^2x^{2C} + B^2x^2 - B(1-2A)x + A^2 - C^2n^2\right]y = 0. \quad (2.26)$$

Examining this equation we note the following:

(1) It is a linear, second order ordinary differential equation with variable coefficients. That is, the coefficients of the dependent variable y and its first and second derivatives are functions of the independent variable x.

(2) A, B, C, D, and n are constants. Their values vary depending on the equation under consideration. Thus, eq. (2.26) represents a class of many Bessel differential equations.

(3) n is called the *order* of the differential equation.

(4) D can be real or imaginary.

2.3.2 Solutions: Bessel Functions

The general solution to eq. (2.26) can be constructed in the form of infinite power series. Since eq. (2.26) is second order, two linearly independent solutions are needed. The form of the solution depends on the constants n and D. There are four possible combinations:

(1) *n* is zero or positive integer, *D* is real. The solution is

$$y(x) = x^A \exp(Bx)[C_1 J_n(Dx^C) + C_2 Y_n(Dx^C)], \qquad (2.27)$$

where

C_1, C_2 = constants of integration

$J_n(Dx^C)$ = *Bessel function of order n of the first kind*

$Y_n(Dx^C)$ = *Bessel function of order n of the second kind*

Note the following:

(i) The term (Dx^C) is the argument of the Bessel function.

(ii) Values of Bessel functions are tabulated. See, for example, Reference [2].

(2) *n* is neither zero nor a positive integer, *D* is real. The solution is

$$y(x) = x^A \exp(Bx)[C_1 J_n(Dx^C) + C_2 J_{-n}(Dx^C)]. \qquad (2.28)$$

(3) *n* is zero or positive integer, *D* is imaginary. The solution is

$$y(x) = x^A \exp(Bx)[C_1 I_n(px^C) + C_2 K_n(px^C)], \qquad (2.29)$$

where

$p = D/i$, where i is imaginary, $i = \sqrt{-1}$

$I_n(px^C)$ = *modified Bessel function of order n of the first kind*

$K_n(px^C)$ = *modified Bessel function of order n of the second kind*

(4) *n* is neither zero nor a positive integer, *D* is imaginary. The solution is

$$y(x) = x^A \exp(Bx)[C_1 I_n(px^C) + C_2 I_{-n}(px^C)]. \qquad (2.30)$$

Example: To illustrate how the above can be applied to obtain solutions to Bessel differential equations, consider the annular fin of Section 2.2.14. The heat equation for this fin is

$$\frac{d^2T}{dr^2} + \frac{1}{r}\frac{dT}{dr} - (2h/kt)(T - T_\infty) = 0. \qquad (2.24)$$

Although this is a second order linear differential equation with variable coefficients, it may not be a Bessel equation, unless it is a special case of eq. (2.26). Thus the first step in seeking a solution is to compare eq. (2.24) with eq. (2.26). Since the comparison is carried out term by term, eq. (2.24) is multiplied through by r^2 so that its first term is identical to that of eq. (2.26). Next, eq. (2.24) is transformed into a homogeneous equation to match eq. (2.26). This is done by introducing a new temperature variable θ, defined as

$$\theta = T - T_\infty . \tag{a}$$

Substituting (a) into eq. (2.24) and multiplying through by r^2

$$r^2 \frac{d^2\theta}{dr^2} + r\frac{d\theta}{dr} - \beta^2 r^2 \theta = 0, \tag{b}$$

where $\beta^2 = (2h/kt)$. Comparing the coefficients of (b) with their counterparts in eq. (2.26) gives

$$A = B = 0, \quad C = 1, \quad D = \beta i, \quad n = 0 .$$

Since $n = 0$ and D is imaginary, the solution is given by eq. (2.29)

$$\theta(r) = T(r) - T_\infty = C_1 I_0(\beta r) + C_2 K_0(\beta r) . \tag{c}$$

2.3.3 Forms of Bessel Functions

The Bessel functions $J_n, Y_n, J_{-n}, I_n, I_{-n}, K_n$ and K_{-n} are symbols representing different infinite power series. The series form depends on n. For example, $J_n(mx)$ represents the following infinite power series

$$J_n(mx) = \sum_{k=0}^{\infty} \frac{(-1)^k (mx/2)^{2k+n}}{k!\Gamma(k+n+1)}, \tag{2.31}$$

where Γ is the *Gamma function*. More details on Bessel functions are found in the literature [2-4].

2.3.4 Special Closed-form Bessel Functions: $n = \dfrac{\text{odd integer}}{2}$

Bessel functions for $n = 1/2$ take the following forms:

$$J_{1/2}(x) = \sqrt{\frac{2}{\pi x}} \sin x, \tag{2.32}$$

$$J_{-1/2}(x) = \sqrt{\frac{2}{\pi x}} \cos x. \tag{2.33}$$

Bessel functions of order 3/2, 5/2, 7/2, ..., are determined from eqs. (2.32) and (2.33), and the following recurrence formula

$$J_{k+1/2}(x) = \frac{2k-1}{x} J_{k-1/2}(x) - J_{k-3/2}(x) \qquad k = 1, 2, 3, ... \tag{2.34}$$

Similarly, modified Bessel functions for $n = 1/2$ take the following forms:

$$I_{1/2}(x) = \sqrt{\frac{2}{\pi x}} \sinh x, \tag{2.35}$$

$$I_{-1/2}(x) = \sqrt{\frac{2}{\pi x}} \cosh x. \tag{2.36}$$

Modified Bessel functions of order 3/2, 5/2, 7/2..., are determined from equations (2.35) and (2.36) and the following recurrence formula

$$I_{k+1/2}(x) = -\frac{2k-1}{x} I_{k-1/2}(x) + I_{k-3/2}(x) \qquad k = 1, 2, 3, ... \tag{2.37}$$

2.3.5   Special Relations for $n = 1, 2, 3, ...$

$$J_{-n}(x) = (-1)^n J_n(x), \tag{2.38a}$$

$$Y_{-n}(x) = (-1)^n Y_n(x), \tag{2.38b}$$

$$I_{-n}(x) = I_n(x), \tag{2.38c}$$

$$K_{-n}(x) = K_n(x). \tag{2.38d}$$

2.3.6 Derivatives and Integrals of Bessel Functions [2,3]

In the following formulas for derivatives and integrals of Bessel functions the symbol Z_n represents certain Bessel functions of order n. Appendix B lists additional integral formulas.

$$\frac{d}{dx}\left[x^n Z_n(mx)\right] = \begin{cases} mx^n Z_{n-1}(mx) & Z = J,Y,I \quad (2.39) \\ -mx^n Z_{n-1}(mx) & Z = K \quad (2.40) \end{cases}$$

$$\frac{d}{dx}\left[x^{-n} Z_n(mx)\right] = \begin{cases} -mx^{-n} Z_{n+1}(mx) & J,Y,K \quad (2.41) \\ mx^{-n} Z_{n+1}(mx) & Z = I \quad (2.42) \end{cases}$$

$$\frac{d}{dx}\left[Z_n(mx)\right] = \begin{cases} mZ_{n-1}(mx) - \dfrac{n}{x}Z_n(mx) & Z = J,Y,I \quad (2.43) \\ -mZ_{n-1}(mx) - \dfrac{n}{x}Z_n(mx) & Z = K \quad (2.44) \end{cases}$$

$$\frac{d}{dx}\left[Z_n(mx)\right] = \begin{cases} -mZ_{n+1}(mx) + \dfrac{n}{x}Z_n(mx) & Z = J,Y,K \quad (2.45) \\ mZ_{n+1}(mx) + \dfrac{n}{x}Z_n(mx) & Z = I \quad (2.46) \end{cases}$$

$$\int x^n Z_{n-1}(mx)\,dx = (1/m)x^n Z_n(mx) \qquad Z = J,Y,I \quad (2.47)$$

$$\int x^{-n} Z_{n+1}(mx)\,dx = -(1/m)x^{-n} Z_n(mx) \qquad Z = J,Y,K \quad (2.48)$$

2.3.7 Tabulation and Graphical Representation of Selected Bessel Functions

Tabulated values of Bessel functions are found in the literature [2]. Appendix C lists values of certain common Bessel functions. Subroutines are available in the IMSL for obtaining numerical values of various Bessel functions. Values for the limiting arguments of $x = 0$ and $x = \infty$ are given in

Table 2.1. Graphical representation of selected Bessel functions are shown in Fig. 2.15.

Table 2.1						
x	$J_0(x)$	$J_n(x)$	$I_0(x)$	$I_n(x)$	$Y_n(x)$	$K_n(x)$
0	1	0	1	0	$-\infty$	∞
∞	0	0	∞	∞	0	0

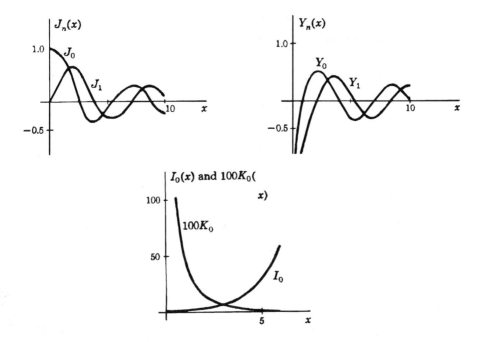

Fig. 2.15 Graphs of selected Bessel functions

2.4 Equidimensional (Euler) Equation

Consider the following second order differential equation with variable coefficients

$$x^2 \frac{d^2 y}{dx^2} + a_1 x \frac{dy}{dx} + a_0 y = 0, \qquad (2.49)$$

where a_0 and a_1 are constant. Note the distinct pattern of the coefficients: x^2 multiplies the second derivative, x multiplies the first derivative and x^0 multiplies the function y. This equation is a special case of a class of equations known as *equidimensional* or *Euler* equation. The more general form of the equidimensional equation has higher order derivatives with coefficients following the pattern of eq. (2.49). For example, the first term of eq. (2.47) takes the form $x^n \dfrac{d^n x}{dx^n}$. The solution to eq. (2.49) depends on the roots of the following equation

$$r_{1,2} = \frac{-(a_1 - 1) \pm \sqrt{(a_1 - 1)^2 - 4a_0}}{2}. \qquad (2.50)$$

There are three possibilities:

(1) If the roots are distinct, the solution takes the form

$$y(x) = C_1 x^{r_1} + C_2 x^{r_2}. \qquad (2.51)$$

(2) If the roots are imaginary as $r_{1,2} = a \pm bi$, where $i = \sqrt{-1}$, the solution is

$$y(x) = x^a \left[C_1 \cos(b \log x) + C_2 \sin(b \log x) \right]. \qquad (2.52)$$

(3) If there is only one root, the solution is

$$y(x) = x^r (C_1 + C_2 \log x). \qquad (2.53)$$

2.5 Graphically Presented Solutions to Fin Heat Transfer Rate [5]

Solutions to many variable area fins have been obtained and published in the literature. Of particular interest is the heat transfer rate q_f for such fins. Mathematical solutions are usually expressed in terms of power series such as Bessel functions. The availability of subroutines in the IMSL makes it convenient to compute q_f. Solutions to common fin geometries have been used to construct dimensionless graphs to determine q_f. However, instead of presenting q_f directly, fin efficiency η_f is plotted. Once η_f is known, q_f can be computed from the definition of η_f in equation (2.18). Fig. 2.16 gives fin efficiency of three types of straight fins and Fig. 2.17 gives fin efficiency of annular fins of constant thickness.

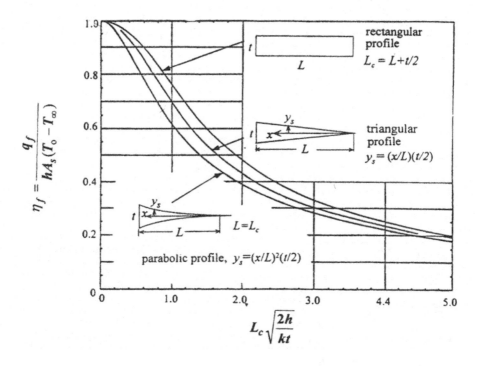

Fig. 2.16 Fin Efficiency of three types of straight fins [5]

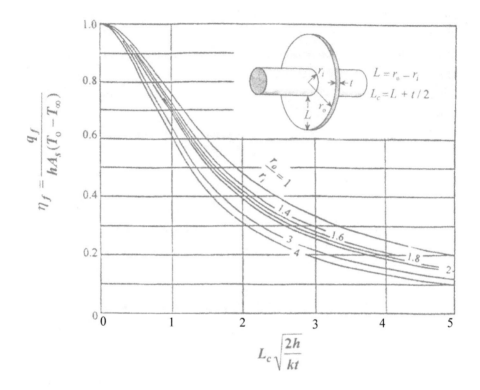

Fig. 2.17 Fin Efficiency of annular fins of constant thickness [5]

REFERENCES

[1] Sherwood, T.K., and Reed, C.E., *Applied Mathematics in Chemical Engineering,* McGraw-Hill, New York, 1939.

[2] Abramowitz, M., and Stegun, I.A., *Handbook of Mathematical Functions with Formulas, Graphs and Mathematical Tables,* U.S. Department of Commerce, National Bureau of Standards, AMS 55, 1964.

[3] Hildebrand, F.B., *Advanced Calculus for Applications*, 2nd edition, Prentice-Hall, Englewood Cliffs, New Jersey, 1976.

[4] McLachlan, N.W., *Bessel Functions for Engineers*, 2nd edition, Clarendon Press, London, 1961.

[5] Gardner, K.A., "Efficiency of Extended Surfaces," *Transactions of ASME*, vol. 67, pp. 621-631, 1945.

PROBLEMS

2.1   Radiation is used to heat a plate of thickness L and thermal conductivity k. The radiation has the effect of volumetric energy generation at a variable rate given by

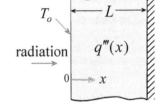

$$q'''(x) = q'''_o\, e^{-bx},$$

where q'''_o and b are constant and x is the distance along the plate. The heated surface at $x = 0$ is maintained at uniform temperature T_o while the opposite surface is insulated. Determine the temperature of the insulated surface.

2.2   Repeat Example 2.3 with the outer plates generating energy at a rate q''' and no energy is generated in the inner plate.

2.3   A plate of thickness L_1 and conductivity k_1 moves with a velocity U over a stationary plate of thickness L_2 and conductivity k_2. The pressure between the two plates is P and the coefficient of friction is μ. The surface of the stationary plate is insulated while that of the moving plate is maintained at constant temperature T_o. Determine the steady state temperature distribution in the two plates.

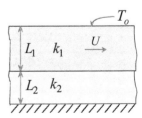

2.4   A very thin electric element is wedged between two plates of
conductivities k_1 and k_2. The element dissipates uniform heat flux
q_o''. The thickness of one plate is L_1 and
that of the other is L_2. One plate is
insulated while the other exchanges heat
by convection. The ambient temperature
is T_∞ and the heat transfer coefficient is
h. Determine the temperature of the
insulated surface for one-dimensional
steady state conduction.

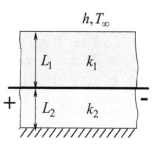

2.5   A thin electric element is sandwiched
between two plates of conductivity k and
thickness L each. The element dissipates a
flux q_o''. Each plate generates energy at a
volumetric rate of q''' and exchanges heat by
convection with an ambient fluid at T_∞.

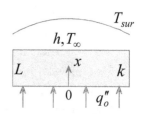

The heat transfer coefficient is h. Determine the temperature of the
electric element.

2.6   One side of a plate is heated with uniform flux q_o'' while the other
side exchanges heat by convection and radiation. The surface
emissivity is ε, heat transfer coefficient h, ambient temperature
T_∞, surroundings temperature T_{sur}, plate thickness L and the
conductivity is k. Assume one-dimensional steady state conduction
and use a simplified radiation model. Determine:

[a] The temperature distribution in the plate.

[b] The two surface temperatures for:

$$h = 27\,\text{W/m}^2 - {}^\circ\text{C}, \, k = 15\,\text{W/m} - {}^\circ\text{C},$$
$$L = 0.08\,\text{m}, \, q_o'' = 19{,}500\,\text{W/m}^2,$$
$$T_{sur} = 18\,^\circ\text{C}, T_\infty = 22\,^\circ\text{C}, \, \varepsilon = 0.95.$$

2.7   A hollow shaft of outer radius R_s and inner
radius R_i rotates inside a sleeve of inner
radius R_s and outer radius R_o. Frictional
heat is generated at the interface at a
flux q_s''.   At the inner shaft surface heat is
added at a flux q_i''. The sleeve is cooled by
convection with a heat transfer coefficient h.
The ambient temperature is T_∞. Determine
the steady state one-dimensional temperature
distribution in the sleeve.

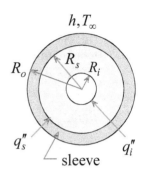

2.8   A hollow cylinder of inner radius r_i and outer radius r_o is heated
with a flux q_i'' at its inner surface. The outside surface exchanges heat
with the ambient and surroundings by convection and radiation. The
heat transfer coefficient is h, ambient temperature T_∞, surroundings
temperature T_{sur} and surface emissivity is ε. Assume steady state
one-dimensional conduction and use a simplified radiation model.
Determine:

[a] The temperature distribution.

[b] Temperature at r_i and r_o , for:

$q_i'' = 35{,}500 \text{ W/m}^2$, $T_\infty = 24 \,^\circ\text{C}$,

$T_{sur} = 14^\circ\text{C}$, $k = 3.8 \text{ W/m}-^\circ\text{C}$,

$r_o = 12 \text{ cm}$, $r_i = 5.5 \text{ cm}$

$\varepsilon = 0.92$ $h = 31.4 \text{ W/m}^2-^\circ\text{C}$.

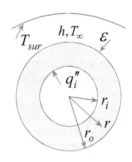

2.9   Radiation is used to heat a hollow sphere of
inner radius r_i,   outer radius r_o and
conductivity k.   Due   to   the   thermal
absorption characteristics of the material
the radiation results in a variable energy
generation rate given by

$$q'''(r) = q_o''' \frac{r^2}{r_o^2},$$

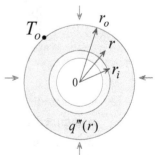

where q_o''' is constant and r is the radial coordinate. The inside surface is insulated and the outside surface is maintained at temperature T_o. Determine the steady state temperature distribution for one-dimensional conduction.

2.10 An electric wire of radius 2 mm and conductivity 398 W/m–°C generates energy at a rate of 1.25×10^5 W/m³. The surroundings and ambient temperatures are 78 °C and 82 °C, respectively. The heat transfer coefficient is 6.5 W/m²–°C and surface emissivity is 0.9. Neglecting axial conduction and assuming steady state, determine surface and centerline temperatures.

2.11 The cross-sectional area of a fin is $A_c(x)$ and its circumference is $C(x)$. Along its upper half surface heat is exchanged by convection with an ambient fluid at T_∞. The heat transfer coefficient is h. Along its lower half surface heat is exchanged by radiation with a surroundings at T_{sur}. Using a simplified radiation model formulate the steady state fin heat equation for this fin.

2.12 A constant area fin of length $2L$ and cross-sectional area A_c generates heat at a volumetric rate q'''. Half the fin is insulated while the other half exchanges heat by convection. The heat transfer coefficient is h and the ambient temperature is T_∞. The base and tip are insulated. Determine the steady state temperature at the mid-section. At what location is the temperature highest?

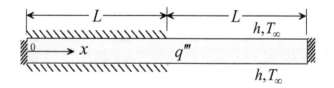

2.13 A constant area fin of length L and cross-sectional area A_c is maintained at T_o at the base and is insulated at the tip. The fin is insulated along a distance b from the base end while heat is exchanged by convection along its remaining surface. The ambient temperature is T_∞ and the heat transfer coefficient is h. Determine the steady state heat transfer rate from the fin.

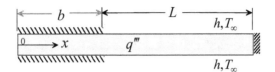

2.14 Heat is removed by convection from an electronic package using a single fin of circular cross section. Of interest is increasing the heat transfer rate without increasing the weight or changing the material of the fin.   It is recommended to use two identical fins of circular cross-sections with a total weight equal to that of the single fin. Evaluate this recommendation using a semi-infinite fin model.

2.15 A plate which generates heat at a volumetric rate q''' is placed inside a tube and cooled by convection. The width of the plate is w and its thickness is δ. The heat transfer coefficient is h and coolant temperature is T_∞. The temperature at the interface between the tube and the plate is T_w. Because of concern that the

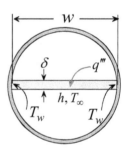

temperature at the center may exceed design limit, you are asked to estimate the steady state center temperature using a simplified fin model. The following data are given:

$$h = 62 \text{ W/m}^2-{}^\circ\text{C}, \quad w = 3 \text{ cm}, \quad \delta = 2.5 \text{ mm}, \quad T_w = 110\,^\circ\text{C},$$
$$T_\infty = 94\,^\circ\text{C}, \quad q''' = 2.5 \times 10^7 \text{ W/m}^3, \quad k = 18 \text{ W/m}-{}^\circ\text{C}.$$

2.16 An electric heater with a capacity P is used to heat air in a spherical chamber. The inside radius is r_i, outside radius r_o and the conductivity is k.   At the inside surface heat is exchanged by convection. The inside heat transfer coefficient is h_i. Heat loss from the outside surface is by

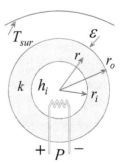

radiation. The surroundings temperature is T_{sur} and the surface emissivity is ε. Assuming one-dimensional steady state conduction, use a simplified radiation model to determine:

[a] The temperature distribution in the spherical wall.

[b] The inside air temperatures for the following conditions:

$$h = 6.5 \text{ W/m}^2-^\circ\text{C}, \ P = 1,500 \text{ W}, \ \varepsilon = 0.81, \ T_{sur} = 18^\circ\text{C},$$

$$k = 2.4 \text{ W/m}-^\circ\text{C}, \quad r_i = 10 \text{ cm}, \quad r_o = 14 \text{ cm}.$$

2.17 A section of a long rotating shaft of radius r_o is buried in a material of very low thermal conductivity. The length of the buried section is L. Frictional heat generated in the buried section can be modeled as surface heat flux. Along the buried surface the radial heat flux is q_r'' and the axial heat flux is q_x''. The exposed surface exchanges heat by convection with the ambient. The heat transfer coefficient is h and the ambient temperature is T_∞. Model the shaft as semi-infinite fin and assume that all frictional heat is conducted through the shaft. Determine the temperature distribution.

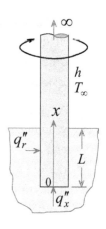

2.18 A conical spine with a length L and base radius R exchanges heat by convection with the ambient. The coefficient of heat transfer is h and the ambient temperature is T_∞. The base is maintained at T_o. Using a fin model, determine the steady state heat transfer rate.

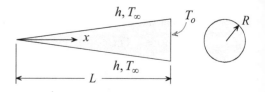

2.19 In many applications it is desirable to reduce the weight of a fin without significantly reducing its heat transfer capacity. Compare the heat transfer rate from two straight fins of identical material. The profile of one fin is rectangular while that of the other is triangular. Both fins have the same length L and base thickness δ. Surface heat transfer is by convection. Base temperature is T_o and ambient temperature is T_∞. The following data are given:

$$h = 28 \text{ W/m}^2-^\circ\text{C}, \ k = 186 \text{ W/m}-^\circ\text{C}, \ L = 18 \text{ cm}, \ \delta = 1\text{cm}$$

2.20   The profile of a straight fin is given by

$$y_s = \delta\,(x/L)^2 ,$$

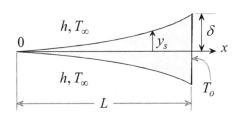

where L is the fin length and δ is half the base thickness. The fin exchanges heat with the ambient by convection. The ambient temperature is T_∞ and the heat transfer coefficient is h. The base temperature is T_o. Assume that $(\delta/L) \ll 1$, determine the steady state fin heat transfer rate.

2.21   A circular disk of radius R_o is mounted on a tube of radius R_i. The profile of the disk is described by $y_s = \delta R_i^2/r^2$, where δ is half the thickness at the tube. The disk exchanges heat with the surroundings by convection. The heat transfer coefficient is h and the ambient temperature is T_∞. The base is maintained at T_o and the tip is insulated. Assume that $4\delta/R_i \ll 1$, use a fin model to determine the steady state heat transfer rate from the disk.

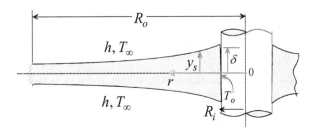

2.22   A very large disk of thickness δ is mounted on a shaft of radius R_i. The shaft rotates with angular velocity ω. The pressure at the interface between the shaft and the disk is P and the coefficient of friction is μ. The disk exchanges heat with the ambient by convection. The heat transfer coefficient is h and the ambient temperature is T_∞. Neglecting heat transfer to the shaft at the interface, use a fin model to determine the interface temperature.

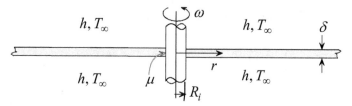

2.23 A circular disk of thickness δ and outer radius R_o is mounted on a shaft of radius R_i. The shaft and disk rotate inside a stationary sleeve with angular velocity ω. Because of friction between the disk and sleeve, heat is generated at a flux q_o''. The disk exchanges heat with the surroundings by convection. The ambient temperature is T_∞. Due to radial variation in the tangential velocity, the heat transfer coefficient varies with radius according to

$$h = h_i (r / R_i)^2 .$$

Assume that no heat is conducted to the sleeve and shaft; use a fin approximation to determine the steady state temperature distribution.

2.24 A specially designed heat exchanger consists of a tube with fins mounted on its inside surface. The fins are circular disks of inner radius R_i and variable thickness given by

$$y_s = \delta (r / R_o)^{3/2} ,$$

where R_o is the tube radius. The disks exchange heat with the ambient fluid by convection. The heat transfer coefficient is h and the fluid temperature is T_∞. The tube surface is maintained at T_o. Modeling each disk as a fin, determine the steady state heat transfer rate. Neglect heat loss from the disk surface at R_i and assume that $(\delta / R_o) \ll 1$.

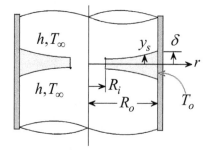

2.25 A disk of radius R_o and thickness δ is mounted on a steam pipe of outer radius R_i. The disk's surface is insulated from $r = R_i$ to $r = R_b$. The remaining surface exchanges heat with the surroundings by convection. The heat transfer coefficient is h and the ambient temperature is T_∞. The temperature of the pipe is T_o and the surface at $r = R_o$ is insulated. Formulate the governing equations and boundary conditions and obtain a solution for the temperature distribution in terms of constants of integration.

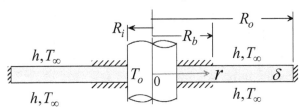

2.26 Consider the moving plastic sheet of Example 2.4. In addition to cooling by convection at the top surface, the sheet is cooled at the bottom at a constant flux q_o''. Determine the temperature distribution in the sheet.

2.27 A cable of radius r_o, conductivity k, density ρ and specific heat c_p moves through a furnace with velocity U and leaves at temperature T_o. Electric energy is dissipated in the cable resulting in a volumetric energy generation rate of q'''. Outside the furnace the cable is cooled by convection. The heat transfer coefficient is h and the ambient temperature is T_∞. Model the cable as a fin and assume that it is infinitely long. Determine the steady state temperature distribution in the cable.

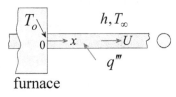

2.28 A cable of radius r_o, conductivity k, density ρ and specific heat c_p moves with velocity U through a furnace and leaves at temperature T_o. Outside the furnace the cable is

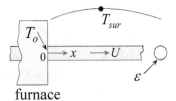

cooled by radiation. The surroundings temperature is T_{sur} and surface emissivity is ε. In certain cases axial conduction can be neglected. Introduce this simplification and assume that the cable is infinitely long. Use a fin model to determine the steady state temperature distribution in the cable.

2.29 A thin wire of diameter d, specific heat c_p, conductivity k and density ρ moves with velocity U through a furnace of length L. Heat is added to the wire in the furnace at a uniform surface flux q_o''. Far away from the inlet of the furnace the wire is at temperature T_i. Assume that no heat is exchanged with the wire before it enters and after it leaves the furnace. Determine the temperature of the wire leaving the furnace.

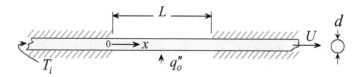

2.30 A wire of diameter d is formed into a circular loop of radius R. The loop rotates with constant angular velocity ω. One half of the loop passes through a furnace where it is heated at a uniform surface flux q_o''. In the remaining half the wire is cooled by convection. The heat transfer coefficient is h and the ambient temperature is T_∞. The specific heat of the wire is c_p, conductivity k and density ρ. Use fin approximation to determine the steady state temperature distribution in the wire.

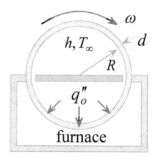

2.31 A wire of radius r_o moves with a velocity U through a furnace of length L. It enters the furnace at T_i where it is heated by convection. The furnace

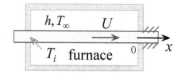

temperature is T_∞ and the heat transfer coefficient is h. Assume that no heat is removed from the wire after it leaves the furnace. Using fin approximation, determine the wire temperature leaving the furnace.

2.32   Hot water at T_o flows downward from a faucet of radius r_o at a rate $\dot{m}$. The water is cooled by convection as it flows. The ambient temperature is T_∞ and the heat transfer coefficient is h. Suggest a model for analyzing the temperature distribution in the water. Formulate the governing equations and boundary conditions.

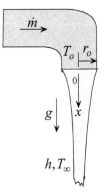

3

TWO-DIMENSIONAL STEADY STATE CONDUCTION

Two-dimensional steady state conduction is governed by a second order partial differential equation. A solution must satisfy the differential equation and four boundary conditions. The method of *separation of variables* [1] will be used to construct solutions. The mathematical tools needed to apply this method will be outlined first. Examples will be presented to illustrate the application of this method to the solution of various problems.

3.1 The Heat Conduction Equation

We begin by examining the governing equation for two-dimensional conduction. Assuming steady state and constant properties, eq. (1.7) becomes

$$\frac{\partial^2 T}{\partial x^2} + \frac{\partial^2 T}{\partial y^2} - \frac{\rho c_p U}{k}\frac{\partial T}{\partial x} + \frac{q'''}{k} = 0. \tag{3.1}$$

Equation (3.1) accounts for the effect of motion and energy generation. For the special case of stationary material and no energy generation, eq. (3.1) reduces to

$$\frac{\partial^2 T}{\partial x^2} + \frac{\partial^2 T}{\partial y^2} = 0. \tag{3.2}$$

The corresponding equation in cylindrical coordinates is

$$\frac{1}{r}\frac{\partial}{\partial r}\left(r\frac{\partial T}{\partial r}\right) + \frac{\partial^2 T}{\partial z^2} = 0. \tag{3.3}$$

Equation (3.1), with $q''' = 0$, eq. (3.2) and eq. (3.3) are special cases of a more general partial differential equation of the form

$$f_2(x)\frac{\partial^2 T}{\partial x^2} + f_1(x)\frac{\partial T}{\partial x} + f_0(x)T + g_2(y)\frac{\partial^2 T}{\partial y^2} + g_1(y)\frac{\partial T}{\partial y} + g_0(y)T = 0.$$

(3.4)

This homogenous, second order partial differential equation with variable coefficients can be solved by the method of *separation of variables.*

3.2 Method of Solution and Limitations

One approach to solving partial differential equations is the *separation of variables* method. The basic idea in this approach is to replace the partial differential equation with sets of ordinary differential equations. The number of such sets is equal to the number of independent variables in the partial differential equation. Thus for steady two-dimensional conduction the heat equation is replaced with two sets of ordinary differential equations. There are two important limitations on this method. (1) It applies to linear equations. An equation is *linear* if the dependent variable and/or its derivatives appear raised to power unity and if their products do not occur. Thus eqs. (3.1)-(3.4) are linear. (2) The geometry of the region must be described by an orthogonal coordinate system. Examples are: squares, rectangles, solid or hollow spheres and cylinders, and sections of spheres or cylinders. This rules out problems involving triangles, trapezoids, and all irregularly shaped objects.

3.3 Homogeneous Differential Equations and Boundary Conditions

Because homogeneity of linear equations plays a crucial role in the application of the method of separation of variables, it is important to understand its meaning. A linear equation is homogenous if it is not altered when the dependent variable in the equation is multiplied by a constant. The same definition holds for boundary conditions. Using this definition to check eq. (3.1), we multiply the dependent variable T by a constant c, take c out of the differentiation signs and divide through by c to obtain

$$\frac{\partial^2 T}{\partial x^2} + \frac{\partial^2 T}{\partial y^2} - \frac{\rho c_p U}{k}\frac{\partial T}{\partial x} + \frac{q'''}{ck} = 0.$$ (a)

Note that the resulting equation (a) is not identical to eq. (3.1). Therefore, eq. (3.1) is non-homogeneous. On the other hand, replacing T by cT in eq. (3.2) does not alter the equation. It follows that eq. (3.2) is homogeneous. To apply this test to boundary conditions, consider the following convection boundary condition

$$-k\frac{\partial T}{\partial x} = h(T - T_\infty).$$ (b)

Replacing T by cT and dividing through by c gives

$$-k\frac{\partial T}{\partial x} = h(T - \frac{T_\infty}{c}).$$ (c)

Since this is different from (b), it follows that boundary condition (b) is non-homogenous. For the special case of $T_\infty = 0$, boundary condition (b) becomes homogeneous.

The separation of variables method can be extended to solve non-homogeneous differential equations and boundary conditions. However, the simplest two-dimensional application is when the equation and three of the four boundary conditions are homogeneous.

3.4 Sturm-Liouville Boundary-Value Problem: Orthogonality [1]

We return to the observation made in Section 3.2 that the basic idea in the separation of variables method is to replace the partial differential equation with sets of ordinary differential equations. One such set belongs to a class of second order ordinary differential equations known as the *Sturm-Liouville boundary-value problem*. The general form of the Sturm-Liouville equation is

$$\frac{d^2\phi_n}{dx^2} + a_1(x)\frac{d\phi_n}{dx} + \left[a_2(x) + \lambda_n^2 a_3(x)\right]\phi_n = 0.$$ (3.5a)

This equation can be written as

$$\frac{d}{dx}\left[p(x)\frac{d\phi_n}{dx}\right] + \left[q(x) + \lambda_n^2 w(x)\right]\phi_n = 0,$$ (3.5b)

where

$$p(x) = e^{\int a_1 dx}, \quad q(x) = a_2 p(x), \quad w(x) = a_3 p(x).$$ (3.6)

The following observations are made regarding eq. (3.5):

(1) The function $w(x)$ plays a special role and is known as the *weighting function.*

(2) Equation (3.5) represents a set of n equations corresponding to n values of λ_n. The corresponding solutions are represented by ϕ_n.

(3) The solutions ϕ_n are known as the *characteristic functions.*

An important property of the Sturm-Liouville problem, which is invoked in the application of the method of separation of variables, is called *orthogonality.* Two functions, $\phi_n(x)$ and $\phi_m(x)$, are orthogonal in the range (a,b) with respect to a weighting function $w(x)$, if

$$\int_a^b \phi_n(x)\,\phi_m(x)\,w(x)\,dx = 0 \qquad n \neq m. \qquad (3.7)$$

The characteristic functions of the Sturm-Liouville problem are orthogonal if the functions $p(x)$, $q(x)$ and $w(x)$ are real, and if the boundary conditions at $x = a$ and $x = b$ are homogenous of the form

$$\phi_n = 0, \qquad (3.8a)$$

$$\frac{d\phi_n}{dx} = 0, \qquad (3.8b)$$

$$\phi_n + \beta \frac{d\phi_n}{dx} = 0, \qquad (3.8c)$$

where β is constant. However, for the special case of $p(x) = 0$ at $x = a$ or $x = b$, these conditions can be extended to include

$$\phi_n(a) = \phi_n(b), \qquad (3.9a)$$

and

$$\frac{d\phi_n(a)}{dx} = \frac{d\phi_n(b)}{dx}. \qquad (3.9b)$$

These conditions are known as *periodic* boundary conditions. In heat transfer problems, boundary conditions (3.8a), (3.8b) and (3.8c) correspond to zero temperature, insulated boundary, and surface convection, respectively. Eqs. (3.9a) and (3.9b) represent continuity of temperature and heat flux. They can occur in cylindrical and spherical problems.

3.5 Procedure for the Application of Separation of Variables Method

To illustrate the use of the separation of variables method, the following example outlines a systematic procedure for solving boundary-value problems in conduction.

Example 3.1: Conduction in a Rectangular Plate

Consider two-dimensional steady state conduction in the plate shown. Three sides are maintained at zero temperature. Along the fourth side the temperature varies spatially.   Determine the temperature distribution T(x,y) in the plate.

(1) Observations. (i) This is a steady state two-dimensional conduction problem. (ii) Four boundary conditions are needed. (iii) Three conditions are homogenous.

(2) Origin and Coordinates.

Anticipating the need to write boundary conditions, an origin and coordinate axes are selected. Although many choices are possible, some are more suitable than others.   In general, it is convenient to select the origin at the intersection of the two

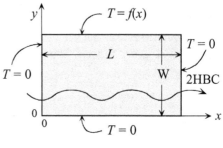

Fig. 3.1

simplest boundary conditions and to orient the coordinate axes such that they are parallel to the boundaries of the region under consideration. This choice results in the simplest solution form and avoids unnecessary algebraic manipulation. Therefore, the origin and coordinate axes are selected as shown in Fig. 3.1.

(3) Formulation.

   (i) Assumptions. (1) Two-dimensional, (2) steady, (3) no energy generation and (4) constant thermal conductivity.

   (ii) Governing Equations. The heat equation for this case is given by eq. (3.2)

$$\frac{\partial^2 T}{\partial x^2} + \frac{\partial^2 T}{\partial y^2} = 0 . \qquad (3.2)$$

(iii) Independent Variable with Two Homogeneous Boundary Conditions. The variable, or direction, having two homogeneous boundary conditions is identified by examining the four boundary conditions. In this example, the boundaries at $x = 0$ and $x = L$ have homogeneous conditions. Thus, the x-variable has two homogeneous boundary conditions. Because this plays a special role in the solution, it is identified with a wavy arrow and marked "2HBC" in Fig. 3.1.

(iv) Boundary Conditions. The required four boundary conditions are listed in the following order: starting with the two boundaries in the variable having two homogeneous conditions and ending with the non-homogeneous condition. Thus, the first three conditions in this order are homogeneous and the fourth is non-homogeneous. Therefore, we write

(1) $T(0,y) = 0$,   homogeneous

(2) $T(L,y) = 0$,   homogeneous

(3) $T(x,0) = 0$,   homogeneous

(4) $T(x,W) = f(x)$,   non-homogeneous

(4) Solution.

(i) Assumed Product Solution. We seek a solution $T(x,y)$ which satisfies eq. (3.2) and the four boundary conditions. We assume that a solution can be constructed in the form of a product of two functions: one which depends on x only, $X(x)$, and one which depends on y only, $Y(y)$. Thus

$$T(x,y) = X(x)\,Y(y). \tag{a}$$

The problem becomes one of finding these two functions. Since a solution must satisfy the differential equation, we substitute (a) into eq. (3.2)

$$\frac{\partial^2[X(x)Y(y)]}{\partial x^2} + \frac{\partial^2[X(x)Y(y)]}{\partial y^2} = 0, \tag{b}$$

or

$$Y(y)\frac{d^2X}{dx^2} + X(x)\frac{d^2Y}{dy^2} = 0.$$

Separating variables and rearranging, we obtain

$$\frac{1}{X(x)}\frac{d^2X}{dx^2} = -\frac{1}{Y(y)}\frac{d^2Y}{dy^2}.$$ (c)

The left side of (c) is a function of x only, $F(x)$, and the right side is a function of y only, $G(y)$. That is

$$F(x) = G(y).$$

Since x and y are independent variables, that is one can be changed independently of the other, the two functions must be equal to a constant. Therefore (c) is rewritten as

$$\frac{1}{X(x)}\frac{d^2X}{dx^2} = -\frac{1}{Y(y)}\frac{d^2Y}{dy^2} = \pm \lambda_n^2,$$ (d)

where λ_n^2 is known as the *separation constant*.   The following observations are made regarding equation (d):

(1) The separation constant is squared. This is for convenience only. As we shall see later, the solution will be expressed in terms of the square root of the constant chosen in (d). Thus by starting with the square of a constant we avoid writing the square root sign later.

(2) The constant has a plus and minus sign. We do this because both possibilities satisfy the condition $F(x) = G(y)$. Although only one sign is correct, at this stage we do not know which one. This must be resolved before we can proceed with the solution.

(3) The constant has a subscript n which is introduced to emphasize that many constants may exist that satisfy the condition $F(x) = G(y)$. The constants $\lambda_1, \lambda_2, ... \lambda_n$, are known as *eigenvalues* or *characteristic values*. The possibility that one of the constants is zero must also be considered.

(4) Equation (d) represents two sets of equations

$$\frac{d^2X_n}{dx^2} \mp \lambda_n^2 X_n = 0,$$ (e)

and

$$\frac{d^2Y_n}{dy^2} \pm \lambda_n^2 Y_n = 0.$$ (f)

In writing equations (e) and (f), care must be exercised in using the correct sign of the λ_n^2 terms. In (e) it is $\mp$ and in (f) it is $\pm$. Since λ_n denotes many constants, each of these equations represents a set of ordinary differential equations corresponding to all values of λ_n. This is why the subscript n is introduced in X_n and Y_n. The functions X_n and Y_n are known as *eigenfunctions* or *characteristic functions*.

 (ii) Selecting the Sign of the λ_n^2 Terms. To proceed we must first select one of the two signs in equations (e) and (f). To do so, we return to the variable, or direction, having two homogeneous boundary conditions. The rule is to select the plus sign in the equation representing this variable. The second equation takes the minus sign. In this example, as was noted previously and shown in Fig. 3.1, the x-variable has two homogeneous boundary conditions. Therefore, (e) and (f) become

$$\frac{d^2 X_n}{dx^2} + \lambda_n^2 X_n = 0, \tag{g}$$

and

$$\frac{d^2 Y_n}{dy^2} - \lambda_n^2 Y_n = 0. \tag{h}$$

If the choice of the sign is reversed, a trivial solution will result. For the important case of $\lambda_n = 0$, equations (g) and (h) become

$$\frac{d^2 X_0}{dx^2} = 0, \tag{i}$$

and

$$\frac{d^2 Y_0}{dy^2} = 0. \tag{j}$$

 (iii) Solutions to the Ordinary Differential Equations. What has been accomplished so far is that instead of solving a partial differential equation, eq. (3.2), we have to solve ordinary differential equations. Solutions to (g) and (h) are (see Appendix A)

$$X_n(x) = A_n \sin \lambda_n x + B_n \cos \lambda_n x, \tag{k}$$

and

$$Y_n(y) = C_n \sinh \lambda_n y + D_n \cosh \lambda_n y. \tag{l}$$

Solutions to (i) and (j) are

$$X_0(x) = A_0 x + B_0,$$ (m)

and

$$Y_0(y) = C_0 y + D_0.$$ (n)

According to (a), each product $X_n Y_n$ is a solution to eq. (3.2). That is

$$T_n(x, y) = X_n(x)Y_n(y),$$ (o)

and

$$T_0(x, y) = X_0(x)Y_0(y).$$ (p)

are solutions. Since eq. (3.2) is linear, it follows that the sum of all solutions is also a solution. Thus, the complete solution is

$$T(x, y) = X_0(x)Y_0(y) + \sum_{n=1}^{\infty} X_n(x)Y_n(y).$$ (q)

(iv) Application of Boundary Conditions. To complete the solution, the constants $A_0, B_0, C_0, D_0, A_n, B_n, C_n, D_n$ and the characteristic values λ_n must be determined. We proceed by applying the boundary conditions in the order listed in step (3) above. Note that temperature boundary conditions will be replaced by conditions on the product $X_n(x)Y_n(y)$. Note further that each product solution, including $X_0(x)Y_0(y)$, must satisfy the boundary conditions. Therefore, boundary condition (1) gives

$$T_n(0, y) = X_n(0)Y_n(y) = 0,$$

and

$$T_0(0, y) = X_0(0)Y_0(y) = 0.$$

Since the variables $Y_n(y)$ and $Y_0(y)$ cannot vanish, it follows that

$$X_n(0) = 0,$$ (r)

and

$$X_0(0) = 0.$$ (s)

Applying (r) and (s) to solutions (k) and (m) gives

$$B_n = B_0 = 0.$$

Similarly, boundary condition (2) gives

$$T_n(L, y) = X_n(L)Y_n(y) = 0,$$

and

$$T_0(L, y) = X_0(L)Y_0(y) = 0.$$

Since the variables $Y_n(y)$ and $Y_0(y)$ cannot be zero, it follows that

$$X_n(L) = 0,$$

and

$$X_0(L) = 0.$$

Applying (k)

$$X_n(L) = A_n \sin \lambda_n L = 0.$$

Since A_n cannot vanish, this result gives

$$\sin \lambda_n L = 0. \tag{t}$$

The solution to (t) gives λ_n

$$\lambda_n = \frac{n\pi}{L}, \qquad n = 1,2,3.... \tag{u}$$

Similarly, using (m) gives

$$X_0(L) = A_0 = 0.$$

With $A_0 = B_0 = 0$, the solution corresponding to $\lambda_n = 0$ vanishes. Equation (u) is known as the *characteristic equation*. All solutions obtained by the method of separation of variables must include a characteristic equation for determining the characteristic values λ_n.

Application of boundary condition (3) gives

$$T_n(x,0) = X_n(x)Y_n(0) = 0.$$

Since the variable $X_n(x)$ cannot vanish, it follows that

$$Y_n(0) = 0.$$

Using (l) gives

$$D_n = 0.$$

The fourth boundary condition is non-homogenous. According to its order, it should be applied last. Before invoking this condition, it is helpful to examine the result obtained so far. With $A_0 = B_0 = B_n = D_n = 0$, the solutions to $X_n(x)$ and $Y_n(y)$ become

$$X_n(x) = A_n \sin \lambda_n x,$$

and

$$Y_n(y) = C_n \sinh \lambda_n y.$$

Thus, according to (q), the temperature solution becomes

$$T(x, y) = \sum_{n=1}^{\infty} a_n (\sin \lambda_n x)(\sinh \lambda_n y), \qquad (3.10)$$

where $a_n = A_n C_n$. The only remaining unknown is the set of constants a_n. At this point boundary condition (4) is applied to eq. (3.10)

$$T(x, W) = f(x) = \sum_{n=1}^{\infty} (a_n \sinh \lambda_n W) \sin \lambda_n x. \qquad (3.11)$$

Eq. (3.11) cannot be solved directly for the constants a_n because of the variable x and the summation sign. To proceed we invoke *orthogonality*, eq. (3.7).

(v) Orthogonality. Eq. (3.7) is used to determine a_n. We note that the function $\sin \lambda_n x$ in eq. (3.11) is the solution to equation (g). Orthogonality can be applied to eq. (3.11) only if (g) is a Sturm-Liouville equation and if the boundary conditions at $x = 0$ and $x = L$ are homogeneous of the type described by eq. (3.8). Comparing (g) with eq. (3.5a) we obtain

$$a_1(x) = a_2(x) = 0 \text{ and } a_3(x) = 1.$$

Eq. (3.6) gives

$$p(x) = w(x) = 1 \text{ and } q(x) = 0.$$

Since the two boundary conditions on $X_n(x) = \sin \lambda_n x$ are homogeneous of type (3.8a), we conclude that the characteristic functions $\phi_n(x) = X_n(x) = \sin \lambda_n x$ are orthogonal with respect to the weighting function $w(x) = 1$. We are now in a position to apply orthogonality to eq. (3.11) to determine a_n. Multiplying both sides of equation (3.11) by $w(x) \sin \lambda_m x \, dx$ and integrating from $x = 0$ to $x = L$

$$\int_0^L f(x)w(x)\sin \lambda_m x \, dx = \int_0^L \left[\sum_{n=1}^{\infty} (a_n \sinh \lambda_n W) \sin \lambda_n x \right] w(x)(\sin \lambda_m x) \, dx.$$

Interchanging the integration and summation signs and noting that $w(x) = 1$, the above is rewritten as

$$\int_0^L f(x)\sin \lambda_m x \, dx = \sum_{n=1}^{\infty} a_n \sinh \lambda_n W \int_0^L (\sin \lambda_n x)(\sin \lambda_m x) \, dx. \quad (3.12)$$

According to orthogonality, eq. (3.7), the integral on the right side of eq. (3.12) vanishes when $m \neq n$. Thus, all integrals under the summation sign vanish except when $m = n$. Equation (3.12) becomes

$$\int_0^L f(x)\sin \lambda_n(x)dx = a_n \sinh \lambda_n W \int_0^L \sin^2 \lambda_n x \, dx.$$

Solving for a_n and noting that the integral on the right side is equal to $L/2$, we obtain

$$a_n = \frac{2}{L \sinh \lambda_n W} \int_0^L f(x)\sin \lambda_n x \, dx. \quad (3.13)$$

(5) Checking. *Dimensional check*: (i) The arguments of sin and sinh must be dimensionless. According to (u), λ_n is measured in units of $1/m$. Thus $\lambda_n x$ and $\lambda_n y$ are dimensionless. (ii) The coefficient a_n in eq. (3.10) must have units of temperature. Since $f(x)$ in boundary condition (4) represents temperature, it follows from eq. (3.13) that a_n is measured in units of temperature.

Limiting check: If $f(x) = 0$, the temperature should be zero throughout. Setting $f(x) = 0$ in eq. (3.13) gives $a_n = 0$. When this is substituted into eq. (3.10) we obtain $T(x,y) = 0$.

(6) Comments. (i) Application of the non-homogeneous boundary condition gives an equation having an infinite series with unknown coefficients. Orthogonality is used to determine the coefficients. (ii) In applying orthogonality, it is important to identify the characteristic and weighting functions of the Sturm-Liouville problem.

3.6 Cartesian Coordinates: Examples

To further illustrate the use of the separation of variables method, we will consider three examples using Cartesian coordinates.   The examples present other aspects that may be encountered in solving conduction problems. The procedure outlined in Section 3.5 will be followed in an abridged form.

Example 3.2: Semi-infinite Plate with Surface Convection

Consider two-dimensional conduction in the semi-infinite plate shown in Fig. 3.2. The plate exchanges heat by convection with an ambient fluid at T_∞. The heat transfer coefficient is h and the plate width is L. Determine the temperature distribution and the total heat transfer rate from the upper and lower surfaces.

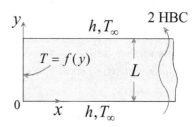

Fig. 3.2

(1) Observations. (i) At $x = \infty$, the plate reaches ambient temperature. (ii) All four boundary conditions are non-homogenous. However, by defining a temperature variable $\theta = T - T_\infty$, three of the four boundary conditions become homogeneous. (iii) For steady state, the rate of heat convected from the two surfaces must equal to the heat conducted at $x = 0$. Thus applying Fourier's law at $x = 0$ gives the heat transfer rate.

(2) Origin and Coordinates. Fig. 3.2 shows the origin and coordinate axes.

(3) Formulation.

(i) **Assumptions.** (1) Two-dimensional, (2) steady, (3) uniform heat transfer coefficient and ambient temperature, (4) constant conductivity and (5) no energy generation.

(ii) **Governing Equations.** Introducing the above assumptions into eq. (1.8) gives

$$\frac{\partial^2 \theta}{\partial x^2} + \frac{\partial^2 \theta}{\partial y^2} = 0, \tag{a}$$

where θ is defined as

$$\theta(x, y) = T(x, y) - T_\infty.$$

(iii) **Independent Variable with Two Homogeneous Boundary Conditions.** In terms of θ, the y-variable has two homogeneous conditions. Thus the y-direction in Fig. 3.2 is marked "2 HBC".

(iv) Boundary Conditions. The order of the four boundary conditions is $\qquad$ *home* $\qquad\qquad\qquad\qquad\qquad\qquad\qquad\qquad$ *hom-*

(1) $k\dfrac{\partial\theta(x,0)}{\partial y} = h\theta(x,0),$ (2) $-k\dfrac{\partial\theta(x,L)}{\partial y} = h\theta(x,L),$

(3) $\theta(\infty, y) = 0,$ *homo*   (4) $\theta(0, y) = f(y) - T_\infty$.   *non-home*

(4) Solution.

(i) Assumed Product Solution. Let

$$\theta(x, y) = X(x)Y(y).$$

Substituting into (a), separating variables and setting the resulting equation equal to constant

$$\frac{1}{X}\frac{d^2 X}{dx^2} = -\frac{1}{Y}\frac{d^2 Y}{dy^2} = \pm\lambda_n^2.$$

Assuming that λ_n is multi-valued, the above gives

$$\frac{d^2 X_n}{dx^2} \mp \lambda_n^2 X_n = 0, \tag{b}$$

and

$$\frac{d^2 Y_n}{dy^2} \pm \lambda_n^2 Y_n = 0. \tag{c}$$

(ii) Selecting the Sign of the λ_n^2 Terms. Since the y-variable has two homogeneous conditions, the $\lambda_n^2 Y_n$ term in (c) takes the positive sign. Thus

$$\frac{d^2 X_n}{dx^2} - \lambda_n^2 X_n = 0, \tag{d}$$

and

$$\frac{d^2 Y_n}{dy^2} + \lambda_n^2 Y_n = 0. \tag{e}$$

For the special case of $\lambda_n = 0$, equations (d) and (e) become

$$\frac{d^2 X_0}{dx^2} = 0, \tag{f}$$

and

$$\frac{d^2 Y_0}{d y^2} = 0. \tag{g}$$

(iii) Solutions to the Ordinary Differential Equations. The solutions to eqs. (d)-(g) are

$$X_n(x) = A_n \exp(-\lambda_n x) + B_n \exp(\lambda_n x), \tag{h}$$

$$Y_n(y) = C_n \sin \lambda_n y + D_n \cos \lambda_n y, \tag{i}$$

$$X_0(x) = A_0 x + B_0, \tag{j}$$

$$Y_0(y) = C_0 y + D_0. \tag{k}$$

Corresponding to each value of λ_n there is a temperature solution $\theta_n(x, y)$. Thus

$$\theta_n(x, y) = X_n(x) Y_n(y), \tag{l}$$

and

$$\theta_0(x, y) = X_0(x) Y_0(y). \tag{m}$$

The complete solution becomes

$$\theta(x, y) = X_0(x) Y_0(y) + \sum_{n=1}^{\infty} X_n(x) Y_n(y). \tag{n}$$

(iv) Application of Boundary Conditions. Boundary condition (1) gives

$$C_n = (h / k \lambda_n) D_n. \tag{o}$$

Applying boundary condition (2)

$$-k \lambda_n [(h / k \lambda_n) D_n \cos \lambda_n L - D_n \sin \lambda_n L] =$$
$$h[(h / k \lambda_n) D_n \sin \lambda_n L + D_n \cos \lambda_n L].$$

This result simplifies to

$$\frac{\lambda_n L}{Bi} - \frac{Bi}{\lambda_n L} = 2 \cot \lambda_n L, \tag{3.14}$$

where $Bi = hL/k$ is the Biot number. Eq. (3.14) is the characteristic equation whose solution gives the characteristic values λ_n. Boundary condition (3) gives

$$A_n \exp(-\infty) + B_n \exp(\infty) = 0.$$

Thus

$$B_n = 0.$$

Boundary condition (3) also requires that $A_0 = B_0 = 0$. Thus, the solution corresponding to $\lambda_n = 0$ vanishes. The temperature solution (n) becomes

$$\theta(x, y) = \sum_{n=1}^{\infty} a_n \exp(-\lambda_n x)\left[(Bi/\lambda_n L)\sin \lambda_n y + \cos \lambda_n y\right]. \quad (3.15)$$

Finally, we apply boundary condition (4) to determine a_n

$$f(y) - T_\infty = \sum_{n=1}^{\infty} a_n \left[(Bi/\lambda_n L)\sin \lambda_n y + \cos \lambda_n y\right]. \qquad (p)$$

(v) **Orthogonality.**   To determine a_n in equation (p) we apply orthogonality. Note that the characteristic functions

$$\phi_n(y) = (Bi/\lambda_n L)\sin \lambda_n y + \cos \lambda_n y,$$

are solutions to equation (e). Comparing (e) with equation (3.5a) shows that it is a Sturm-Liouville equation with

$$a_1 = a_2 = 0 \text{ and } a_3 = 1,$$

Thus eq. (3.6) gives

$$p = w = 1 \text{ and } q = 0. \qquad (q)$$

Since the boundary conditions at $y = 0$ and $y = L$ are homogeneous, it follows that the characteristic functions are orthogonal with respect to $w = 1$. Multiplying both sides of (p) by

$$\phi_m = \left[(Bi/\lambda_m L)\sin \lambda_m y + \cos \lambda_m y\right]dy,$$

integrating from $y = 0$ to $y = L$ and invoking orthogonality, eq. (3.7), we obtain

$$\int_0^L [f(y) - T_\infty][(Bi/\lambda_n L)\sin \lambda_n y + \cos \lambda_n y]dy$$

$$= a_n \int_0^L [(Bi/\lambda_n L)\sin \lambda_n y + \cos \lambda_n y]^2 \, dy.$$

Evaluating the integral on the right side, using eq. (3.14) and solving for a_n gives

$$a_n = \frac{\int_0^L [f(y) - T_\infty][(Bi/\lambda_n L)\sin \lambda_n y + \cos \lambda_n y]dy}{(L/2)\left[1 + (Bi/\lambda_n L)^2 + 2Bi/(\lambda_n L)^2\right]}. \tag{r}$$

To determine the heat transfer rate q, we apply Fourier's law at $x = 0$. Since temperature gradient is two-dimensional, it is necessary to integrate along y. Using the temperature solution, eq. (3.15), and Fourier's law, we obtain

$$q = k\sum_{n=1}^\infty a_n \lambda_n \int_0^L [(Bi/\lambda_n L)\sin \lambda_n y + \cos \lambda_n y]dy \; .$$

Evaluating the integral gives

$$q = k\sum_{n=1}^\infty a_n \left[(Bi/\lambda_n L)(1 - \cos \lambda_n L) + \sin \lambda_n L\right]. \tag{3.16}$$

(5) Checking. *Dimensional check*: Units of q in eq. (3.16) should be W/m. Since Bi and $\lambda_n L$ are dimensionless and a_n is measured in $^\circ C$, it follows that units of ka_n are W/m.

Limiting check: if $f(y) = T_\infty$, no heat transfer takes place within the plate and consequently the entire plate will be at T_∞. Setting $f(y) = T_\infty$ in (r) gives $a_n = 0$. When this result is substituted into eq. (3.15) we obtain $T(x, y) = T_\infty$.

(6) Comments. By introducing the definition $\theta = T - T_\infty$, it was possible to transform three of the four boundary conditions from non-homogeneous to homogeneous.

Example 3.3: Moving Plate with Surface Convection

A semi-infinite plate of thickness 2L moves through a furnace with velocity U and leaves at temperature T_o. The plate is cooled outside the furnace by convection. The heat transfer coefficient is h and the ambient temperature is T_∞. Determine the two-dimensional temperature distribution in the plate.

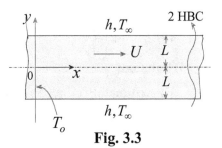

Fig. 3.3

(1) Observations. (i) Temperature distribution is symmetrical about the x-axis. (ii) At $x = \infty$ the plate reaches ambient temperature. (iii) All four boundary conditions are non-homogenous. However, by defining a new variable, $\theta = T - T_\infty$, three of the four boundary conditions become homogeneous.

(2) Origin and Coordinates. Fig. 3.3 shows the origin and coordinate axes.

(3) Formulation.

   (i) Assumptions. (1) Two-dimensional, (2) steady, (3) constant properties, (4) uniform velocity, (5) $q''' = 0$ and (6) uniform h and T_∞.

   (ii) Governing Equations. Defining $\theta(x, y) = T(x, y) - T_\infty$, eq. (1.7) gives

$$\frac{\partial^2 \theta}{\partial x^2} + \frac{\partial^2 \theta}{\partial y^2} - 2\beta \frac{\partial \theta}{\partial x} = 0, \tag{a}$$

where β is defined as

$$\beta = \frac{\rho c_p U}{2k}. \tag{b}$$

   (iii) Independent Variable with Two Homogeneous Boundary Conditions. Only the y-direction can have two homogeneous conditions. Thus it is marked "2 HBC" in Fig. 3.3.

   (iv) Boundary Conditions. Noting symmetry, the four boundary conditions are

(1) $\dfrac{\partial \theta(x,0)}{\partial y} = 0,$ ~hom~ (2) $-k\dfrac{\partial \theta(x,L)}{\partial y} = h\theta(x,L),$ ~home~

(3) $\theta(\infty, y) = 0,$ ~home~ (4) $\theta(0, y) = T_o - T_\infty.$ ~non~

(4) Solution.

(i) Assumed Product Solution.

Assume a product solution $\theta = X(x)Y(y)$. Substituting into (a), separating variables and setting the resulting equation equal to constant gives

$$\frac{d^2 X_n}{dx^2} - 2\beta \frac{dX_n}{dx} \pm \lambda_n^2 X_n = 0, \tag{c}$$

and

$$\frac{d^2 Y_n}{dy^2} \mp \lambda_n^2 Y_n = 0. \tag{d}$$

(ii) Selecting the Sign of the λ_n^2 Terms.
Since the y-variable has two homogeneous conditions, the correct signs in (c) and (d) are

$$\frac{d^2 X_n}{dx^2} - 2\beta \frac{dX_n}{dx} - \lambda_n^2 X_n = 0, \tag{e}$$

and

$$\frac{d^2 Y_n}{dy^2} + \lambda_n^2 Y_n = 0. \tag{f}$$

For $\lambda_n = 0$, equations (e) and (f) become

$$\frac{d^2 X_0}{dx^2} - 2\beta \frac{dX_0}{dx} = 0, \tag{g}$$

and

$$\frac{d^2 Y_0}{dy^2} = 0. \tag{h}$$

(iii) Solutions to the Ordinary Differential Equations.
The solutions to equations (e)-(h) are

$$X_n(x) = [A_n \exp(\beta x + \sqrt{\beta^2 + \lambda_n^2}\, x) + B_n \exp(\beta x - \sqrt{\beta^2 + \lambda_n^2}\, x)], \quad \text{(i)}$$

$$Y_n(y) = C_n \sin \lambda_n y + D_n \cos \lambda_n y, \tag{j}$$

$$X_0(x) = A_0 \exp(2\beta x) + B_0, \tag{k}$$

and

$$Y_0(y) = C_0 + D_0 y. \tag{l}$$

The complete solution becomes

$$\theta(x, y) = X_0(x)Y_0(y) + \sum_{n=1}^{\infty} X_n(x)Y_n(y). \tag{m}$$

(iv) **Application of Boundary Conditions.**   The four boundary conditions will be used to determine the constants of integration and the characteristic values λ_n. Condition (1) gives

$$C_n = D_0 = 0.$$

Boundary condition (2) gives

$$\lambda_n L \tan \lambda_n L = Bi. \tag{n}$$

where Bi is the Biot number. This is the characteristic equation whose roots give λ_n. Boundary condition (2) gives

$$C_0 = 0.$$

Condition (3) gives

$$A_n = 0.$$

With $C_0 = D_0 = 0$, it follows that $X_0 Y_0 = 0$. Thus equation (m) becomes

$$\theta(x, y) = \sum_{n=1}^{\infty} a_n \left[\exp(\beta - \sqrt{\beta^2 + \lambda_n^2}\,)x\right]\cos \lambda_n y. \tag{3.17}$$

Finally, we apply boundary condition (4) to determine a_n

$$T_o - T_\infty = \sum_{n=1}^{\infty} a_n \cos \lambda_n y. \tag{o}$$

(v) Orthogonality. Note that the characteristic functions $\phi_n = \cos \lambda_n y$ are solutions to equation (f). Comparing (f) with equation (3.5a) shows that it is a Sturm-Liouville equation with

$$a_1 = a_2 = 0 \text{ and } a_3 = 1.$$

Thus eq. (3.6) gives

$$p = w = 1 \text{ and } q = 0. \tag{p}$$

Since the boundary conditions at $y = 0$ and $y = L$ are homogeneous, it follows that the characteristic functions $\phi_n = \cos \lambda_n y$ are orthogonal with respect to $w = 1$. Multiplying both sides of (o) by $\cos \lambda_m y dy$, integrating from $y = 0$ to $y = L$ and invoking orthogonality, eq. (3.7), we obtain a_n

$$a_n = \frac{(T_o - T_\infty) \displaystyle\int_0^L \cos \lambda_n y dy}{\displaystyle\int_0^L \cos^2 \lambda_n y dy}.$$

Evaluating the integrals gives

$$a_n = \frac{2(T_o - T_\infty)\sin \lambda_n L}{\lambda_n L + \sin \lambda_n L \cos \lambda_n L}. \tag{q}$$

(5) Checking. *Dimensional check*: Eq. (3.17) requires that units of β be the same as that of λ_n. According to (d), λ_n is measured in (1/m). From the definition of β in (b) we have

$$\beta = \frac{\rho(\text{kg/m}^3)c_p(\text{J/kg-}^\circ\text{C})U(\text{m/s})}{k(\text{W/m-}^\circ\text{C})} = (1/\text{m}).$$

Limiting check: If $T_\infty = T_o$, no heat transfer takes place in the plate and consequently the temperature remains uniform. Setting $T_\infty = T_o$ in (q) gives $a_n = 0$. When this result is substituted into eq. (3.17) we obtain $T(x, y) = T_\infty = T_o$.

(6) Comments. For a stationary plate ($U = \beta = 0$), the solution to this special case should be identical to that of Example 3.2 with $f(y) = T_o$. However, before comparing the two solutions, coordinate transformation must be made since the origin of the coordinate y is not the same for both problems. Also, the definitions of both L and Bi differ by a factor of two.

Example 3.4: Plate with Two insulated Surfaces

Consider two-dimensional conduction in the rectangular plate shown in Fig. 3.4. Two opposite sides are insulated and the remaining two sides are at specified temperatures. Determine the temperature distribution in the plate.

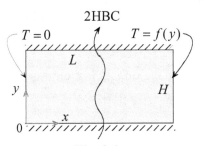

2HBC

Fig. 3.4

(1) Observations. (i) Only one of the four boundary conditions is non-homogeneous. (ii) The y-variable has two homogeneous conditions.

(2) Origin and Coordinates. Fig. 3.4 shows the origin and coordinate axes.

(3) Formulation.

(i) Assumptions. (1) Two-dimensional, (2) steady, (3) constant thermal conductivity and (4) no energy generation.

(ii) Governing Equations. Equation (1.8) gives

$$\frac{\partial^2 T}{\partial x^2} + \frac{\partial^2 T}{\partial y^2} = 0 . \qquad (a)$$

(iii) Independent Variable with Two Homogeneous Boundary Conditions. The y-direction has two homogeneous conditions as indicated in Fig. 3.4.

(iv) Boundary Conditions. The order of the four boundary conditions is

(1) $\dfrac{\partial T(x,0)}{\partial y} = 0$, (2) $\dfrac{\partial T(x,H)}{\partial y} = 0$, (3) $T(0,y) = 0$, (4) $T(L,y) = f(y)$.

(4) Solution.

(i) Assumed Product Solution. Let

$$T(x,y) = X(x)Y(y) .$$

Substituting into (a), separating variables and setting the resulting equation equal to constant

$$\frac{1}{X}\frac{d^2X}{dx^2} = -\frac{1}{Y}\frac{d^2Y}{dy^2} = \pm\lambda_n^2 .$$

Assuming that λ_n is multi-valued, the above gives

$$\frac{d^2X_n}{dx^2} \mp \lambda_n^2 X_n = 0 , \tag{b}$$

and

$$\frac{d^2Y_n}{dy^2} \pm \lambda_n^2 Y_n = 0 . \tag{c}$$

(ii) Selecting the Sign of the λ_n^2 Terms. Since the y-variable has two homogeneous conditions, the $\lambda_n^2 Y_n$ term in (c) takes the positive sign. Thus

$$\frac{d^2X_n}{dx^2} - \lambda_n^2 X_n = 0 , \tag{d}$$

and

$$\frac{d^2Y_n}{dy^2} + \lambda_n^2 Y_n = 0 . \tag{e}$$

For the special case of $\lambda_n = 0$, equations (d) and (e) become

$$\frac{d^2X_0}{dx^2} = 0 , \tag{f}$$

and

$$\frac{d^2Y_0}{dy^2} = 0 . \tag{g}$$

(iii) Solutions to the Ordinary Differential Equations. The solutions to eqs. (d)-(g) are

$$X_n(x) = A_n \sinh\lambda_n x + B_n \cosh\lambda_n x , \tag{h}$$

$$Y_n(y) = C_n \sin\lambda_n y + D_n \cos\lambda_n y , \tag{i}$$

$$X_0(x) = A_0 x + B_0,$$ (j)

and

$$Y_0(y) = C_0 y + D_0.$$ (k)

Corresponding to each value of λ_n there is a temperature solution $T_n(x, y)$. Thus

$$T_n(x, y) = X_n(x) Y_n(y),$$ (l)

and

$$T_0(x, y) = X_0(x) Y_0(y).$$ (m)

The complete solution becomes

$$T(x, y) = X_0(x) Y_0(y) + \sum_{n=1}^{\infty} X_n(x) Y_n(y).$$ (n)

(iv) Application of Boundary Conditions. Boundary condition (1) gives

$$C_n = C_0 = 0.$$

Boundary condition (2) gives the characteristic equation

$$\sin \lambda_n H = 0.$$ (o)

Thus

$$\lambda_n = n\pi / H, \qquad n = 1, 2, ...$$ (p)

Boundary condition (3) gives

$$B_n = B_0 = 0.$$ (q)

The temperature solution (n) becomes

$$T(x, y) = a_0 x + \sum_{n=1}^{\infty} a_n \sinh \lambda_n x \, \cos \lambda_n y.$$ (3.18)

where $a_0 = A_0 D_0$ and $a_n = A_n D_n$. Boundary condition (4) gives

$$f(y) = a_0 L + \sum_{n=1}^{\infty} a_n \sinh \lambda_n L \, \cos \lambda_n y.$$ (r)

(v) Orthogonality. To determine a_0 and a_n using equation (r), we apply orthogonality. Note that the characteristic functions $\cos \lambda_n y$ are solutions to (e). Comparing (e) with equation (3.5a) shows that it is a Sturm-Liouville equation with

$$a_1 = a_2 = 0 \text{ and } a_3 = 1.$$

Thus eq. (3.6) gives

$$p = w = 1 \text{ and } q = 0.$$

Since the boundary conditions at $y = 0$ and $y = H$ are homogeneous, it follows that the characteristic functions are orthogonal with respect to $w(y) = 1$. Multiplying both sides of (r) by $w(y)\cos\lambda_m y dy$, integrating from $y = 0$ to $y = H$ and invoking orthogonality, eq. (3.7), we obtain

$$\int_0^H f(y)\cos\lambda_n y dy = a_0 L \int_0^H \cos\lambda_n y dy + a_n \sinh\lambda_n L \int_0^H \cos^2\lambda_n y\, dy. \quad \text{(s)}$$

However

$$\int_0^H \cos\lambda_n y dy = 0. \quad \text{(t)}$$

It is important to understand why the integral in (t) vanishes. The integrand in (t) represents the product of two characteristic functions: $\cos\lambda_n y$, and 1 which corresponds to $\lambda_n = 0$. Thus, by orthogonality, the integral must vanish. This can also be verified by direct integration and application of (p). Evaluating the last integral in (s) and using (t), equation (s) gives a_n

$$a_n = \frac{2\displaystyle\int_0^H f(y)\cos\lambda_n y dy}{H\sinh\lambda_n L}. \quad \text{(u)}$$

The same procedure is followed in determining a_0. Multiplying both sides of (r) by unity, integrating from $y = 0$ to $y = H$ and applying orthogonality gives a_0

$$a_0 = \frac{1}{HL}\int_0^H f(y)dy. \quad \text{(v)}$$

(5) Checking. *Dimensional check*: Units of a_n in (u) are in $^\circ C$ and units of a_0 in (v) are in $^\circ C/m$. Thus each term in eq. (3.18) is expressed in units of temperature.

Limiting check: if $f(y) = \text{constant} = T_o$, the problem becomes one-dimensional and the temperature distribution linear. The integral in (u) for this case becomes

$$\int_0^H f(y)\cos \lambda_n y\,dy = T_o \int_0^H \cos \lambda_n y\,dy = (T_o / \lambda_n)\sin \lambda_n H = 0.$$

Substituting this result into (u) gives $a_n = 0$. Evaluating the integral in (v) gives $a_0 = T_o / L$. Thus eq. (3.18) reduces to

$$T(x) = (T_o / L)x. \tag{w}$$

This is the one-dimensional temperature distribution in the plate.

(6) Comments. This example shows that the solution corresponding to $\lambda_n = 0$ does not always vanish. Here it represents the steady state one-dimensional solution.

3.7 Cylindrical Coordinates: Examples

Steady state conduction in cylindrical coordinates depends on the variables r, θ, z. In two-dimensional problems the temperature is a function of two of the three variables. We will consider an example in which temperature varies with r and z.

Example 3.5: Radial and Axial Conduction in a Cylinder [2]

Two identical solid cylinders of radius r_o and length L are pressed co-axially with a force F and rotated in opposite directions. The angular velocity is ω and the interface coefficient of friction is μ. The cylindrical surfaces exchange heat by convection. Determine the interface temperature.

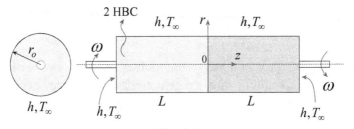

Fig. 3.5

(1) Observations. (i) Temperature distribution is symmetrical with respect to the interface plane. (ii) Interface frictional heat depends on the tangential velocity and tangential force. Since the tangential velocity varies with radius, it follows that the interface heat flux is non-uniform. Thus the interface boundary condition represents a specified heat flux which is a function of radius. (iii) Convection boundary conditions can be made homogeneous by defining a new temperature variable, $\theta = T - T_\infty$.

(2) Origin and Coordinates. Because of axial symmetry, the origin is selected at the interface as shown in Fig. 3.5. Thus, only one cylinder is analyzed.

(3) Formulation.

(i) Assumptions. (1) Two-dimensional, (2) steady, (3) constant conductivity, (4) uniform interface pressure, (5) uniform ambient temperature, (6) uniform heat transfer coefficient, (7) no energy generation and (8) the radius of rod holding each cylinder is small compared to cylinder radius.

(ii) Governing Equations. Defining $\theta = T - T_\infty$, eq. (1.11) gives

$$\frac{1}{r}\frac{\partial}{\partial r}\left(r\frac{\partial \theta}{\partial r}\right) + \frac{\partial^2 \theta}{\partial z^2} = 0. \tag{3.19}$$

(iii) Independent Variable with Two Homogeneous Boundary Conditions. Only the r-variable can have two homogeneous conditions. Fig. 3.5 is marked accordingly.

(iv) Boundary Conditions.

(1) $\theta(0,z)$ = finite,

(2) $-k\dfrac{\partial \theta(r_o,z)}{\partial r} = h\,\theta(r_o,z)$,

(3) $-k\dfrac{\partial \theta(r,L)}{\partial z} = h\,\theta(r,L)$

(4) $-k\dfrac{\partial \theta(r,0)}{\partial z} = \mu\dfrac{F}{\pi r_o^2}\omega r \equiv f(r)$.

Note that the interface boundary condition is obtained by considering an element dr, determining the tangential friction force acting on it, multiplying it by the tangential velocity $2\omega r$ and applying Fourier's law.

(4) Solution.

(i) Assumed Product Solution. Let

$$\theta(r,z) = R(r)Z(z).$$
(a)

Substitute (a) into eq. (3.19), separate variables and set the resulting equation equal to $\pm \lambda_k^2$ gives

$$\frac{1}{r}\frac{d}{dr}\left(r\frac{dR_k}{dr}\right) \pm \lambda_k^2 R_k = 0,$$
(b)

and

$$\frac{d^2 Z_k}{d z^2} \mp \lambda_k^2 Z_k = 0.$$
(c)

(ii) Selecting the Sign of the λ_k^2 Terms. Since the r-variable has two homogenous boundary conditions, $+\lambda_k^2$ is selected in (b). Thus equations (b) and (c) become

$$r^2 \frac{d^2 R_k}{dr^2} + r\frac{dR_k}{dr} + \lambda_k^2 r^2 R_k = 0,$$
(d)

and

$$\frac{d^2 Z_k}{d z^2} - \lambda_k^2 Z_k = 0.$$
(e)

For the special case of $\lambda_k = 0$, the above equations take the form

$$r\frac{d^2 R_0}{dr^2} + \frac{dR_0}{dr} = 0,$$
(f)

and

$$\frac{d^2 Z_0}{d z^2} = 0.$$
(g)

(iii) Solutions to the Ordinary Differential Equations. Equation (d) is a second order ordinary differential equation with variable coefficients. Comparing (d) with Bessel equation (2.26), gives

$$A = B = 0, \quad C = 1, \quad D = \lambda_k, \quad n = 0.$$

Since $n = 0$ and D is real, the solution to (d) is

$$R_k(r) = A_k J_0(\lambda_k r) + B_k Y_0(\lambda_k r).$$
(h)

The solutions to (e), (f) and (g) are

$$Z_k(z) = C_k \sinh \lambda_k z + D_k \cosh \lambda_k z, \qquad (i)$$

$$R_0(r) = A_0 \ln r + B_0, \qquad (j)$$

and

$$Z_0(z) = C_0 z + D_0. \qquad (k)$$

The complete solution is

$$\theta(r,z) = R_0(r)Z_0(z) + \sum_{k=1}^{\infty} R_k(r)Z_k(z). \qquad (l)$$

(iv) Application of Boundary Conditions.   Since $Y_0(0) = -\infty$ and $\ln 0 = -\infty$, boundary condition (1) gives

$$B_k = A_0 = 0.$$

Boundary condition (2) gives

$$-k \frac{d}{dr}[J_0(\lambda_k r)]_{r_o,z} = h J_0(\lambda_k r_o),$$

or

$$(\lambda_k r_o) \frac{J_1(\lambda_k r_o)}{J_0(\lambda_k r_o)} = Bi, \qquad (m)$$

where the Bi is the Biot number, defined as

$$Bi = hr_o / k. \qquad (n)$$

Equation (m) is the characteristic equation whose roots give λ_k. Boundary condition (2) also gives

$$B_0 = 0.$$

Since $A_0 = B_0 = 0$, the solution corresponding to $\lambda_k = 0$ vanishes. Application of boundary condition (3) gives

$$D_k = -C_k \frac{\cosh \lambda_k L + (Bi / \lambda_k L) \sinh \lambda_k L}{\sinh \lambda_k L + (Bi / \lambda_k L) \cosh \lambda_k L}.$$

Substituting into (l)

$$\theta(r,z) = \sum_{k=1}^{\infty} a_k \left[\sinh \lambda_k z - \frac{\cosh \lambda_k L + (Bi / \lambda_k L) \sinh \lambda_k L}{\sinh \lambda_k L + (Bi / \lambda_k L) \cosh \lambda_k L} \cosh \lambda_k z \right] J_0(\lambda_k r).$$

(3.20)

Finally, boundary condition (4) gives

$$f(r) = -k \sum_{k=1}^{\infty} a_k \lambda_k J_0(\lambda_k r).$$

(o)

(v) Orthogonality. Note that $J_0(\lambda_k r)$ is a solution to equation (d) which has two homogeneous boundary conditions in the r-variable. Comparing (d) with eq. (3.5a) shows that it is a Sturm-Liouville equation with

$$a_1 = 1/r, \quad a_2 = 0, \quad a_3 = 1.$$

Thus eq. (3.6) gives

$$p = w = r, \quad q = 0.$$

(p)

It follows that $J_0(\lambda_k r)$ and $J_0(\lambda_i r)$ are orthogonal with respect to the weighting function $w(r) = r$. Multiplying both sides of (o) by $r J_0(\lambda_i r) dr$, integrating from $r = 0$ to $r = r_o$ and applying orthogonality, eq. (3.7), gives a_k

$$a_k = -\frac{\int_0^{r_o} f(r) r J_0(\lambda_k r) \, dr}{k \lambda_k \int_0^{r_o} r J_0^2(\lambda_k r) \, dr} = \frac{2\lambda_k \int_0^{r_o} f(r) r J_0(\lambda_k r) \, dr}{k[(\lambda_k r_o)^2 + (hr_o / k)^2] J_0^2(\lambda_k r_o)}.$$

(q)

The integral in the denominator of equation (q) is obtained from Table 3.1 of Section 3.8. Interface temperature is obtained by setting $z = 0$ in eq. (3.20)

$$\theta(r,0) = T(r,0) - T_{\infty} = -\sum_{k=1}^{\infty} a_k \left[\frac{\cosh \lambda_k L + (Bi / \lambda_k L) \sinh \lambda_k L}{\sinh \lambda_k L + (Bi / \lambda_k L) \cosh \lambda_k L} \right] J_0(\lambda_k r).$$

(r)

(5) Checking. *Dimensional check*: Units of a_k in (q) must be in $^\circ$C. From the definition of $f(r)$ in boundary condition (4) we conclude that $f(r)$ has units of W/m^2. A dimensional check of (q) confirms that a_k has units of $^\circ$C.

Limiting check: If the coefficient of friction is zero, no heat will be generated at the interface and the cylinder will be at the ambient temperature T_∞. Setting $\mu = 0$ in boundary condition (4) gives $f(r) = 0$. Substituting $f(r) = 0$ into (q) gives $a_k = 0$. Thus eq. (3.20) gives $\theta = 0$ or $T(r,z) = T_\infty$. The same check holds for $\omega = 0$.

(6) Comments. (i) It is important to determine the weighing function by comparison with the Sturm-Liouville problem and not assume that it is equal to unity. (ii) The subscript k in the characteristic values and summation index is used instead of n to avoid confusion with the order of Bessel functions which often occurs in cylindrical solutions.

3.8 Integrals of Bessel Functions

The application of orthogonality leads to integrals involving the characteristic functions of the problem. A common integral which occurs when the characteristic functions are Bessel functions takes the form

$$N_n = \int_0^{r_o} r J_n^2(\lambda_k r)\, dr. \qquad (3.21)$$

where N_n is called the *normalization integral*. The value of this integral depends on the form of the homogeneous boundary condition that leads to the characteristic equation. Table 3.1 gives N_n for solid cylinders, corresponding to the three boundary conditions at $r = r_o$ given in eq. (3.8). The Biot number in Table 3.1 is defined as $Bi = hr_o / k$.

Table 3.1 Normalizing integrals for solid cylinders [3]

Boundary condition at r_o	$N_n = \int_0^{r_o} r J_n^2(\lambda_k r)\, dr$
$J_n(\lambda_k r_o) = 0$	$\dfrac{r_o^2}{2\lambda_k^2}\left[\dfrac{dJ_n(\lambda_k r_o)}{dr}\right]^2$
$\dfrac{dJ_n(\lambda_k r_o)}{dr} = 0$	$\dfrac{1}{2\lambda_k^2}\left[(\lambda_k r_o)^2 - n^2\right]J_n^2(\lambda_k r_o)$
$-k\dfrac{dJ_n(\lambda_k r_o)}{dr} = hJ_n(\lambda_k r_o)$	$\dfrac{1}{2\lambda_k^2}\left[(Bi)^2 + (\lambda_k r_o)^2 - n^2\right]J_n^2(\lambda_k r_o)$

3.9 Non-homogeneous Differential Equations

The method of separation of variables can not be directly applied to solve non-homogeneous differential equations. However, a simple modification in the solution procedure makes it possible to extend the separation of variables method to non-homogeneous equations. The following example illustrates the modified approach.

Example 3.6: Cylinder with Energy Generation.

A solid cylinder of radius r_o and length L generates heat at a volumetric rate of q'''. One plane surface is maintained at T_o while the other is insulated. The cylindrical surface is at temperature T_a. Determine the temperature distribution at steady state.

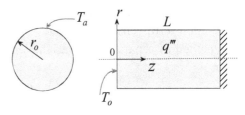

Fig. 3.5

(1) Observations. (i) The energy generation term makes the differential equation non-homogeneous. (ii) Defining a new temperature variable, $\theta = T - T_a$, makes the boundary condition at the cylindrical surface homogeneous. (iii) A cylindrical coordinate system should be used.

(2) Origin and Coordinates.

Fig. 3.5 shows the origin and coordinate axes.

(3) Formulation.

  (i) Assumptions. (1) Two-dimensional, (2) steady, (3) uniform energy generation and (4) constant conductivity.

  (ii) Governing Equations. Defining $\theta = T - T_a$ and applying the above assumptions to eq. (1.11) gives

$$\frac{1}{r}\frac{\partial}{\partial r}\left(r\frac{\partial\theta}{\partial r}\right) + \frac{\partial^2\theta}{\partial z^2} + \frac{q'''}{k} = 0. \tag{3.22}$$

  (iii) Independent Variable with Two Homogeneous Boundary Conditions. The r-variable has two homogeneous conditions. However, this must be re-examined depending on the method used to deal with the non-homogeneous nature of eq. (3.22).

(iv) Boundary Conditions.

(1) $\theta(0,z) = $ finite,   (2) $\theta(r_o,z) = 0$,

(3) $\dfrac{\partial \theta(r,L)}{\partial z} = 0$,   (4) $\theta(r,0) = T_o - T_a$.

(4) Solution.   Equation (3.22) is non-homogeneous due to the heat generation term. Thus we can not proceed directly with the application of the method of separation of variables. Instead, we assume a solution of the form

$$\theta(r,z) = \psi(r,z) + \phi(z). \tag{a}$$

Note that $\psi(r,z)$ depends on two variables while $\phi(z)$ depends on a single variable. Substituting (a) into eq. (3.22)

$$\frac{1}{r}\frac{\partial}{\partial r}\left(r\frac{\partial \psi}{\partial r}\right) + \frac{\partial^2 \psi}{\partial z^2} + \frac{d^2 \phi}{dz^2} + \frac{q'''}{k} = 0. \tag{b}$$

The next step is to split (b) into two equations, one for $\psi(r,z)$ and the other for $\phi(z)$. We select

$$\frac{1}{r}\frac{\partial}{\partial r}\left(r\frac{\partial \psi}{\partial r}\right) + \frac{\partial^2 \psi}{\partial z^2} = 0. \tag{c}$$

Thus, the second part must be

$$\frac{d^2 \phi}{dz^2} + \frac{q'''}{k} = 0. \tag{d}$$

Note the following regarding this procedure:

(1) Equation (c) is a homogeneous partial differential equation.
(2) Equation (d) is a non-homogeneous ordinary differential equation.
(3) The guideline in splitting (b) is to exclude the term q'''/k from the part governed by the partial differential equation, as is done in (c).

Boundary conditions on $\psi(r,z)$ and $\phi(z)$ are obtained by substituting (a) into the four boundary conditions on θ. Thus (a) and condition (1) give

$$\psi(r,z) = \text{finite.} \tag{c-1}$$

Boundary condition (2) gives

$$\psi(r_o,z) = -\phi(z). \tag{c-2}$$

Boundary condition (3) yields

$$\frac{\partial \psi(r,L)}{\partial z} + \frac{d\phi(L)}{dz} = 0.$$

Let

$$\frac{\partial \psi(r,L)}{\partial z} = 0. \tag{c-3}$$

Thus

$$\frac{d\phi(L)}{dz} = 0. \tag{d-1}$$

Finally, boundary condition (4) gives

$$\psi(r,0) + \phi(0) = T_o - T_a.$$

Let

$$\psi(r,0) = 0. \tag{c-4}$$

Thus

$$\phi(0) = T_o - T_a. \tag{d-2}$$

Note that splitting the boundary conditions is guided by the rule that, whenever possible, conditions on the partial differential equation are selected such that they are homogeneous.

Thus the non-homogeneous equation (3.22) is replaced by two equations: (c), which is a homogeneous partial differential equation and (d), which is a non-homogeneous ordinary differential equation. Equation (c) has three homogeneous conditions, (c-1), (c-3) and (c-4) and one non-homogeneous condition, (c-2). The solution to (d) is

$$\phi(z) = -\frac{q'''}{2k}z^2 + Ez + F, \tag{e}$$

where E and F are constants of integration.

(i) **Assumed Product Solution.** To solve partial differential equation (c) we assume a product solution. Let

$$\psi(r,z) = R(r)Z(z). \tag{f}$$

Substituting (f) into (c), separating variables and setting the resulting two equations equal to a constant, $\pm \lambda_k^2$, we obtain

$$\frac{d^2 Z_k}{d z^2} \pm \lambda_k^2 Z_k = 0, \tag{g}$$

and

$$r^2 \frac{d^2 R_k}{dr^2} + r\frac{dR_k}{dr} \mp \lambda_k^2 r^2 R_k = 0. \tag{h}$$

(ii) Selecting the Sign of the λ_k^2 Terms. Since the z-variable has two homogeneous boundary conditions, the plus sign is selected in (g). Thus (g) and (h) become

$$\frac{d^2 Z_k}{d z^2} + \lambda_k^2 Z_k = 0, \tag{i}$$

and

$$r^2 \frac{d^2 R_k}{dr^2} + r\frac{dR_k}{dr} - \lambda_k^2 r^2 R_k = 0. \tag{j}$$

For the special case of $\lambda_k = 0$ the above equations take the form

$$\frac{d^2 Z_0}{d z^2} = 0, \tag{k}$$

and

$$r\frac{d^2 R_0}{dr^2} + \frac{dR_0}{dr} = 0. \tag{l}$$

(iii) Solutions to the Ordinary Differential Equations. The solution to (i) is

$$Z_k(z) = A_k \sin \lambda_k z + B_k \cos \lambda_k z. \tag{m}$$

Equation (j) is a Bessel differential equation with $A = B = n = 0$, $C = 1$, $D = \lambda_n i$. Thus

$$R_k = C_k I_0(\lambda_k r) + D_k K_0(\lambda_k r). \tag{n}$$

The solutions to (k) and (l) are

$$Z_0(z) = A_0 z + B_0, \tag{o}$$

and

$$R_0(r) = C_0 \ln r + D_0. \tag{p}$$

The complete solution to $\theta(r, z)$ becomes

$$\theta(r,z) = \phi(z) + R_0(r)Z_0(z) + \sum_{k=1}^{\infty} R_k(r)Z_k(z). \tag{q}$$

(iv) Application of Boundary Conditions.   Applying boundary condition (c-1) to equation (n) and (p) gives

$$D_k = C_0 = 0.$$

Condition (c-4) applied to (m) and (o) gives

$$B_k = B_0 = 0.$$

Condition (c-3) applied to (m) and (o) gives

$$A_0 = 0,$$

and the characteristic equation for λ_k is given by

$$\cos \lambda_k L = 0,$$

or

$$\lambda_k L = \frac{(2k-1)\pi}{2}, \quad k = 1,2,.... \tag{r}$$

With $A_0 = B_0 = 0$, the solution $R_0 Z_0$ vanishes. The solution to $\psi(r,z)$ becomes

$$\psi(r,z) = \sum_{k=1}^{\infty} a_k I_0(\lambda_k r)\sin \lambda_k z. \tag{s}$$

Returning now to the function $\phi(z)$, conditions (d-1) and (d-2) give the constants E and F

$$E = q'''L/k,$$

and

$$F = T_o - T_a.$$

Substituting into (e) gives

$$\phi(z) = \frac{q''' L^2}{2k} \left[2(z/L) - (z/L)^2 \right] + (T_o - T_a).$$ (t)

Application of condition (c-2) gives

$$-\frac{q''' L^2}{2k} \left[2(z/L) - (z/L)^2 \right] - (T_o - T_a) = \sum_{k=1}^{\infty} a_k I_0(\lambda_k r_o) \sin \lambda_k z.$$ (u)

(v) Orthogonality. Note that the characteristic functions, $\sin \lambda_k z$, in (u) are solutions to equation (i). Comparing (i) with eq. (3.5a) shows that it is a Sturm-Liouville equation with

$$a_1 = a_2 = 0 \text{ and } a_3 = 1.$$

Thus eq. (3.6) gives

$$p = w = 1 \text{ and } q = 0.$$

Since the boundary conditions at $z = 0$ and $z = L$ are homogeneous, it follows that the characteristic functions $\sin \lambda_k z$ are orthogonal with respect to $w = 1$. Multiplying both sides of (u) by $\sin \lambda_i z \, dz$, integrating from $z = 0$ to $z = L$ and invoking orthogonality, eq. (3.7), we obtain

$$\int_0^L \left\{ -\frac{q''' L^2}{2k} \left[2(z/L) - (z/L)^2 \right] - (T_o - T_a) \right\} \sin \lambda_k z \, dz =$$

$$a_k I_0(\lambda_k r_o) \int_0^L \sin^2 \lambda_k z \, dz.$$

Evaluating the integrals and solving for a_k

$$a_k = \frac{-2}{(\lambda_k L) I_0(\lambda_k r_o)} \left[(T_o - T_a) + q''' / k\lambda_k^2) \right].$$ (v)

Thus the temperature solution $\theta(r,z)$ becomes

$$\theta(r,z) = T(r,z) - T_a =$$

$$(T_o - T_a) + \frac{q''' L^2}{2k} \left[2(z/L) - (z/L)^2 \right] + \sum_{k=1}^{\infty} a_k I_0(\lambda_k r) \sin \lambda_k z.$$ (w)

(5) Checking. *Dimensional checks:* Each term in (w) has units of temperature.

Limiting check: If $q''' = 0$ and $T_a = T_o$, physical consideration requires that the cylinder be at uniform temperature. Under this condition equation (v) gives $a_k = 0$. Substituting into (w) gives $T(r,z) = T_a$.

(6) Comments. An alternate solution is to replace (a) with the following:

$$\theta(r,z) = \psi(r,z) + \phi(r).$$

This leads to two homogeneous boundary conditions in the *r*-variable instead of the *z*-variable.

3.10 Non-homogeneous Boundary Conditions:
The Method of Superposition

The separation of variables method can be applied to solve problems with non-homogeneous boundary conditions using the principle of superposition. In this approach, a problem is decomposed into as many simpler problems as the number of non-homogeneous boundary conditions. Each simple problem is assigned a single non-homogeneous boundary condition such that when the simpler problems are added they satisfy the conditions on the original problem. To illustrate this method, consider steady state two-dimensional conduction in the plate shown in Fig. 3.7. Since all four boundary conditions are non-homogeneous, the solutions to four simpler problems, each having a single non-homogeneous condition, are superimposed. The heat equation for constant conductivity is

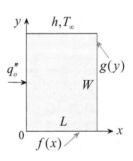

Fig. 3.7

$$\frac{\partial^2 T}{\partial x^2} + \frac{\partial^2 T}{\partial y^2} = 0.$$

(a)

The boundary conditions are

(1) $-k\dfrac{\partial T(0,y)}{\partial x} = q_o''$,   (2) $T(L,y) = g(y)$,

(3) $T(x,0) = f(x)$,   (4) $-k\dfrac{\partial T(x,W)}{\partial y} = h[T(x,W) - T_\infty]$.

We assume that the solution $T(x,y)$ is the sum of the solutions to four problems, given by

$$T(x,y) = T_1(x,y) + T_2(x,y) + T_3(x,y) + T_4(x,y). \qquad \text{(b)}$$

The four solutions, $T_n(x,y)$, $n = 1,2,3,4$, must satisfy equation (a) and the four boundary conditions. Substituting (b) into (a) gives four identical equations

$$\frac{\partial^2 T_n}{\partial x^2} + \frac{\partial^2 T_n}{\partial y^2} = 0 \quad n = 1,2,3,4 \qquad \text{(c)}$$

Substituting (b) into condition (1)

$$-k\frac{\partial T_1(0,y)}{\partial x} - k\frac{\partial T_2(0,y)}{\partial x} - k\frac{\partial T_3(0,y)}{\partial x} - k\frac{\partial T_4(0,y)}{\partial x} = q_o''.$$

We assign the non-homogeneous part of this equation to $T_1(x,y)$, leaving homogeneous conditions for the remaining three problems. Thus

$$-k\frac{\partial T_1(0,y)}{\partial x} = q_o'', \quad \frac{\partial T_2(0,y)}{\partial x} = 0, \quad \frac{\partial T_3(0,y)}{\partial x} = 0, \quad \frac{\partial T_4(o,y)}{\partial x} = 0. \quad \text{(d-1)}$$

Similarly, the remaining three conditions are subdivided to give

$$T_1(L,y) = 0, \quad T_2(L,y) = g(y), \quad T_3(L,y) = 0, \quad T_4(L,y) = 0, \quad \text{(d-2)}$$

$$T_1(x,0) = 0, \quad T_2(x,0) = 0, \quad T_3(x,0) = f(x), \quad T_4(x,0) = 0, \quad \text{(d-3)}$$

$$-k\frac{\partial T_1(x,W)}{\partial y} = hT_1(x,W), \quad -k\frac{\partial T_2(x,W)}{\partial y} = hT_2(x,W),$$

$$-k\frac{\partial T_3(x,W)}{\partial y} = hT_3(x,W), \quad -k\frac{\partial T_4(x,W)}{\partial y} = h[T_4(x,W) - T_\infty].\text{(d-4)}$$

Note that each of the four problems has one non-homogeneous condition, as shown in Fig. 3.8. In solving the simpler problems it may be more convenient to change the location of the origin and/or the direction of coordinates. However, all solutions must be transformed to a common origin and coordinates before they are added together.

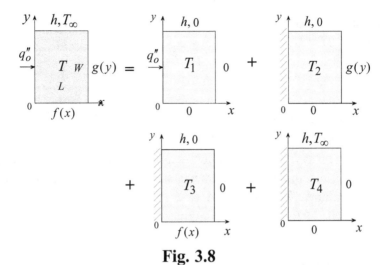

Fig. 3.8

REFERENCES

[1] Hildebrand, F.B., *Advanced Calculus for Applications*, 2nd edition, Prentice-Hall, Englewood Cliffs, New Jersey, 1976.

[2] Arpaci, V.S., *Conduction Heat Transfer*, Addison-Wesley, Reading, Massachusetts, 1966.

[3] Ozisik, M.N., *Heat Conduction*, Wiley, New York, 1979.

PROBLEMS

3.1   A rectangular plate of length L and width W is insulated on sides $x = L$ and $y = W$. The plate is heated with variable flux along $x = 0$. The fourth side is at temperature T_1. Determine the steady state two-dimensional temperature distribution.

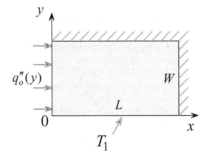

3.2   Heat is generated in a rectangular plate $3L \times H$ at a uniform volumetric rate of q'''. The surface at $(0,y)$ is maintained at T_o. Along surface $(3L,y)$ the plate exchanges heat by convection. The heat transfer coefficient is h and the ambient temperature is T_∞. Surface (x,H) is divided into three equal segments which are maintained at uniform temperatures T_1, T_2, and T_3, respectively. Surface $(x,0)$ is insulated. Determine the two-dimensional steady state temperature distribution.

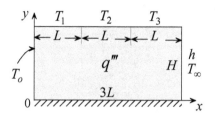

3.3   A rectangular plate $L \times H$ is maintained at uniform temperature T_o along three sides. Half the fourth side is insulated while the other half is heated at uniform flux q_o''. Determine the steady state heat transfer rate through surface $(0,y)$.

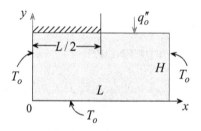

3.4   A semi-infinite plate of width $2L$ is maintained at uniform temperature T_o along its two semi-infinite sides. Half the surface $(0,y)$ is insulated while the remaining half is heated at a flux q_o''. Determine the two-dimensional steady state temperature distribution.

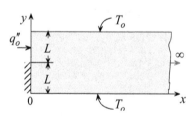

3.5   Consider two-dimensional steady state conduction in a rectangular plate $L \times H$. The sides $(0,y)$ and (x,H) exchange heat by convection. The heat transfer coefficient is h and ambient temperature is T_∞. Side $(x,0)$ is insulated. Half the surface at $x = L$ is cooled at a flux q_1'' while the other half is heated

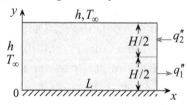

at a flux q_2''. Determine the heat transfer rate along $(0,y)$.

3.6   Radiation strikes one side of a rectangular element $L \times H$ and is partially absorbed. The absorbed energy results in a non-uniform volumetric energy generation given by

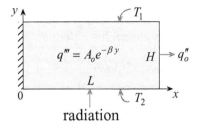

$$q''' = A_o \, e^{-\beta y},$$

where A_o and β are constant. The surface which is exposed to radiation, $(x,0)$, is maintained at T_2 while the opposite surface (x,H) is at T_1. The element is insulated along surface $(0,y)$ and cooled along (L,y) at a uniform flux q_o''. Determine the steady state temperature of the insulated surface.

3.7   An electric strip heater of width $2b$ dissipates energy at a flux q_o''. The heater is symmetrically mounted on one side of a plate of height H and width $2L$. Heat is exchanged by convection along surfaces $(0,y)$ and $(2L,y)$. The ambient temperature is T_∞ and the heat transfer coefficient is h. Surfaces $(x,0)$ and (x,H) are insulated. Determine the temperature of the plate-heater interface for two-dimensional steady state conduction.

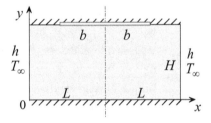

3.8   The cross section of a long prism is a right angle isosceles triangle of side L. One side is heated with uniform flux q_o'' while the second side is maintained at zero temperature. The hypotenuse is insulated. Determine the temperature T_c at the midpoint of the hypotenuse.

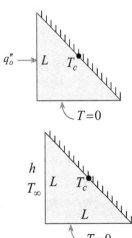

3.9   The cross section of a long prism is a right angle isosceles triangle of side L. One side exchanges heat by convection with the surroundings while the second side is maintained at zero temperature. The heat transfer coefficient is h and ambient temperature is T_∞. The hypotenuse is insulated. Determine the temperature T_c at the mid-point of the hypotenuse.

3.10 By neglecting lateral temperature variation in the analysis of fins, two-dimensional conduction is modeled as a one-dimensional problem. To examine this approximation, consider a semi-infinite plate of thickness $2H$. The base is maintained at uniform temperature T_o. The plate exchanges heat by convection at its semi-infinite surfaces. The heat transfer coefficient is h and the ambient temperature is T_∞. Determine the heat transfer rate at the base.

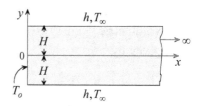

3.11 A very long plate of thickness $2H$ leaves a furnace at temperature T_f. The plate is cooled as it moves with velocity U through a liquid tank. Because of the high heat transfer coefficient, the surface of the plate is maintained at temperature T_o. Determine the steady state two-dimensional temperature distribution in the plate.

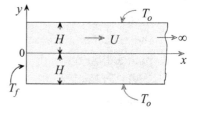

3.12 Liquid metal flows through a cylindrical channel of width H and inner radius R. It enters at T_i and is cooled by convection along the outer surface to T_o. The ambient temperature is T_∞ and heat transfer coefficient is h. The inner surface is insulated. Assume that the liquid metal flows with a uniform velocity U and neglect curvature effect $(H/R \ll 1)$, determine the steady state temperature of the insulated surface.

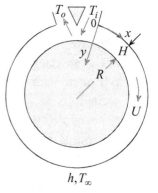

3.13 A solid cylinder of radius r_o and length L is maintained at T_1 at one end. The other end is maintained at T_2 and the cylindrical surface at T_3. Determine the steady state two-dimensional temperature distribution in the cylinder.

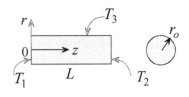

3.14 A solid cylinder of radius r_o and length L is maintained at T_1 at one end and is cooled with uniform flux q_o'' at the other end. The cylindrical surface is maintained at a uniform temperature T_2. Determine the steady state two-dimensional temperature distribution in the cylinder.

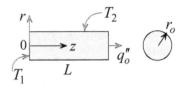

3.15 The volumetric heat generation rate in a solid cylinder of radius r_o and length L varies along the radius according to

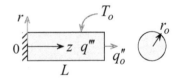

$$q''' = q_o''' r,$$

where q_o''' is constant. One end is insulated while the other end is cooled with uniform flux q_o''. The cylindrical surface is maintained at uniform temperature T_o. Determine the steady state temperature of the insulated surface.

3.16 Heat is added at a uniform flux q_o'' over a circular area of radius b at one end of a solid cylinder of radius r_o and length L. The remaining surface of the heated end and the opposite surface are insulated. The cylindrical surface is maintained at uniform temperature T_o. Determine the steady state two-dimensional temperature distribution.

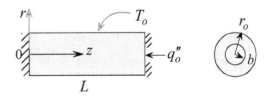

3.17 A disk of radius $r_o / 2$ is pressed co-axially with a force F against one end of a solid cylinder of radius r_o and length L. The disk

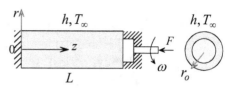

rotates with angular velocity ω. The coefficient of friction between the disk and cylinder is μ. The two ends and the exposed surface of the disk are insulated. Heat is exchanged by convection along the cylindrical surface. The heat transfer coefficient is h and the ambient temperature is T_∞. Determine the temperature of cylinder-disk interface.

3.18 A shaft of radius r_o and length L is maintained at T_1 at one end and T_2 at the other end. The shaft rotates inside a sleeve of length b. Heat is generated at the interface between the

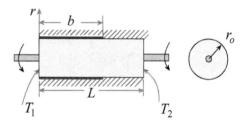

sleeve and the shaft at uniform flux q_o''. The cylindrical surface of the shaft is well insulated. Determine the steady state interface temperature.

3.19 A glass rod of radius r_i rotates co-axially with angular velocity ω inside a hollow cylinder of inner radius r_i, outer radius r_o and length L. Interface

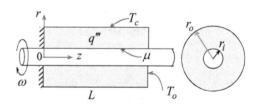

pressure is P and the coefficient of friction is μ. The cylinder generates heat at a volumetric rate of q'''. One end of the cylinder is insulated while the other end is maintained at T_o. The cylindrical surface is at T_c. Neglecting heat conduction through the rod, determine the interface temperature.

3.20 Consider steady state two-dimensional conduction in a hollow half disk. The inner radius is r_i, outer radius r_o and the thickness is δ.

2222222The disk exchanges heat by convection along its two plane surfaces. The heat transfer coefficient is h and the ambient temperature is T_∞. The cylindrical surface at r_i is maintained at T_i and that at r_o at T_o. The two end surfaces at $\theta = 0$ and $\theta = \pi$ are insulated. Determine the steady state temperature distribution in the disk.

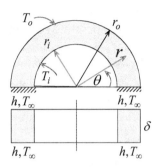

3.21 A hollow cylinder has an inner radius r_i, outer radius r_o and length L. One end is maintained at T_i while the other end is insulated. The cylinder exchanges heat by convection at its inside

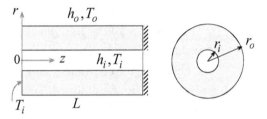

and outside surfaces. The inside ambient temperature is T_i and the inside heat transfer coefficient is h_i. The outside ambient temperature is T_o and the outside heat transfer coefficient is h_o. Determine the two–dimensional steady state temperature distribution in the cylinder.

3.22 In the annular fin model, lateral temperature variation is neglected. Thus, two-dimensional temperature distribution is modeled as one-dimensional conduction.

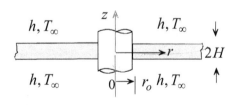

To examine the accuracy of this approximation, consider a large plate of thickness $2H$ which is mounted on a tube of radius r_o. The tube is maintained at a uniform temperature T_o. Heat transfer at the two surfaces is by convection. The ambient temperature is T_∞ and the heat transfer coefficient is h. Determine the two-dimensional steady state temperature distribution and the heat transfer rate from the plate.

3.20 A rod of radius r_o moves through a furnace with velocity U and leaves at temperature T_o. The rod is cooled

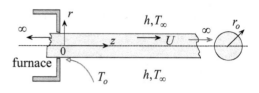

outside the furnace by convection. The ambient temperature is T_∞ and the heat transfer coefficient is h. Assume steady state conduction, determine the temperature distribution.

3.21 Heat us generated in an electric wire of radius r_o at a volumetric rate q_o'''. The wire moves with velocity U through a large chamber where it exchanges heat by convection. The heat transfer coefficient is h and chamber temperature is T_∞. The wire enters the chamber at temperature T_i. Determine the steady state two-dimensional temperature distribution.

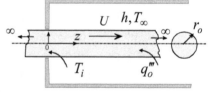

TRANSIENT CONDUCTION

In transient conduction the temperature at any location in a region changes with time. This is a common condition which is encountered in many engineering applications. In this chapter we will first present a simplified model for solving certain transient problems in which spatial variation of temperature is neglected, and then treat transient conduction in one spatial coordinate. The use of the method of separation of variables to solve transient problems will be presented. Time dependent boundary conditions will be treated using *Duhamel's superposition integral*. The *similarity transformation* method will be introduced and applied to the solution of transient conduction in semi-infinite regions.

4.1 Simplified Model: Lumped-Capacity Method

In the simplest approach to solving transient problems, spatial temperature variation is neglected and thus the temperature is assumed to vary with time only. That is

$$T = T(t). \tag{4.1}$$

This idealization is not always justified but can be made under certain conditions discussed in the following section.

4.1.1 Criterion for Neglecting Spatial Temperature Variation

Consider a thin copper wire which is heated in an oven and then removed and allowed to cool by convection. Heat is conducted through the interior and removed from the surface. Thus, there is a temperature drop ΔT across the radius of the wire. It is this temperature drop that is neglected in the *lumped-capacity* model. Factors influencing this drop are: (1) the radius of the wire r_o, (2) thermal conductivity k, and (3) the heat transfer coefficient h. We expect ΔT to be small for a small radius and for a large conductivity. The role of h is not as obvious. A small heat transfer coefficient can be thought of as an insulation layer restricting heat from leaving the wire and therefore forcing a more uniform temperature within. How these three factors combine to form a parameter that gives a measure of the temperature drop can be established by dimensional analysis of the governing equation and boundary conditions. This parameter is found to be the Biot number Bi which is defined as

$$Bi = \frac{h\,\delta}{k}, \qquad (4.2)$$

where δ is a length scale equal to the distance across which the temperature drop ΔT is to be neglected. In this example $\delta = r_o$. Based on our observation, we conclude that the smaller the Biot number, the smaller ΔT is. The next question is: how small must the Biot number be for the temperature drop to be negligible? The answer is obtained by comparing approximate transient solutions in which spatial temperature variation is neglected with exact solutions. Results indicate that for $Bi \le 0.1$ the temperature drop ΔT is less than 5% of the temperature difference between the center and the ambient fluid. Thus the criterion for neglecting spatial temperature variation and the justification for using the lumped-capacity method is

$$Bi = \frac{h\,\delta}{k} < 0.1. \qquad (4.3)$$

Care must be exercised in identifying the length scale δ in eq. (4.3). For a long cylinder, $\delta = r_o$. For a large plate of thickness L which is heated or cooled along both its surfaces, symmetry suggests that $\delta = L/2$. However, for a plate which is insulated on one side, $\delta = L$. For an irregularly shaped body, or one in which heat is transferred from more than one surface, a reasonable definition of the length scale in the Biot number is determined by dividing the volume of the object, V, by its surface area, A_s. That is

$$\delta = \frac{V}{A_s}. \qquad (4.4)$$

The physical significance of the Biot number is revealed if eq. (4.2) is rearranged as

$$Bi = \frac{\delta/k}{1/h} .$$

This indicates that the Biot number is the ratio of the conduction resistance (internal) to the convection resistance (external). Thus, a small Biot number implies a small internal resistance compared to the external resistance.

4.1.2 Lumped-Capacity Analysis

Consider a region of surface area A_s and volume V which is initially at temperature T_i. Energy is suddenly generated volumetrically at a rate q''' while the region is allowed to exchange heat with the surroundings by convection. The heat transfer coefficient is h and the ambient temperature is T_∞. Of interest is the determination of the transient temperature. The first step is the formulation of an appropriate governing equation for temperature behavior. Assuming that $Bi < 0.1$, the lumped-capacity model can be used. That is

$$T = T(t) . \tag{a}$$

Temperature variation is governed by conservation of energy

$$\dot{E}_{in} + \dot{E}_g - \dot{E}_{out} = \dot{E} , \tag{1.6}$$

where

$\dot{E}$ = rate of energy change within the region

$\dot{E}_{in}$ = rate of energy added

$\dot{E}_g$ = rate of energy generated

$\dot{E}_{out}$ = rate of energy removed

We can proceed by pretending that the region either loses or gains heat. Both options yield the same result. Assuming that heat is removed ($\dot{E}_{in} = 0$), eq. (1.6) becomes

$$\dot{E}_g - \dot{E}_{out} = \dot{E} . \tag{b}$$

Neglecting radiation and assuming that heat is removed by convection, Newton's law of cooling gives

$$\dot{E}_{out} = hA_s (T - T_\infty) . \tag{c}$$

To formulate $\dot{E}_g$ we assume that energy is generated uniformly throughout at a rate q''' per unit volume. Thus

$$\dot{E}_g = Vq''' .$$ (d)

For incompressible material $\dot{E}$ is given by

$$\dot{E} = \rho c_p V \frac{dT}{dt} ,$$ (e)

where ρ is density and c_p is specific heat. Substituting (c), (d) and (e) into (b) gives

$$Vq''' - hA_s(T - T_\infty) = \rho c_p V \frac{dT}{dt} .$$

Separating variables and rearranging

$$\frac{dT}{(hA_s / q'''V)(T - T_\infty) - 1} = -\frac{q'''}{\rho c_p} dt .$$ (4.5)

Equation (4.5) is the *lumped-capacity* governing equation for all bodies that exchange heat by convection and generate energy volumetrically. It is based on the following assumptions: (1) Energy is generated uniformly, (2) negligible radiation, (3) incompressible material and (4) $Bi < 0.1$. The initial condition for this first order ordinary differential equation is

$$T(0) = T_i .$$ (4.6)

The solution to eq. (4.5) is obtained by direct integration. Assuming constant h, T_∞ and q''', and using, eq. (4.6), the result is

$$\frac{T - T_\infty}{T_i - T_\infty} = [1 - \frac{q'''V}{hA_s(T_i - T_\infty)}] \exp[-\frac{hA_s}{\rho c_p V}t] + \frac{q'''V}{hA_s(T_i - T_\infty)} .$$ (4.7)

Units for the quantities in (4.7) are

A_s = surface area, m^2

c_p = specific heat, J/kg-°C

h = heat transfer coefficient, W/m^2-°C

q''' = volumetric energy generation, W/m^3

t = time, s

T = temperature, °C

V = volume, m^3

ρ = density, kg/m^3

The following observations are made regarding eq. (4.7): (1) Temperature decay is exponential for all lumped-capacity solutions when heat is exchanged by surface convection. (2) The thermal conductivity does not enter into this solution since neglecting spatial temperature variation implies infinite conductivity. (3) Steady state temperature is determined by setting $t = \infty$ in eq. (4.7) to obtain

$$T(\infty) = T_\infty + \frac{q'''V}{hA_s}. \qquad (4.8)$$

Note that this result represents a balance between energy generated $q'''V$ and energy convected $hA_s[T(\infty) - T_\infty]$. Equation (4.7) can be applied to the following special cases:

Case (i): No energy generation. Set $q''' = 0$ in eq. (4.7)

$$\frac{T - T_\infty}{T_i - T_\infty} = \exp\left[-\frac{hA_s}{\rho c_p V}t\right]. \qquad (4.9)$$

The steady state temperature for this case is $T(\infty) = T_\infty$.

Case (ii): No convection. Set $h = 0$ in eq. (4.7). Expanding the exponential term in eq. (4.7) for small h and then setting $h = 0$, gives

$$T = T_i + \frac{q'''}{\rho c_p}t. \qquad (4.10)$$

This represents a balance between energy generated and energy stored. Since in this case no energy leaves, it follows that a steady state does not exist. This is confirmed by eq. (4.10).

Case (iii) Initial and ambient temperatures are the same. Set $T_i = T_\infty$ in eq. (4.7)

$$T = T_\infty + \frac{q'''V}{hA_s}\left[1 - \exp\left(-\frac{hA_s}{\rho c_p V}t\right)\right]. \qquad (4.11)$$

The steady state for this case is obtained by setting $t = \infty$. The result is given by eq. (4.8).

4.2 Transient Conduction in Plates

For problems in which $Bi > 0.1$, spatial temperature variation becomes important and thus must be taken into consideration. Conduction in this case is governed by a partial differential equation. The method of separation of variables can be applied to the solution of such transient problems. As with steady state two-dimensional conduction, the application of this method is simplest when the differential equation is homogeneous and the spatial variable has two homogeneous boundary conditions. However, separation of variables is also applicable to non-homogeneous equations and/or boundary conditions using the procedure presented in Section 3.10.

In addition to boundary conditions, an initial condition must be specified for all transient problems. This condition describes the temperature distribution at $t = 0$.

Example 4.1: Plate with Surface Convection

Consider one-dimensional transient conduction in a plate of width 2L which is initially at a specified temperature distribution given by $T_i = T(x,0) = f(x)$. The plate is suddenly allowed to exchange heat by convection with an ambient fluid at T_∞. The convection coefficient is h and the thermal diffusivity is α. Assume that f(x) is symmetrical about the center plane; determine the transient temperature of the plate.

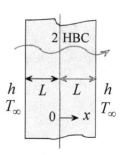

Fig. 4.1

(1) Observations. (1) Temperature distribution is symmetrical about the center plane.   (2) A convection boundary condition can be made homogeneous by defining a new temperature variable $\theta = T - T_\infty$.

(2) Origin and Coordinates. Due to symmetry, the origin is taken at the center plane as shown in Fig. 4.1.

(3) Formulation

(i) **Assumptions.**   (1) One-dimensional conduction, (2) constant conductivity, (3) constant diffusivity and (4) constant heat transfer coefficient and ambient temperature.

(ii) **Governing Equations.** To make the convection boundary condition homogeneous, we introduce the following temperature variable

$$\theta(x,t) = T(x,t) - T_\infty. \tag{a}$$

Based on the above assumptions, eq. (1.8) gives

$$\frac{\partial^2 \theta}{\partial x^2} = \frac{1}{\alpha}\frac{\partial \theta}{\partial t}. \tag{4.12}$$

(iii) **Independent Variable with Two Homogeneous Boundary Conditions.** Only the x-variable can have two homogenous conditions. Thus Fig. 4.1 is identified accordingly.

(iv) **Boundary and Initial Conditions.** The boundary conditions at $x = 0$ and $x = L$ are

(1) $\dfrac{\partial \theta(0,t)}{\partial x} = 0$

(2) $-k\dfrac{\partial \theta(L,t)}{\partial x} = h\theta(L,t)$

The initial condition is

(3) $\theta(x,0) = f(x) - T_\infty$

(4) Solution.

(i) **Assumed Product Solution.** The method of separation of variables is applied to solve this problem. Assume a product solution of the form

$$\theta(x,t) = X(x)\,\tau(t). \tag{b}$$

Substituting (b) into eq. (4.12), separating variables and setting the resulting equation equal to the separation constant $\pm\lambda_n^2$, gives

$$\frac{d^2 X_n}{dX^2} \mp \lambda_n^2 X_n = 0, \tag{c}$$

and

$$\frac{d\tau_n}{dt} \mp \alpha \lambda_n^2 \, \tau_n = 0 \,. \tag{d}$$

(ii) Selecting the Sign of the λ_n^2 Terms. Since the x-variable has two homogeneous conditions, the plus sign must be selected in (c). Thus (c) and (d) become

$$\frac{d^2 X_n}{dx^2} + \lambda_n^2 X_n = 0 \,, \tag{e}$$

and

$$\frac{d\tau_n}{dt} + \alpha \lambda_n^2 \, \tau_n = 0 \,. \tag{f}$$

It can be shown that the corresponding equations for $\lambda_n^2 = 0$ yield a trivial solution for $X_0\tau_0$ and will not be detailed here.

(iii) Solutions to the Ordinary Differential Equations. The solutions to (e) and (f) are

$$X_n(x) = A_n \sin \lambda_n x + B_n \cos \lambda_n x \,, \tag{g}$$

and

$$\tau_n(t) = C_n \exp(-\alpha \lambda_n^2 t) \,. \tag{h}$$

(iv) Application of Boundary and Initial Conditions. Condition (1) gives $A_n = 0$, thus

$$X_n(x) = B_n \cos \lambda_n x \,. \tag{i}$$

Condition (2) gives the characteristic equation for λ_n

$$\lambda_n L \tan \lambda_n L = Bi \,, \tag{4.13}$$

where Bi is the Biot number defined as

$$Bi = hL / k \,. \tag{j}$$

Substituting into (b) and summing all solutions

$$\theta(x,t) = T(x,t) - T_\infty = \sum_{n=0}^{\infty} a_n \exp(-\alpha \lambda_n^2 t) \cos \lambda_n x \,. \tag{4.14}$$

Application of the non-homogeneous initial condition (3) yields

$$f(x) - T_\infty = \sum_{n=0}^{\infty} a_n \cos\lambda_n x . \tag{k}$$

(v) **Orthogonality.** The characteristic functions $\cos\lambda_n x$ in equation (k) are solutions to (e). Comparing (e) with eq. (3.5b) shows that it is a Sturm-Liouville equation with

$$a_1 = a_2 = 0 \text{ and } a_3 = 1.$$

Thus eq. (3.6) gives

$$p = w = 1 \text{ and } q = 0.$$

Since the conditions at $x = 0$ and $x = L$ are homogeneous, it follows that $\cos\lambda_n x$ are orthogonal with respect to $w(x) = 1$. Multiplying both sides of (k) by $\cos\lambda_m x\, dx$, integrating from $x = 0$ to $x = L$ and applying orthogonality, eq.(3.7), gives a_n

$$a_n = \frac{2\lambda_n \int_0^L [f(x) - T_\infty]\cos \lambda_n x\, dx}{\lambda_n L + (\sin \lambda_n L)(\cos \lambda_n L)} . \tag{4.15}$$

For the special case of uniform initial temperature T_i, we set $f(x) = T_i$ in eq.(4.15) and evaluate the integral to obtain

$$a_n = \frac{2\,(T_i - T_\infty)\sin \lambda_n L}{\lambda_n L + (\sin \lambda_n L)(\cos \lambda_n L)} . \tag{4.16}$$

(5) **Checking.** *Dimensional check*: The exponent of the exponential in eq.(4.14) must be dimensionless. Since λ_n has units of 1/m, units of the exponent are

$$\alpha(\mathrm{m}^2\,/\,\mathrm{s})\lambda_n^2(1/\mathrm{m}^2)t(\mathrm{s}) = \text{unity}$$

Limiting check: If the plate is initially at the ambient temperature T_∞, no heat transfer takes place and consequently the temperature remains constant throughout. Setting $f(x) = T_\infty$ in eq.(4.15) gives $a_n = 0$. Substituting this result into eq.(4.14) gives $T(x,t) = T_\infty$.

(6) **Comments.** (i) Unlike steady state two-dimensional conduction, the sign of λ_n^2 term is positive in the two separated ordinary differential equations. (ii) The assumption that the initial temperature distribution is symmetrical about $x = 0$ is made for convenience only.

4.3 Non-homogeneous Equations and Boundary Conditions

The approach used in solving two-dimensional steady state non-homogeneous problems is followed in solving transient problems.

Example 4.2: Plate with Energy Generation and Specified Surface Temperatures

A plate of thickness L is initially at a uniform temperature T_i. Electricity is suddenly passed through the plate resulting in a volumetric heat generation rate of q'''. Simultaneously one surface is maintained at uniform temperature T_1 and the other at T_2. Determine the transient temperature distribution.

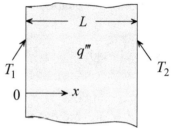

Fig. 4.2

(1) Observations. (i) Temperature distribution is asymmetric. (ii) The x-variable has two non-homogeneous conditions. (iii) The heat equation is non-homogeneous due to the heat generation term q'''.

(2) Origin and Coordinates. Fig. 4.2 shows the origin and coordinate x.

(3) Formulation

(i) **Assumptions.**   (1) One-dimensional conduction, (2) constant conductivity and (3) constant diffusivity.

(ii) **Governing Equations.** Based on the above assumptions, eq. (1.8) gives

$$\frac{\partial^2 T}{\partial x^2} + \frac{q'''}{k} = \frac{1}{\alpha}\frac{\partial T}{\partial t}. \tag{4.17}$$

(iii) **Independent Variable with Two Homogeneous Boundary Conditions.** Only the x-variable can have two homogeneous conditions.

(iv) **Boundary and Initial Conditions.** The boundary conditions at $x = 0$ and $x = L$ are

(1) $T(0,t) = T_1$
(2) $T(L,t) = T_2$

The initial condition is

(3) $T(x,0) = T_i$

(4) Solution. Since the differential equation and boundary conditions are non-homogeneous, the method of separation of variables can not be applied directly. We assume a solution of the form

$$T(x,t) = \psi(x,t) + \phi(x). \tag{a}$$

The function $\phi(x)$ is introduced to account for the energy generation term and to satisfy the non-homogeneous boundary conditions. Substituting (a) into eq. (4.17)

$$\frac{\partial^2 \psi}{\partial x^2} + \frac{d^2 \phi}{dx^2} + \frac{q'''}{k} = \frac{1}{\alpha}\frac{\partial \psi}{\partial t}. \tag{b}$$

Equation (b) will be split in two parts, one for $\psi(x,t)$ and the other for $\phi(x)$. Let

$$\frac{\partial^2 \psi}{\partial x^2} = \frac{1}{\alpha}\frac{\partial \psi}{\partial t}. \tag{c}$$

Thus,

$$\frac{d^2 \phi}{dx^2} + \frac{q'''}{k} = 0. \tag{d}$$

To solve equations (c) and (d) requires two boundary conditions for each and an initial condition for (c). These conditions are obtained by requiring (a) to satisfy boundary conditions (1) and (2) and initial condition (3). Substituting (a) into condition (1)

$$\psi(0,t) + \phi(0) = T_1.$$

Let

$$\psi(0,t) = 0. \tag{c-1}$$

Therefore,

$$\phi(0) = T_1. \tag{d-1}$$

Similarly, condition (2) gives

$$\psi(L,t) = 0, \tag{c-2}$$

and

$$\phi(L) = T_2. \tag{d-2}$$

The initial condition gives

$$\psi(x,0) = T_i - \phi(x).$$
(c-3)

The solution to equation (d) is obtained by direct integration

$$\phi(x) = -\frac{q'''}{2k}x^2 + C_1 x + C_2.$$
(e)

(i) Assumed Product Solution. Partial differential equation (c) has two homogeneous conditions in the x-variable and can be solved by the method of separation of variables. Assume a product solution of the form

$$\psi(x,t) = X(x)\tau(t).$$
(f)

Substituting (f) into (c), separating variables and setting the resulting equation equal to a constant $\pm \lambda_n^2$ gives

$$\frac{d^2 X_n}{dx^2} \mp \lambda_n^2 X_n = 0,$$
(g)

and

$$\frac{d\tau_n}{dt} \mp \alpha \; \lambda_n^2 \, \tau_n = 0.$$
(h)

(ii) Selecting the Sign of the Two λ_n^2 Terms. The plus sign must be selected in (g) because the x-variable has two homogeneous conditions. Thus (g) and (h) become

$$\frac{d^2 X_n}{dx^2} + \lambda_n^2 X_n(x) = 0,$$
(i)

and

$$\frac{d\tau_n}{dt} + \alpha \, \lambda_n^2 \, \tau_n = 0.$$
(j)

It can be shown that the corresponding equations for $\lambda_n^2 = 0$ yield a trivial solution for $X_0 \tau_0$ and will not be detailed here.

(iii) Solutions to the Ordinary Differential Equations. The solutions to (i) and (j) are

$$X_n(x) = A_n \sin \lambda_n x + B_n \cos \lambda_n x, \qquad .$$
(k)

and

$$\tau_n(t) = C_n \exp(-\alpha \lambda_n^2 t).$$
(l)

(iv) Application of Boundary and Initial Conditions. Boundary condition (c-1) gives

$$B_n = 0.$$

Condition (c-2) gives the characteristic equation

$$\sin \lambda_n L = 0, \quad \lambda_n L = n\pi, \quad n = 1,2,3... \tag{m}$$

Substituting into (f) and summing all solutions gives

$$\psi(x,t) = \sum_{n=1}^{\infty} a_n [\exp(-\alpha\lambda_n^2 t)] \sin \lambda_n x. \tag{n}$$

Returning to the function $\phi(x)$, boundary conditions (d-1) and (d-2) give the constants of integration C_1 and C_2. The solution to $\phi(x)$ becomes

$$\phi(x) = T_1 + (T_2 - T_1)\frac{x}{L} + \frac{q'''}{2k}(L-x)x. \tag{o}$$

Application of the non-homogeneous initial condition (c-3) yields

$$T_i - \phi(x) = \sum_{n=1}^{\infty} a_n \sin\lambda_n x. \tag{p}$$

(v) Orthogonality. The characteristic functions $\sin \lambda_n x$ appearing in (p) are solutions to equation (i). Comparing (i) with eq. (3.5a) shows that it is a Sturm-Liouville equation with $w(x) = 1$. Multiplying both sides of (p) by $\sin \lambda_m x \, dx$, integrating from $x = 0$ to $x = L$ and applying orthogonality, eq. (3.7), gives a_n

$$a_n = \frac{\displaystyle\int_0^L [T_i - \phi(x)]\sin \lambda_n x \, dx}{\displaystyle\int_0^L \sin^2 \lambda_n x \, dx}. \tag{q}$$

Substituting (o) into (q), evaluating the integrals and using (m) gives

$$a_n = 2\frac{1-(-1)^n}{n\pi}(T_i - T_1) + \frac{2(-1)^n}{n\pi}\left[(T_2 - T_1) + (q'''L^2/2k)\right] - $$
$$\frac{\left[(\pi^2 n^2 - 2)(-1)^n + 2\right]q'''L^2}{(n\pi)^3}\frac{}{k}. \tag{r}$$

The complete solution to the transient temperature is

$$\frac{T(x,t)-T_1}{T_2-T_1}=\frac{x}{L}+\frac{q'''L^2}{2k(T_2-T_1)}\left[1-(x/L)^2\right](x/L)+$$

$$\sum_{n=1}^{\infty}\frac{a_n}{(T_2-T_1)}\left[\exp(-\alpha n^2\pi^2 t/L^2)\right]\sin(n\pi x/L).$$

(s)

(5) Checking. *Dimensional check*: Each term in (s) is dimensionless. According to (q), a_n has units of temperature. Thus the term $a_n/(T_2-T_1)$ is dimensionless. Checking the units of $q''L^2/k(T_2-T_1)$ shows that this is a dimensionless group.

Limiting Check: At steady state, $t=\infty$, the temperature distribution should be that of a plate with energy generation and specified temperature at $x=0$ and $x=L$. Setting $t=\infty$ in (s) gives the expected solution.

(6) Comments. (i) If the initial temperature is not uniform, T_i in equation (q) is replaced by $T_i(x)$. (ii) The approach used to solve this problem can also be used to deal with other non-homogeneous boundary conditions. (iii) The solution is governed by two dimensionless parameters: $(T_i-T_1)/(T_2-T_1)$ and $q'''L^2/k(T_2-T_1)$.

4.4 Transient Conduction in Cylinders

As with steady state two-dimensional conduction, the method of separation of variables can be used to solve transient conduction in cylindrical coordinates. The following example illustrates the application of this method to the solution of non-homogeneous heat equations.

Example 4.3: Cylinder with Energy Generation

A long solid cylinder of radius r_o is initially at a uniform temperature T_i. Electricity is suddenly passed through the cylinder resulting in volumetric heat generation rate of q'''. The cylinder is cooled at its surface by convection. The heat transfer coefficient is h and the ambient temperature is T_∞. Determine the transient temperature of the cylinder.

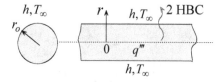

Fig. 4.3

(1) Observations. (i) The heat generation term makes the governing equation non-homogeneous. (ii) A convection boundary condition can be made homogeneous by defining a new temperature variable, $\theta = T - T_\infty$. With this transformation the r-variable will have two homogeneous conditions.

(2) Origin and Coordinates. Fig. 4.3 shows the origin and coordinate r.

(3) Formulation

(i) **Assumptions.** (1) One-dimensional conduction, (2) uniform h and T_∞, (3) constant conductivity, (4) constant diffusivity and (5) negligible end effect.

(ii) **Governing Equations.** To make the convection boundary condition homogeneous, we introduce the following temperature variable

$$\theta(r,t) = T(r,t) - T_\infty .$$

Based on the above assumptions, eq. (1.11) gives

$$\frac{\partial^2 \theta}{\partial r^2} + \frac{1}{r}\frac{\partial \theta}{\partial r} + \frac{q'''}{k} = \frac{1}{\alpha}\frac{\partial \theta}{\partial t} . \tag{4.18}$$

(iii) **Independent Variable with Two Homogeneous Boundary Conditions.** Only the r-variable can have two homogeneous conditions. Thus Fig. 4.3 is marked accordingly.

(iv) **Boundary Conditions.** The boundary conditions at $r = 0$ and $r = r_o$ are

(1) $\dfrac{\partial \theta(0,t)}{\partial r} = 0$ or $\theta(0,t) = \text{finite}$

(2) $-k\dfrac{\partial \theta(r_o,t)}{\partial r} = h\theta(r_o,t)$

The initial condition is

(3) $\theta(r,0) = T_i - T_\infty$

(4) Solution.   Since the differential equation is non-homogeneous, we assume a solution of the form

$$\theta(r,t) = \psi(r,t) + \phi(r). \tag{a}$$

Note that $\psi(r,t)$ depends on two variables while $\phi(r)$ depends on one variable. Substituting (a) into eq. (4.18)

$$\frac{\partial^2\psi}{\partial r^2} + \frac{1}{r}\frac{\partial\psi}{\partial r} + \frac{d^2\phi}{dr^2} + \frac{1}{r}\frac{d\phi}{dr} + \frac{q'''}{k} = \frac{1}{\alpha}\frac{\partial\psi}{\partial t}. \tag{b}$$

The next step is to split (b) into two equations, one for $\psi(r,t)$ and the other for $\phi(r)$. We let

$$\frac{\partial^2\psi}{\partial r^2} + \frac{1}{r}\frac{\partial\psi}{\partial r} = \frac{1}{\alpha}\frac{\partial\psi}{\partial t}. \tag{c}$$

Thus

$$\frac{d^2\phi}{dr^2} + \frac{1}{r}\frac{d\phi}{dr} + \frac{q'''}{k} = 0. \tag{d}$$

To solve equations (c) and (d) we need two boundary conditions for each and an initial condition for (c). Substituting (a) into boundary condition (1)

$$\frac{\partial\psi(0,t)}{\partial r} + \frac{d\phi(0)}{dr} = 0.$$

Let

$$\frac{\partial\psi(0,t)}{\partial r} = 0 \quad \text{or} \quad \psi(0,t) = \text{finite}. \tag{c-1}$$

Thus

$$\frac{d\phi(0)}{dr} = 0. \tag{d-1}$$

Similarly, condition (2) gives

$$-k\frac{\partial\psi(r_o,t)}{\partial r} = h\psi(r_o,t), \tag{c-2}$$

and

$$-k\frac{d\phi(r_o)}{dr} = h\phi(r_o). \tag{d-2}$$

The initial condition gives

$$\psi(r,0) = (T_i - T_\infty) - \phi(r). \tag{c-3}$$

Integrating (d) gives

$$\phi(r) = -\frac{q'''}{4k}r^2 + C_1 \ln r + C_2 \,. \tag{e}$$

(i) Assumed Product Solution. Equation (c) is solved by the method of separation of variables. Assume a product solution of the form

$$\psi(r,t) = R(r)\tau(t)\,. \tag{f}$$

Substituting (f) into (c), separating variables and setting the resulting equation equal to a constant, $\pm \lambda_k^2$, gives

$$\frac{d^2 R_k}{dr^2} + \frac{1}{r}\frac{dR_k}{dr} \mp \lambda_k^2 R_k = 0\,, \tag{g}$$

and

$$\frac{d\tau_k}{dt} \mp \lambda_k^2 \alpha \tau_k = 0\,. \tag{h}$$

(ii) Selecting the Sign of the λ_k^2 Terms. Since the r-variable has two homogeneous conditions, the plus sign must be selected in (g). Equations (g) and (h) become

$$\frac{d^2 R_k}{dr^2} + \frac{1}{r}\frac{dR_k}{dr} + \lambda_k^2 R_k = 0\,, \tag{i}$$

and

$$\frac{d\tau_k}{dt} + \lambda_k^2 \alpha \tau_k = 0\,. \tag{j}$$

It can be shown that the corresponding equations for $\lambda_k^2 = 0$ yield a trivial solution for $R_0 \tau_0$ and will not be detailed here.

(iii) Solutions to the Ordinary Differential Equations. Solutions to (i) and (j) are

$$R_k(r) = A_k J_0(\lambda_k r) + B_k Y_0(\lambda_k r)\,, \tag{k}$$

and

$$\tau_k(t) = C_k \exp(-\lambda_k^2 \alpha t)\,. \tag{l}$$

(iv) Application of Boundary and Initial Conditions. Conditions (c-1) and (c-2) give

$$B_k = 0\,,$$

and

$$Bi J_0(\lambda_k r_o) = (\lambda_k r_o)\, J_1(\lambda_k r_o)\,. \tag{m}$$

where Bi is the Biot number defined as $Bi = hr_o / k$. The roots of (m) give the constants λ_k. Substituting (k) and (l) into (f) and summing all solutions

$$\psi(r,t) = \sum_{k=1}^{\infty} a_k \exp(-\lambda_k^2 \alpha t) J_0(\lambda_k r). \qquad \text{(n)}$$

Returning to solution (e) for $\phi(r)$, boundary conditions (d-1) and (d-2) give the constants C_1 and C_2. Thus (e) becomes

$$\phi(r) = \frac{q'''}{4k}(r_o^2 - r^2) + \frac{q''' r_o}{2h}. \qquad \text{(o)}$$

Application of the non-homogeneous initial condition (c-3) yields

$$(T_i - T_\infty) - \phi(r) = \sum_{k=0}^{\infty} a_k J_0(\lambda_k r). \qquad \text{(p)}$$

(v) Orthogonality.

The characteristic functions $J_0(\lambda_k r)$ in equation (p) are solutions to (i). Comparing (i) with eq. (3.5a) shows that it is a Sturm-Liouville equation with

$$a_1 = 1/r, \ a_2 = 0 \text{ and } a_3 = 1.$$

Thus eq. (3.6) gives

$$p = w = r \text{ and } q = 0.$$

Since the boundary conditions at $r = 0$ and $r = r_o$ are homogeneous, it follows that $J_0(\lambda_k r)$ are orthogonal with respect to $w(r) = r$. Multiplying both sides of (p) by $J_0(\lambda_i r)r\,dr$, integrating from $r = 0$ to $r = r_o$ and invoking orthogonality, eq. (3.7), gives a_k

$$a_k = \frac{2\lambda_k \int_0^{r_o} [(T_i - T_\infty) - \phi(r)] J_0(\lambda_k r) r\,dr}{\int_0^{r_o} J_0^2(\lambda_k r) r\,dr}. \qquad \text{(q)}$$

Substituting (o) into (q), using Appendix B to evaluate the integral in the numerator and Table 3.1 to evaluate the integral in the denominator, gives

$$a_k = \frac{2(\lambda_k r_o)(T_i - T_\infty)}{(Bi^2 + \lambda_k^2 r_o^2)\, J_0^2(\lambda_k r_o)} \left\{ \left[1 - \frac{q''' r_o^2}{k(T_i - T_\infty)}\left(\frac{1}{2Bi} + \frac{1}{\lambda_k^2 r_o^2} \right) \right] J_1(\lambda_k r_o) \right.$$

$$\left. + \frac{q''' r_o^2}{k(T_i - T_\infty)}\, \frac{J_0(\lambda_k r_o)}{(\lambda_k r_o)} \right\}. \qquad (r)$$

The complete solution, expressed in dimensionless form, is

$$\frac{T(r,t) - T_\infty}{T_i - T_\infty} = \frac{q''' r_o^2}{4k(T_i - T_\infty)}\left[1 - \frac{r^2}{r_o^2} + \frac{2}{Bi} \right] +$$

$$\frac{1}{(T_i - T_\infty)} \sum_{k=1}^{\infty} a_k \exp(-\lambda_k^2 \alpha t)\, J_0(\lambda_k r). \qquad (4.19)$$

(5) Checking. *Dimensional check*: (i) The exponent of the exponential in eq. (4.19) must be dimensionless. Since λ_n has units of 1/m, units of the exponent are

$$\alpha(m^2/s)\lambda_n^2(1/m^2)t(s) = \text{unity}$$

(ii) Each term in eq. (4.19) is dimensionless

$$\frac{q'''(W/m^3)r_o^2(m^2)}{k(W/m-{}^\circ C)(T_i - T_\infty)({}^\circ C)}\left[1 - (r^2/r_o^2)(m^2/m^2) \right] = \text{unity}$$

Limiting check: (i) If the cylinder is initially at the ambient temperature T_∞ and there is no heat generation, no heat transfer takes place and consequently the temperature remains uniform throughout. For this case $a_k = 0$. Setting $a_k = q''' = 0$ in eq. (4.19) gives $T(r,t) = T_\infty$.

(ii) At steady state, $t = \infty$, the temperature should be that of a cylinder with energy generation and convection at the surface. Setting $t = \infty$ eq. (4.19) gives the correct result.

(6) Comments. (i) The more general problem of a non-uniform initial temperature, $T(r,0) = f(r)$, can be treated in the same manner without additional complication. (ii) The solution to this problem is expressed in terms of two parameters: the Biot number Bi and a heat generation parameter $(q'''r_o^2)/k(T_i - T_\infty)$. (iii) The solution for a cylinder with no energy generation is obtained by setting $q''' = 0$ in equations (r) and (4.19).

4.5 Transient Conduction in Spheres

The following example illustrates the application of the method of separation of variables to the solution of one-dimensional transient conduction in spheres.

Example 4.4: Sphere with Surface Convection

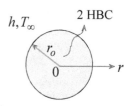

A solid sphere of radius r_o is initially at a uniform temperature T_i. At time $t > 0$ it is allowed to exchange heat with the surroundings by convection. The heat transfer coefficient is h and the ambient temperature is T_∞. Determine the transient temperature of the sphere.

Fig. 4.4

(1) Observations. (i) If h and T_∞ are uniform, the temperature distribution in the sphere will depend on the radial distance only. (ii) A convection boundary condition can be made homogeneous by defining a new temperature variable, $\theta = T - T_\infty$. With this transformation the r-variable has two homogeneous conditions.

(2) Origin and Coordinates. Fig. 4.4 shows the origin and coordinate r.

(3) Formulation

   (i) Assumptions. (1) One-dimensional conduction, (2) constant diffusivity, (3) constant conductivity and (4) uniform heat transfer coefficient and ambient temperature.

   (ii) Governing Equations. To make the convection boundary condition homogeneous, we introduce the following temperature variable

$$\theta(r,t) = T(r,t) - T_\infty .$$
(a)

Based on the above assumptions, eq. (1.13) gives

$$\frac{1}{r^2}\frac{\partial}{\partial r}\left(r^2 \frac{\partial \theta}{\partial r}\right) = \frac{1}{\alpha}\frac{\partial \theta}{\partial t} .$$
(4.20)

(iii) **Independent Variable with Two Homogeneous Boundary Conditions.** Only the r-variable can have two homogeneous conditions. Thus Fig. 4.4 is marked accordingly.

(iv) **Boundary and Initial Conditions.** The boundary conditions at $r = 0$ and $r = r_o$ are

(1) $\dfrac{\partial\theta(0,t)}{\partial r} = 0$ or $\theta(0,t) = \text{finite}$

(2) $-k\dfrac{\partial\theta(r_o,t)}{\partial r} = h\,\theta(r_o,t)$

The initial condition is

(3) $\theta(r,0) = T_i - T_\infty$

(4) Solution.

(i) **Assumed Product Solution.** Equation (4.20) is solved by the method of separation of variables. Assume a product solution of the form

$$\theta(r,t) = R(r)\,\tau(t).\tag{b}$$

Substituting (b) into eq. (4.20), separating variables and setting the resulting equation equal to a constant, $\pm\lambda_k^2$, gives

$$r^2\frac{d^2 R_k}{dr^2} + 2r\frac{dR_k}{dr} \mp \lambda_k^2 r^2 R_k = 0,\tag{c}$$

and

$$\frac{d\tau_k}{dt} \mp \lambda_k^2\alpha\tau_k = 0.\tag{d}$$

(ii) **Selecting the Sign of the λ_k^2 Terms.** Since the r-variable has two homogeneous conditions, the plus sign must be selected in (c). Equations (c) and (d) become

$$r^2\frac{d^2 R_k}{dr^2} + 2r\frac{dR_k}{dr} + \lambda_k^2 r^2 R_k = 0,\tag{e}$$

$$\frac{d\tau_k}{dt} + \lambda_k^2\alpha\tau_k = 0.\tag{f}$$

It can be shown that the corresponding equations for $\lambda_k^2 = 0$ give a trivial solution for $R_o \tau_o$.

(iii) Solutions to the Ordinary Differential Equations. Eq.(e) is a Bessel equation whose solution is

$$R_k(r) = r^{-1/2} \left[A_k J_{1/2}(\lambda_k r) + B_k J_{-1/2}(\lambda_k r) \right].$$

Using equations (2.32) and (2.33), the above becomes

$$R_k(r) = \frac{1}{r}(A_k \sin \lambda_k r + B_k \cos \lambda_k r). \tag{g}$$

The solution to (f) is

$$\tau_k(t) = C_k \exp(-\alpha \lambda_k^2 t). \tag{h}$$

(iv) Application of Boundary and Initial Conditions. Condition (1) gives

$$B_k = 0.$$

Condition (2) gives

$$(1 - Bi)\tan \lambda_k r_o = \lambda_k r_o, \quad k = 1, 2, \ldots \tag{i}$$

where Bi is the Biot number, defined as $Bi = hr_o / k$. The roots of equation (i) give the constants λ_k. Substituting (g) and (h) into (b) and summing all solutions

$$\theta(r,t) = T(r,t) - T_\infty = \sum_{k=1}^{\infty} a_k \exp(-\alpha \lambda_k^2 t) \frac{\sin \lambda_k r}{r}. \tag{4.21}$$

Application of initial condition (3) gives

$$T_i - T_\infty = \sum_{k=1}^{\infty} a_k \frac{\sin \lambda_k r}{r}. \tag{j}$$

(v) Orthogonality. The characteristic functions $(1/r) \sin \lambda_k r$ in equation (j) are solutions to (e). Comparing (e) with eq. (3.5a) shows that it is a Sturm-Liouville equation with

$$a_1 = 2/r, \quad a_2 = 0, \quad a_3 = 1,$$

$$p = r^2, \quad q = 0, \quad w = r^2.$$

Since the boundary conditions at $r = 0$ and $r = r_o$ are homogeneous, it follows that the functions $(1/r)\sin \lambda_k r$ are orthogonal with respect to $w(r) = r^2$. Multiplying both sides of (j) by $(1/r)(\sin \lambda_i r) r^2 dr$, integrating from $r = 0$ to $r = r_o$ and invoking orthogonality, eq. (3.7), gives a_k

$$a_k = \frac{(T_i - T_\infty) \int_0^{r_o} r \sin \lambda_k r \, dr}{\int_0^{r_o} \sin^2 \lambda_k r \, dr} = 2(T_i - T_\infty) \frac{\sin \lambda_k r_o - \lambda_k r_o \cos \lambda_k r_o}{\lambda_k [\lambda_k r_o - \sin \lambda_k r_o \cos \lambda_k r_o]}.$$

(k)

(5) Checking. *Dimensional check*: The exponent of the exponential in eq. (4.21) must be dimensionless. Since λ_n has units of $1/m$, units of the exponent are

$$\alpha(m^2/s)\lambda_k^2(1/m^2)t(s) = \text{unity}$$

Limiting check: (i) If the sphere is initially at the ambient temperature T_∞, no heat transfer takes place and consequently the temperature remains constant throughout. Setting $T_i = T_\infty$ in (k) gives $a_k = 0$. When this result is substituted into eq. (4.21) gives $T(r,t) = T_\infty$. (ii) At steady state, $t = \infty$, the sphere should be at equilibrium with the ambient temperature. Setting $t = \infty$ in (4.21) gives $T(r, \infty) = T_\infty$.

(6) Comments. (i) The solution is expressed in terms of a single parameter which is the Biot number *Bi*. (ii) The more general problem of a non-uniform initial temperature, $T(r,0) = f(r)$, can be treated in the same manner without additional complication.

4.6 Time Dependent Boundary Conditions: Duhamel's Superposition Integral

There are many practical applications where boundary conditions vary with time. Examples include solar heating, re-entry aerodynamic heating, reciprocating surface friction and periodic oscillation of temperature or flux. One of the methods used to solve such problems is based on *Duhamel's superposition integral*. In this method the solution of an auxiliary problem with a constant boundary condition is used to construct the solution to the same problem with a time dependent condition. Since

Duhamel's method is based on superposition of solutions, it follows that it is limited to linear equations.

4.6.1 Formulation of Duhamel's Integral [1]

To illustrate how solutions can be superimposed to construct a solution for a time dependent boundary condition, consider transient conduction in a plate which is initially at a uniform temperature equal to zero. At time $t > 0$ one boundary is maintained at temperature T_{01}. At time $t = \tau_1$, the temperature is changed to T_{02}, as shown in Fig. 4.5. We wish to determine the transient temperature $T(x,t)$ for this time dependent boundary condition. Assuming that the heat equation is linear, the problem can be decomposed into two problems: one which starts $t = 0$ and a second which starts at $t = \tau_1$. Note that each problem has a constant temperature boundary condition. Thus

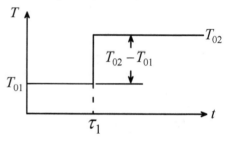

Fig. 4.5

$$T(x,t) = T_1(x,t) + T_2(x,t - \tau_1). \tag{a}$$

Let $\overline{T}(x,t)$ be the solution to an auxiliary problem corresponding to constant surface temperature of magnitude unity. Thus the solution for T_1 is

$$T_1(x,t) = T_{01} \overline{T}(x,t). \tag{b}$$

At $t = \tau_1$ the temperature is increased by $(T_{02} - T_{01})$. The solution to this problem is

$$T_2(x,t) = (T_{02} - T_{01})\overline{T}(x,t - \tau_1). \tag{c}$$

Adding (b) and (c) gives the solution to the problem with the time dependent boundary condition

$$T(x,t) = T_{01} \overline{T}(x,t) + (T_{02} - T_{01}) \overline{T}(x,t - \tau_1). \tag{d}$$

If at time τ_2 another step change in temperature takes place at the boundary, say $(T_{03} - T_{02})$, a third term, $(T_{03} - T_{02})\overline{T}(x,t - \tau_2)$ must be added to solution (d).

We now generalize this superposition scheme to obtain solutions to problems with boundary conditions which change arbitrarily with time according to $F(t)$, as shown in Fig. 4.6. Note that $F(t)$ represents any time dependent boundary condition and that it is not limited to a specified temperature. It could, for example, represent time dependent heat flux, ambient temperature or surface temperature. This problem is solved by superimposing the solutions to many problems, each having a small step change in the boundary condition. Thus the function $F(t)$ is approximated by n consecutive steps of ΔF corresponding to time steps Δt, as shown in Fig. 4.6. The first problem starts at time $t = 0$ and has a finite step function of magnitude $F(0)$. The contribution of this problem to the solution is

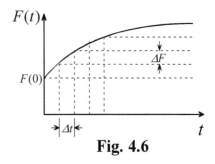

Fig. 4.6

$$T_0(x,t) = F(0)\,\overline{T}(x,t). \tag{e}$$

The contribution of the ith problem, with input $\Delta F(\tau_i)$, to the solution is

$$T_i(x,t) = \Delta F(\tau_i)\,\overline{T}(x,t - \tau_i).$$

Adding all solutions gives

$$T(x,t) = F(0)\,\overline{T}(x,t) + \sum_{i=1}^{n} \Delta F(\tau_i)\overline{T}(x,t - \tau_i). \tag{f}$$

This result can be written as

$$T(x,t) = F(0)\overline{T}(x,t) + \sum_{i=1}^{n} \frac{\Delta F(\tau_i)}{\Delta \tau_i}\,\overline{T}(x,t - \tau_i)\,\Delta \tau_i.$$

In the limit as $n \to \infty$, $\Delta \tau_i \to d\tau$, $\dfrac{\Delta F(\tau_i)}{\Delta \tau_i} \to \dfrac{dF(\tau)}{d\tau}$ and the summation is replaced by integration, the above becomes

$$T(x,t) = F(0)\,\overline{T}(x,t) + \int_0^t \frac{dF(\tau)}{d\tau}\,\overline{T}(x,t-\tau)\,d\tau . \qquad (4.28)$$

Using integration by parts and recalling that $\overline{T}(x,0) = 0$, eq. (4.28) can be written as

$$T(x,t) = \int_0^t F(\tau)\,\frac{\partial \overline{T}(x,t-\tau)}{\partial t}\,d\tau . \qquad (4.29)$$

The following should be noted regarding Duhamel's method: (1) It applies to linear equations. (2) The initial temperature T_i is assumed to be zero. If it is not, a new variable must be defined as $(T - T_i)$. (3) The auxiliary solution $\overline{T}(x,t)$ is based on a constant boundary condition of magnitude unity. (4) Equations (4.28) and (4.29) apply to any coordinate system. (5) The integrals in equations (4.28) and (4.29) are with respect to the dummy variable τ. The variable t is treated as constant in these integrals. (6) The method applies to problems with time dependent heat generation. It can also be applied to lumped-capacity models in which either the heat transfer coefficient or the ambient temperature varies with time.

4.6.2 Extension to Discontinuous Boundary Conditions

If the time dependent boundary condition $F(t)$ is discontinuous, Duhamel's integral must be modified accordingly.   Consider a time dependent boundary condition which is described by two functions $F_1(t)$ and $F_2(t)$ having a discontinuity at time t_a as shown in Fig. 4.7. Thus

$$F(t) = F_1(t) \quad 0 \le t \le t_a ,$$

and

$$F(t) = F_2(t) \quad t \ge t_a .$$

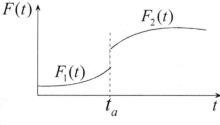

Fig. 4.7

Equations (4.28) and (4.29) can be used with $F(t) = F_1(t)$ to obtain a solution for $t < t_a$. However, for $t > t_a$ equations (4.28) and (4.29) must be modified. Modification of eq. (4.28) requires the addition of another solution, $T_a(x,t)$, which accounts for the step change in F at $t = t_a$. This solution is given by

$$T_a(x,t) = [F_2(t_a) - F_1(t_a)]\,\overline{T}(x,t-t_a).$$

Thus the modified form of equations (4.28) for $t > t_a$ becomes

$$T(x,t) = F_1(0)\,\overline{T}(x,t) + [F_2(t_a) - F_1(t_a)]\,\overline{T}(x,t - t_a)$$
$$+ \int_0^{t_a} \frac{dF_1(\tau)}{d\tau}\,\overline{T}(x,t - \tau)\,d\tau + \int_{t_a}^{t} \frac{dF_2(\tau)}{d\tau}\,\overline{T}(x,t - \tau)\,d\tau. \tag{4.30}$$

Evaluating the integrals in eq. (4.30) by parts and recalling that $\overline{T}(x,0) = 0$, gives

$$T(x,t) = [F_2(t_a) - F_1(t_a)]\overline{T}(x,t - t_a) +$$
$$\int_0^{t_a} F_1(\tau)\,\frac{\partial \overline{T}(x,t - \tau)}{\partial t}\,d\tau + \int_{t_a}^{t} F_2(\tau)\,\frac{\partial \overline{T}(x,t - \tau)}{\partial t}\,d\tau. \tag{4.31}$$

Equations (4.30) and (4.31) can be extended to boundary conditions having more than one discontinuity.

4.6.3 Applications

Example 4.5: Plate with Time Dependent Surface Temperature

A plate of thickness L is initially at zero temperature. One side is maintained at zero while the temperature of the other side is allowed to vary with time according to $T(L,t) = At$, where A is constant. Determine the transient temperature of the plate.

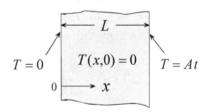

Fig. 4.8

(1) Observations. (i) One of the boundary conditions is time dependent. (ii) Since the initial temperature is zero, Duhamel's method can be used directly without defining a new temperature variable.

(2) Origin and Coordinates. Fig. 4.8 shows the origin and coordinate x.

(3) Formulation.

(i) **Assumptions.** (1) One-dimensional transient conduction, and (2) constant diffusivity.

(ii) Governing equations. Based on the above assumptions, eq. (1.8) gives

$$\frac{\partial^2 T}{\partial x^2} = \frac{1}{\alpha}\frac{\partial T}{\partial t}. \tag{a}$$

(iii) Boundary conditions. The boundary conditions at $x = 0$ and $x = L$ are

(1) $T(0,t) = 0$

(2) $T(L,t) = A\,t$

The initial condition is

(3) $T(x,0) = 0$

(4) Solution. Since eq. (a) is linear, Duhamel's integral, eq. (4.28) or eq. (4.29), can be applied to obtain a solution to the problem. Applying eq. (4.28)

$$T(x,t) = F(0)\,\overline{T}(x,t) + \int_0^t \frac{dF(\tau)}{d\tau}\,\overline{T}(x,t-\tau)\,d\tau, \tag{4.28}$$

where $\overline{T}(x,t)$ is the solution to the same problem with boundary condition (2) replaced by $T(L,t) = 1$. This solution can be obtained from the results of Example 4.2 by setting $q''' = 0$, $T_1 = T_i = 0$ and $T_2 = 1$ to give

$$\overline{T}(x,t) = \frac{x}{L} + \sum_{n=1}^{\infty} a_n \exp[-\alpha(n\pi/L)^2 t]\sin(n\pi x/L), \tag{b}$$

where

$$a_n = \frac{2}{n\pi}(-1)^n. \tag{c}$$

$F(0)$ and $dF(\tau)/d\tau$ appearing in eq. (4.28) are obtained from the time dependent boundary condition. Thus

$$F(\tau) = A\tau, \qquad \frac{dF}{d\tau} = A. \tag{d}$$

Substituting (b) and (d) into eq. (4.28) and noting that $F(0) = 0$

$$T(x,t) = A\frac{x}{L} \int_0^t d\tau + A\sum_{n=1}^{\infty} a_n \sin(n\pi x/L) \int_0^t \exp[-\alpha(n\pi/L)^2(t-\tau)]\, d\tau,$$

$$(e)$$

Evaluating the integrals in (e) and using (c) gives

$$T(x,t) = A\frac{x}{L}t + A\frac{2L^2}{\alpha\pi^3}\sum_{n=1}^{\infty}\frac{(-1)^n}{n^3}\sin(n\pi x/L)\,[1 - \exp(-\alpha n^2\pi^2/L^2)t].$$

$$(f)$$

(5) Checking. *Dimensional check*: Each term in (f) must have units of temperature ($^\circ$C). The constant A has units of $^\circ$C/s and α is measured in m^2/s. Thus units of the first and second terms are in $^\circ$C. The exponent of the exponential term is dimensionless.

Limiting checks: (i) if $A = 0$, the entire plate should remain at $T = 0$ at all times. Setting $A = 0$ in (f) gives $T(x,t) = 0$. (ii) At $t = \infty$, plate temperature should be infinite. Setting $t = \infty$ in (f) gives $T(x,\infty) = \infty$.

(6) Comments. The same procedure can be used for the more general case of a surface temperature at an arbitrary function of time, $T(L,t) = F(t)$.

Example 4.6: Lumped-Capacity Method with Time Dependent Ambient Temperature

A metal foil at T_i is placed in an oven which is also at T_i. When the oven is turned on its temperature rises with time. The variation of oven temperature is approximated by

$$T_\infty(t) = T_o[1 - \exp(-\beta t)] + T_i,$$

where T_o and β are constant. Determine the transient temperature of the foil using a lumped-capacity model and Duhamel's integral.

(1) Observations. (i) Oven temperature is time dependent. (ii) Oven temperature in the lumped-capacity method appears in the heat equation and is not a boundary condition. (iii) If the heat equation for the lumped-capacity model is linear, Duhamel's method can be used to determine the transient temperature.

(2) Origin and Coordinates. Since spatial temperature variation is neglected in the lumped-capacity model, no spatial coordinates are needed.

(3) Formulation.

(i) Assumptions. (1) $Bi < 0.1$, (2) constant density, (3) constant specific heat and (4) constant heat transfer coefficient.

(ii) Governing Equations. Setting $q''' = 0$ in eq. (4.5) gives the governing equation for the transient temperature

$$\frac{dT}{T - T_\infty(t)} = -\frac{hA_s}{\rho c_p V}\, dt\,. \tag{a}$$

Since this equation is linear, Duhamel's method can be applied to obtain a solution. However, application of Duhamel's integral requires that the initial temperature be zero. We therefore define the following temperature variable

$$\theta(t) = T(t) - T_i\,. \tag{b}$$

The oven temperature expressed in terms of this variable becomes

$$\theta_\infty(t) = T_\infty(t) - T_i\,. \tag{c}$$

Substituting into (a)

$$\frac{d\theta}{\theta(t) - \theta_\infty(t)} = -\frac{hA_s}{\rho c_p V}\, dt\,. \tag{d}$$

The initial condition becomes

$$\theta(0) = 0\,. \tag{e}$$

(4) Solution. Apply Duhamel's integral, eq. (4.28)

$$\theta(t) = F(0)\,\overline{\theta}(t) + \int_0^t \frac{dF(\tau)}{d\tau}\,\overline{\theta}(t - \tau)\, d\tau\,. \tag{4.28}$$

where $\overline{\theta}$ is the solution to the auxiliary problem of constant oven temperature equal to unity and zero initial temperature. $F(t)$ is the time dependent oven temperature, given by

$$F(t) = \theta_\infty(t) = T_\infty(t) - T_i = T_o[1 - \exp(-\beta t)]\,. \tag{f}$$

Constant oven temperature equal to unity is given by

$$\overline{\theta}_\infty = T_\infty - T_i = 1. \tag{g}$$

Applying (d) to the auxiliary problem gives

$$\frac{d\overline{\theta}}{\overline{\theta} - 1} = -\frac{hA_s}{\rho c_p V} dt. \tag{h}$$

Initial condition (e) becomes

$$\overline{\theta}(0) = 0. \tag{i}$$

Integrating (h) and using initial condition (i) gives

$$\overline{\theta}(t) = 1 - \exp(-\frac{hA_s}{\rho c_p V} t). \tag{j}$$

To determine $dF(\tau)/d\tau$, replace t in (f) with τ and differentiate to obtain

$$\frac{dF}{d\tau} = \beta T_o \exp(-\beta \tau). \tag{k}$$

Substituting (j) and (k) into eq. (4.28) and noting that $F(0) = 0$, gives

$$\theta(t) = T(t) - T_i = \beta T_o \int_0^t \exp(-\beta \tau)\left\{1 - \exp[-\frac{hA_s(t-\tau)}{\rho c_p V}]\right\} d\tau. \tag{l}$$

Performing the integration and rearranging the result yields

$$T(t) = T_i + T_o + \frac{T_o \beta}{\beta - (hA_s / \rho c_p V)}\left[\exp(-\beta t) - \exp(-hA_s t / \rho c_p V)\right] -$$
$$T_o \exp(-\beta t). \tag{m}$$

(5) Checking. *Dimensional check:* β and $(hA_s / \rho c_p V)$ have units of (1/s). Thus the exponents of the exponentials are dimensionless and each term in (j) has units of temperature.

Limiting checks: (i) Solution (m) should satisfy the initial condition $T(0)$ $= T_i$. Setting $t = 0$ in (m) gives $T(0) = T_i$. (ii) At $t = \infty$ the foil should be at the steady state temperature of the oven. Setting $t = \infty$ in (d) gives the steady state oven temperature as $T_\infty(\infty) = T_o + T_i$. Setting $t = \infty$ in solution (m) gives $T(\infty) = T_o + T_i$.

Governing equation check: Direct substitution of solution (m) into (a) shows that (a) is satisfied.

(6) Comments. (i) The definition of θ in equation (b) meets the requirement of an initial temperature equal to zero. (ii) Direct integration of (a) gives the same solution (m) obtained using Duhamel's method.

4.7 Conduction in Semi-infinite Regions: The Similarity Method

The application of the method of separation of variables to the solution of partial differential equations requires two homogeneous boundary conditions in one spatial variable. A further limitation is that the two boundaries with homogeneous conditions must be separated by a finite distance. Examples include boundary conditions at the center and surface of a cylinder of radius R or on two sides of a plate of thickness L. However, the use of separation of variables in solving conduction problems in semi-infinite regions fails even if there are two homogeneous boundary conditions in one variable. Thus, alternate methods must be used to solve such problems. One such approach is based on the *similarity transformation* method. The basic idea in this method is to combine two independent variables into one. Thus a partial differential equation with two independent variables is transformed into an ordinary differential equation. One of the limitations of this method is that the region must not be finite. In addition, there are limitations on boundary and initial conditions.

To illustrate the use of this method, consider a semi-infinite plate which is initially at a uniform temperature T_i. The plate surface is suddenly maintained at temperature T_o as shown in Fig. 4.9. For constant conductivity, the heat equation for transient one-dimensional conduction is

$$\frac{\partial^2 T}{\partial x^2} = \frac{1}{\alpha} \frac{\partial T}{\partial t}. \qquad \text{(a)}$$

The boundary conditions are

(1) $T(0,t) = T_o$

(2) $T(\infty,t) = T_i$

The initial condition is

(3) $T(x,0) = T_i$

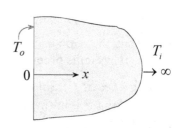

Fig. 4.9

Introducing the definition of a dimensionless temperature variable θ

$$\theta = \frac{T - T_i}{T_o - T_i}. \tag{b}$$

Substituting into (a)

$$\frac{\partial^2 \theta}{\partial x^2} = \frac{1}{\alpha} \frac{\partial \theta}{\partial t}. \tag{c}$$

The boundary and initial conditions become

(1) $\theta(0,t) = 1$

(2) $\theta(\infty,t) = 0$

(3) $\theta(x,0) = 0$

To proceed with the similarity method we assume that the two independent variables x and t can be combined into a single variable $\eta = \eta(x,t)$ called the *similarity variable*. Thus we write

$$\theta(x,t) = \theta(\eta). \tag{d}$$

The key to this method is the determination of the form of the similarity variable. The combination of x and t forming η must be such that the governing equation and boundary and initial conditions be transformed in terms of η only. Formal methods for selecting η for a given problem are available [2]. The applicable similarity variable for this problem is

$$\eta = \frac{x}{\sqrt{4\alpha t}}. \tag{4.32}$$

Using eq. (4.32) to construct the derivatives in (c), we obtain

$$\frac{\partial \theta}{\partial x} = \frac{d\theta}{d\eta} \frac{\partial \eta}{\partial x} = \frac{d\theta}{d\eta} \frac{1}{\sqrt{4\alpha t}},$$

$$\frac{\partial^2 \theta}{\partial x^2} = \frac{d}{d\eta}\left[\frac{d\theta}{d\eta} \frac{1}{\sqrt{4\alpha t}}\right]\frac{\partial \eta}{\partial x} = \frac{1}{4\alpha t}\frac{d^2\theta}{d\eta^2},$$

$$\frac{\partial \theta}{\partial t} = \frac{d\theta}{d\eta} \frac{\partial \eta}{\partial t} = \frac{d\theta}{d\eta}\left[-\frac{x}{2\sqrt{4\alpha}}t^{-3/2}\right] = -\frac{x}{2\sqrt{4\alpha t}}\frac{1}{t}\frac{d\theta}{d\eta} = -\frac{\eta}{2t}\frac{d\theta}{d\eta}.$$

Substituting these results into (c) yields

$$\frac{d^2\theta}{d\eta^2} + 2\eta\frac{d\theta}{d\eta} = 0.$$ (4.33)

The following should be noted regarding this result:

(1) The variables x and t have been successfully eliminated and replaced by a single independent variable η. This confirms that eq. (4.32) is the correct choice for the similarity variable η.

(2) While three conditions must be satisfied by the solution to (c), eq. (4.33) can satisfy two conditions only.

The three conditions will now be transformed in terms of η. From eq. (4.32) we obtain

$x = 0, t = t$ transforms to $\eta = 0$

$x = \infty, t = t$ transforms to $\eta = \infty$

$x = x, t = 0$ transforms to $\eta = \infty$

Thus the two boundary conditions and the initial condition transform to

(1) $\theta(0,t) = \theta(0) = 1$, (2) $\theta(\infty) = 0$, (3) $\theta(\infty) = 0$

Note that two of the three conditions coalesce into one. Thus the transformed problem has the required number of conditions that must be satisfied by a solution to eq. (4.33). Separating variables in eq. (4.33)

$$\frac{d(d\theta/d\eta)}{d\theta/d\eta} = -2\eta d\eta.$$

Integrating gives

$$\ln\frac{d\theta}{d\eta} = -\eta^2 + \ln A,$$

or

$$d\theta = A e^{-\eta^2} d\eta,$$

where A is constant. Integrating from $\eta = 0$ to η

$$\int_0^\theta d\theta = A \int_0^\eta e^{-\eta^2} d\eta,$$

or

$$\theta - \theta(0) = A \int_0^\eta e^{-\eta^2} d\eta .$$ (e)

Using boundary condition (1) and rewriting (e)

$$\theta = 1 + A \frac{\sqrt{\pi}}{2} \left[\frac{2}{\pi} \int_0^\eta e^{-\eta^2} d\eta \right].$$ (f)

The integral in the bracket of (f) is known as the *error function*, defined as

$$\mathrm{erf}\,\eta = \frac{2}{\sqrt{\pi}} \int_0^\eta e^{-\eta^2} d\eta .$$ (g)

Values of the error function are given in Table 4.1. Note that erf 0 = 0 and erf∞ = 1.0. Using (g), equation (f) becomes

$$\theta = 1 + \frac{A\sqrt{\pi}}{2} \mathrm{erf}\,\eta .$$ (h)

Boundary condition (2) gives $A = \dfrac{-2}{\sqrt{\pi}}$. Solution (h) becomes

$$\theta(\eta) = 1 - \mathrm{erf}\,\eta ,$$ (4.33a)

or

$$\theta(x,t) = 1 - \mathrm{erf}\,\frac{x}{\sqrt{4\alpha t}} .$$ (4.33b)

Note that the derivative of erf η is

$$\frac{d}{d\eta}(\mathrm{erf}\,\eta) = \frac{2}{\sqrt{\pi}} e^{-\eta^2} ,$$ (4.34)

and

$$\mathrm{erf}(-\eta) = -\mathrm{erf}\,\eta .$$ (4.35)

Table 4.1

η	erf η	η	erf η
0.0	0.00000	1.6	0.97635
0.1	0.11246	1.7	0.98379
0.2	0.22270	1.8	0.98909
0.3	0.32863	1.9	0.99279
0.4	0.42839	2.0	0.99432
0.5	0.52050	2.1	0.99702
0.6	0.60386	2.2	0.99814
0.7	0.67780	2.3	0.99886
0.8	0.74210	2.4	0.99931
0.9	0.79691	2.5	0.99959
1.0	0.84270	2.6	0.99976
1.1	0.88021	2.7	0.99987
1.2	0.91031	2.8	0.99992
1.3	0.93401	2.9	0.99996
1.4	0.95229	3.0	0.99998
1.5	0.96611		

REFERENCES

[1]   Hildebrand, F.B., *Advanced Calculus for Applications*, 2nd edition, Prentice-Hall,   Englewood Cliffs, New Jersey, 1976.

[2]   Hansen, A.G., *Similarity Analysis of Boundary Value Problems in Engineering*, Prentice-Hall, Englewood Cliffs, New Jersey, 1976.

PROBLEMS

4.1   A thin foil of surface area A_s is initially at a uniform temperature T_i. The foil is cooled by radiation and convection. The ambient and surroundings temperatures are at $T_\infty = 0$ kelvin. The heat transfer coefficient is h and surface emissivity is ε. Use lumped-capacity analysis to determine the transient temperature of the foil.

4.2   Consider a wire of radius r_o which is initially at temperature T_∞. Current is suddenly passed through the wire causing energy generation at a volumetric rate q'''. The wire exchanges heat with the surroundings by convection. The ambient temperature is T_∞ and the heat transfer coefficient is h. Assume that the Biot number is small compared to unity determine the transient temperature of the wire. What is the steady state temperature?

4.3   A coin of radius r_o and thickness δ rests on an inclined plane. Initially it is at the ambient temperature T_∞. The coin is released and begins to slide down the plane. The frictional force F_f is assumed constant and the velocity changes with time according to $V = ct$, where c is constant. Due to changes in velocity, the heat transfer coefficient h varies with time according to

$$h = \beta t,$$

where β is constant. Assume that the Biot number is small compared to unity and neglect heat loss to the plane, determine the transient temperature of the coin.

4.4   Consider a penny and a wire of the same material. The diameter of the wire is the same as the thickness of the penny. The two are heated in an oven by convection. Initially both are at the same temperature. Assume that the heat transfer coefficient is the same for both and that the Biot number is small compared to unity. Which object will be heated faster?

4.5   A copper wire of diameter 0.01 mm is initially at $220\,^{\circ}\mathrm{C}$. The wire is suddenly cooled by an air jet at $18\,^{\circ}\mathrm{C}$. The heat transfer coefficient is $138\ \mathrm{W/m^2-^{\circ}C}$. What will the wire temperature be after 0.5 seconds? Properties of copper are: $c_p = 385$ J/kg–$^{\circ}$C, $k = 398.4$ W/m–$^{\circ}$C and $\rho = 8933\ \mathrm{kg/m^3}$.

4.6   Small glass balls of radius 1.1 mm are cooled in an oil bath at $22\,^{\circ}\mathrm{C}$. The balls enter the bath at $180\,^{\circ}\mathrm{C}$ and are moved through on a conveyor belt. The heat transfer coefficient is $75\ \mathrm{W/m^2-^{\circ}C}$. The bath is 2.5 m long. What should the conveyor speed be for the balls to leave at $40\,^{\circ}\mathrm{C}$? Properties of glass are: $c_p = 810$ J/kg–$^{\circ}$C, $k = 3.83$ W/m–$^{\circ}$C and $\rho = 2600\ \mathrm{kg/m^3}$.

4.7   Consider one-dimensional transient conduction in a plate of thickness L which is initially at uniform temperature T_i. At time $t \geq 0$ one side exchanges heat by convection with an ambient fluid at T_{∞} while the other side is maintained at T_i. The conductivity is k, heat transfer coefficient is h and the thermal diffusivity is α. Determine the transient temperature.

4.8   A plate of thickness $2L$ is initially at temperature T_i. Electricity is suddenly passed through the plate resulting in a volumetric heat generation rate of q'''. Simultaneously, the two sides begin to exchange heat by convection with an ambient fluid at T_{∞}. The thermal conductivity is k, heat transfer coefficient h and the thermal

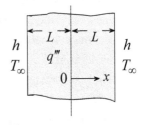

diffusivity is α. Determine the one-dimensional transient temperature.

4.9 The temperature of a plate of thickness L is initially linearly distributed according to

$$T_i = T(x,0) = T_1 + (T_2 - T_1)(x/L),$$

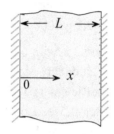

where T_1 and T_2 are constant. At $t \geq 0$ the plate is insulated along its surfaces. Determine the one-dimensional transient temperature distribution. What is the steady state temperature?

4.10 A bar of rectangular cross section $L \times H$ is initially at uniform temperature T_i. At time $t = 0$ the bar starts to slide down an inclined surface with a constant velocity V. The pressure and coefficient of friction at the interface are P and μ, respectively. On the opposite side the bar exchanges heat by convection with the surroundings. The heat transfer coefficient is h and the ambient temperature is T_∞. The two other surfaces are insulated. Assume that the Biot number is large compared to unity and neglect heat transfer in the direction normal to the $L \times H$ plane, determine the transient temperature distribution.

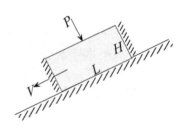

4.11 A plate of thickness L_1 is initially at uniform temperature T_1. A second plate of the same material of thickness L_2 is initially at uniform temperature T_2. At $t \geq 0$ the two plates are fastened together with a perfect contact at the interface. Simultaneously, the surface of one plate is heated with uniform flux q_o'' and the opposite surface begins to exchange heat by convection with the surroundings. The heat

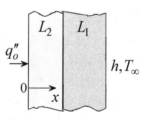

transfer coefficient is h and the ambient temperature is T_∞. Determine the one-dimensional transient temperature distribution in the two plates.

4.12 Consider a constant area fin of length L, cross section area A_c and circumference C. Initially the fin is at uniform temperature T_i. At time $t \geq 0$ one end is insulated while heat is added at the other end at a constant flux q_o''. The fin exchanges heat by convection with the surroundings. The heat transfer coefficient is h and the ambient temperature is T_∞. Determine the heat transfer rate from the fin to the surroundings.

4.13 A tube of inner radius r_i, outer radius r_o and length L is initially at uniform temperature T_i. At $t \geq 0$ one end is maintained at T_o while the other end is insulated. Simultaneously, the tube begins to exchange heat by convection along its inner and outer surfaces. The inside and outside heat transfer coefficients are h_i and h_o, respectively. The inside and outside ambient temperature is T_∞. Neglecting temperature variation in the r-direction, determine the one-dimensional transient temperature distribution in the tube.

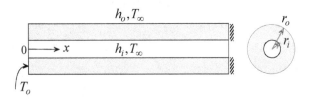

4.14 A wire of radius r_o and length L is initially at uniform temperature T_i. At $t \geq 0$ current is passed through the wire resulting in a uniform volumetric heat generation q'''. The two wire ends are maintained at the initial temperature T_i. The wire exchanges heat by convection with the surroundings. The heat transfer coefficient is h and the

ambient temperature is T_∞. Neglecting temperature variation in the
r-direction, determine the one-dimensional transient temperature
distribution in the wire. What is the steady state temperature of the
mid-section?

4.15 Consider a long solid cylinder
of radius r_o which is initially
at uniform temperature T_i.
Electricity is suddenly passed
through the cylinder resulting

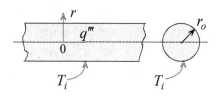

in a volumetric heat generation rate q'''. The cylindrical surface is
maintained at the initial temperature T_i.   Determine the one-
dimensional transient temperature.

4.16 A hollow cylinder of outer radius r_o and inner radius r_i is initially at
uniform temperature T_1. A second cylinder of the same material
which is solid of radius r_i is initially at uniform temperature T_2. At
$t \geq 0$ the solid cylinder is forced inside the hollow cylinder resulting
in a perfect interface contact. The
outer   surface   of   the   hollow
cylinder is maintained at T_1.
Determine   the   one-dimensional
transient temperature distribution
in the two cylinders.

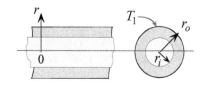

4.17 The initial temperature distribution in a solid cylinder of radius r_o is
given by

$$T(r,0) = T_1 + (T_2 - T_1)(r/r_o)^2,$$

where T_1 and T_2 are constant. At time $t \geq 0$ the cylinder is insulated
along its surface. Neglecting axial conduction, determine the
transient temperature distribution.

4.18 A sphere of radius r_o is initially at temperature $T_i = f(r)$. At time
$t \geq 0$ the   surface   is   maintained   at   uniform   temperature T_o.
Determine the one-dimensional transient temperature distribution.

4.19   A roast at temperature T_i is placed in an oven at temperature T_∞. The roast is cooked by convection and radiation. The heat transfer coefficient is h. Model the roast as a sphere of radius r_o and assume that the radiation flux is constant equal to q_o''. How long will it take for the center temperature to reach a specified level T_o?

4.20   A plate of thickness L is initially at zero temperature. One side is insulated while the other side suddenly exchanges heat by convection. The heat transfer coefficient is h. The ambient temperature $T_\infty(t)$ varies with time according to

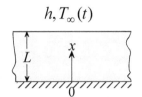

$$T_\infty(t) = T_o e^{-\beta t},$$

where T_o and β are constant. Use Duhamel's superposition integral to determine the one-dimensional transient temperature.

4.21   Consider a plate of thickness L which is initially at zero temperature. At $t \geq 0$ one side is insulated while the other side is maintained at a time dependent temperature given by

$$T = A_1 t, \quad 0 \leq t \leq t_a,$$
$$T = A_2 t, \quad\quad t \geq t_a.$$

Use Duhamel's integral to determine the one-dimensional transient temperature.

4.22   Aerodynamic heating of a space vehicle during entry into a planetary atmosphere varies with time due to variation in altitude and vehicle speed. A simplified heating flux during entry consists of two linear segments described by

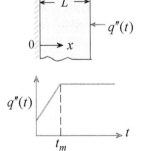

$$q_1''(t) = q_o'' + c_1 t, \quad 0 \leq t \leq t_m,$$

$$q_2''(t) = q_o'' + c_1 t_m, \quad t \geq t_m.$$

Model the skin of a vehicle as a plate of thickness L which is insulated on its back side and assume zero initial temperature, use Duhamel's integral to determine the one-dimensional transient temperature of the skin. For constant heat flux of magnitude unity, the temperature solution to the auxiliary problem for this model is

$$\bar{T}(x,t) = \frac{\alpha}{kL}t + \frac{L}{k}\left[\frac{3x^2 - L^2}{6L^2} - \frac{2}{\pi^2}\sum_{n=1}^{\infty}\frac{(-1)^n}{n^2}e^{-\alpha(n\pi/L)^2 t}\cos(n\pi x/L)\right],$$

where k is thermal conductivity and α is diffusivity.

4.23 A solid cylinder of radius r_o is initially at zero temperature. The cylinder is placed in a convection oven. The oven is turned on and its temperature begins to rise according to

$$T_\infty(t) = T_o\sqrt{t} ,$$

where T_o is constant. The heat transfer coefficient is h, thermal conductivity k and diffusivity α. Use Duhamel's integral to determine the one-dimensional transient temperature.

4.24 A cylinder is heated by convection in an oven. When the oven is turned on its temperature varies with time according to

$$T_\infty(t) = T_o\, e^{\beta t},$$

where T_o and β are constant. The initial temperature is zero. Use Duhamel's integral to determine the one-dimensional transient temperature.

4.25 A solid sphere of radius r_o is initially at zero temperature. At time $t > 0$ the sphere is heated by convection. The heat transfer coefficient is h. The ambient temperature varies with time according to

$$T_\infty(t) = T_o(1 + At).$$

Use Duhamel's integral to determine the one-dimensional transient temperature.

4.26 The surface of a semi-infinite plate which is initially at uniform temperature T_i is suddenly heated with a time dependent flux given by

$$q_o'' = \frac{C}{\sqrt{t}}.$$

where C is constant. Use similarity method to determine the one-dimensional transient temperature.

4.27 In 1864, Kelvin used a simplified transient conduction model to estimate the age of the Earth. Neglecting curvature effect, he modeled the Earth as a semi-infinite region at a uniform initial temperature of $4000\,^\circ C$, which is the melting temperature of rock. He further assumed that the earth surface at $x = 0$ remained at $0\,^\circ C$ as the interior cooled off. Presently, measurements of temperature variation with depth near the Earth's surface show that the gradient is approximately $0.037\,^\circ C/m$. Use Kelvin's model to estimate the age of the Earth.

4.28 Transient conduction in a semi-infinite region is used to determine the thermal diffusivity of material. The procedure is to experimentally heat the surface of a thick plate which is initially at uniform temperature such that the surface is maintained at constant temperature. Measurement of surface temperature and the temperature history at a fixed location in the region provides the necessary data to compute the thermal diffusivity.

[a] Show how the transient solution in a semi-infinite region can be used to determine the thermal diffusivity.

[b] Justify using the solution for a semi-infinite region as a model for a finite thickness plate.

[c] Compute the thermal diffusivity for the following data:

T_i = initial temperature = $20\,^\circ C$,
T_o = surface temperature = $131\,^\circ C$,
$T(x,t) = 64\,^\circ C$ = temperature at location x and time t
x = distance from surface = 1.2 cm
t = time = 25 s

4.29 Two long bars, 1 and 2, are initially at uniform temperatures T_{01} and T_{02}, respectively. The two bars are brought into perfect contact at their ends and allowed to exchange heat by conduction. Assuming that the bars are insulated along their surfaces, determine:

[a] The transient temperature of the bars.

[b] The interface temperature.

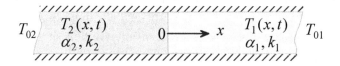

5

CONDUCTION IN POROUS MEDIA

5.1 Examples of Conduction in Porous Media

The flow of fluid in porous media plays an important role in temperature distribution and heat transfer rate. The term *transpiration cooling* is used to describe heat removal associated with fluid flow through porous media. The idea is to provide intimate contact between the coolant and the material to be cooled. Thus heat transfer in porous media is a volumetric process rather than surface action. Fig. 5.1 illustrates several examples of conduction in porous media. In Fig. 5.1a, a porous ring is placed at the throat of a rocket nozzle to protect it from high temperature damage. Air or helium is injected through the porous ring to provide an efficient cooling mechanism. Fig. 5.1b shows a porous heat shield designed to protect a space vehicle during atmospheric entry. Cooling of porous turbine blades is shown in Fig. 5.1c. Transpiration cooling has been proposed as an

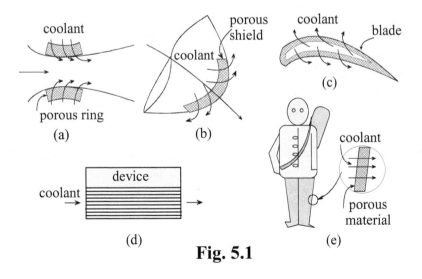

Fig. 5.1

efficient method for protecting microelectronic components. Fig. 5.1d represents an electronic device with a microchannel sink. Coolant flow through the channels provides effective thermal protection. Fig. 5.1e shows a porous fireman's suit for use in a hot environment. Other examples include heat transfer in blood-perfused tissue and geothermal applications.

5.2 Simplified Heat Transfer Model

A rigorous analysis of heat transfer in porous media involves solving two coupled energy equations, one for the porous material and one for the fluid. Thus the solution is represented by two temperature functions. The analysis can be significantly simplified by assuming that at any point in the solid-fluid porous matrix the two temperatures are identical [1]. This important approximation is made throughout this chapter.

5.2.1 Porosity

We define porosity P as the ratio of pore or fluid volume, V_f, to total volume V. Thus

$$P = \frac{V_f}{V}.   (5.1)$$

For a porous wall of face area A and thickness L, the volume is given by

$$V = AL.   (a)$$

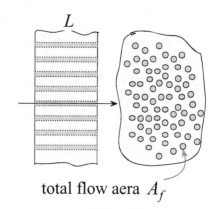

total flow aera A_f

Fig. 5.2

If we model the pores as straight channels, as shown in Fig. 5.2, the fluid volume is given by

$$V_f = A_f L,   (b)$$

where A_f is the total fluid flow area.   Substituting (a) and (b) into eq. (5.1) gives

$$P = \frac{A_f}{A},   (c)$$

or

$$A_f = PA.   (5.2)$$

It follows that the area of the solid material, A_s, is given by

$$A_s = (1-P)A. \tag{5.3}$$

Note that porosity is a property of the porous material. Although its value changes with temperature, it is assumed constant in our simplified model.

5.2.2 Heat Conduction Equation: Cartesian Coordinates

Consider one-dimensional transient conduction in a porous wall shown in Fig. 5.3. The face area is A and the porosity is P. Fluid flows through the wall at a rate $\dot{m}$. We assume that the wall generates energy at a rate q''' per unit volume. The following assumptions are made:

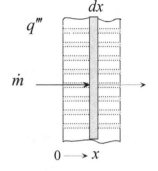

Fig. 5.3

(1) Constant porosity.
(2) Constant properties of both solid and fluid.
(3) Solid and fluid temperatures are identical at each location.
(4) Constant flow rate.
(5) Negligible changes in kinetic and potential energy.

Consider an element dx shown in Fig. 5.3. Energy is exchanged with the element by conduction and convection. Conduction takes place through both the solid material and the fluid. Conservation of energy gives

$$\dot{E}_{in} + \dot{E}_g - \dot{E}_{out} = \dot{E}, \tag{d}$$

where

$\dot{E}$ = rate of energy change within the element
$\dot{E}_{in}$ = rate of energy added to element
$\dot{E}_g$ = rate of energy generated within element
$\dot{E}_{out}$ = rate of energy removed from element

Using Fourier's law to determine conduction through the solid and fluid and accounting for energy added by fluid motion, $\dot{E}_{in}$ is given by

$$\dot{E}_{in} = -k_s(1-P)A\frac{\partial T}{\partial x} - k_f PA\frac{\partial T}{\partial x} + \dot{m}c_{pf}T, \tag{e}$$

where

$A =$ wall area normal to flow direction

$c_{pf} =$ specific heat of fluid

$k_f =$ thermal conductivity of fluid

$k_s =$ thermal conductivity of solid material

$\dot{m} =$ coolant mass flow rate

$T =$ temperature

$x =$ distance

Defining the thermal conductivity of the solid-fluid matrix, $\bar{k}$, as

$$\bar{k} = (1 - P)k_s + Pk_f,$$ (5.4)

equation (e) becomes

$$\dot{E}_{in} = -\bar{k}A\frac{\partial T}{\partial x} + \dot{m}c_{pf}T.$$ (f)

Equation (f) is used to formulate $\dot{E}_{out}$ as

$$\dot{E}_{out} = -\bar{k}A\frac{\partial T}{\partial x} - \bar{k}A\frac{\partial^2 T}{\partial x^2}dx + \dot{m}c_{pf}(T + \frac{\partial T}{\partial x}dx).$$ (g)

Energy generated in the element, $\dot{E}_g$, is given by

$$\dot{E}_g = q'''A\,dx.$$ (h)

The rate of energy change within the element, $\dot{E}$, represents changes in the energy of the solid and of the fluid. Each component is proportional to its mass within the element. Thus

$$\dot{E} = \rho_s c_{ps}(1 - P)A\frac{\partial T}{\partial t}dx + \rho_f c_{pf}PA\frac{\partial T}{\partial t}dx.$$ (i)

Defining the heat capacity of the solid-fluid matrix, $\overline{\rho c_p}$, as

$$\overline{\rho c_p} = (1 - P)\rho_s c_{ps} + P\rho_f c_{pf},$$ (5.5)

equation (i) becomes

$$\dot{E} = \overline{\rho c_p}A\frac{\partial T}{\partial t}dx.$$ (j)

Substituting (f), (g), (h) and (j) into (d) gives

$$\frac{\partial^2 T}{\partial x^2} - \frac{\dot{m} c_{pf}}{A \bar{k}} \frac{\partial T}{\partial x} + \frac{q'''}{\bar{k}} = \frac{1}{\bar{\alpha}} \frac{\partial T}{\partial t}, \tag{5.6}$$

where $\bar{\alpha}$ is the thermal diffusivity of the solid-fluid matrix, defined as

$$\bar{\alpha} = \frac{\bar{k}}{\rho c_p}. \tag{5.7}$$

5.2.3 Boundary Conditions

As with other conduction problems, boundary conditions in porous media can take on various forms. The following are typical examples.

(i) Specified temperature. This condition may be imposed at the inlet or outlet and is expressed as

$$T(0,t) = T_1 \text{ or } T(L,t) = T_2. \tag{5.8}$$

(ii) Convection at outlet boundary. Conservation of energy at the solid part of the outlet plane gives

$$-k_s \frac{\partial T(L,t)}{\partial x} = h[T(L,t) - T_\infty], \tag{5.9}$$

where h is the heat transfer coefficient and T_∞ is the ambient temperature. Note that the fluid plays no role in this boundary condition since energy carried by the fluid to the outlet plane is exactly equal to the energy removed.

(iii) Inlet supply reservoir. Consider a reservoir at temperature T_o supplying fluid to a porous wall shown in Fig. 5.4. Application of conservation of energy to the fluid between the outlet of the reservoir and the inlet to the wall ($x = 0$) gives

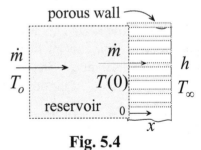

Fig. 5.4

$$\dot{m} c_{pf} [T_o - T(0,t)] = -\bar{k} A \frac{\partial T(0,t)}{\partial x}. \tag{5.10}$$

This boundary condition neglects transient changes in the fluid between the reservoir and wall inlet.

5.2.4 Heat Conduction Equation: Cylindrical Coordinates

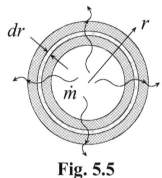

dr

r

$\dot{m}$

Fig. 5.5

Consider radial flow through a porous cylinder of length L. The cylinder generates energy at a volumetric rate of q'''. Application of conservation of energy and Fourier's law of conduction to the element dr shown in Fig. 5.5 gives

$$\frac{\partial^2 T}{\partial r^2} + \left(1 - \frac{\dot{m}c_{pf}}{2\pi \overline{k}L}\right)\frac{1}{r}\frac{\partial T}{\partial r} + \frac{q'''}{\overline{k}} = \frac{1}{\overline{\alpha}}\frac{\partial T}{\partial t}. \tag{5.11}$$

5.3 Applications

Example 5.1: Steady State Conduction in a Porous Plate

A porous plate of thickness L is heated on one side by convection while supplied by a coolant at a flow rate of $\dot{m}$ on the other side. Coolant temperature in the supply reservoir is T_o. The heat transfer coefficient is h and the ambient temperature is T_∞. Assume steady state. Determine the temperature distribution in the plate and the coolant rate per unit surface area needed to maintain the hot side at design temperature T_d.

(1) Observations. (i) This is a one-dimensional conduction problem in a porous plate with coolant flow. (ii) Temperature distribution depends on the coolant flow rate. The higher the coolant flow rate, the lower the surface temperature of the hot side.

(2) Origin and Coordinates. Fig. 5.4 shows the origin and coordinate axis.

(3) Formulation.

   (i) Assumptions. (1) Solid and fluid temperatures are the same at any given location, (2) constant porosity, (3) constant properties of both solid and fluid, (4) constant flow rate, (5) negligible changes in kinetic and potential energy, (6) no energy generation, and (7) steady state.

(ii) Governing Equations. Based on the above assumptions, eq. (5.6) simplifies to

$$\frac{d^2T}{dx^2} - \frac{\dot{m}c_{pf}}{A\overline{k}}\frac{dT}{dx} = 0,$$

or

$$\frac{d^2T}{dx^2} - \frac{\beta}{L}\frac{dT}{dx} = 0, \tag{a}$$

where β is a dimensionless coolant flow rate parameter defined as

$$\beta = \frac{\dot{m}c_{pf}L}{A\overline{k}}. \tag{b}$$

(iii) Boundary Conditions. The supply reservoir boundary condition is obtained from eq. (5.10). Using the definition of β, eq. (5.10) becomes

$$\frac{\beta}{L}[T_o - T(0)] = -\frac{dT(0)}{dx}. \tag{c}$$

The second boundary condition, convection at $x = L$, is given by eq. (5.9)

$$-k_s\frac{dT(L)}{dx} = h[T(L) - T_\infty], \tag{d}$$

where k_s is the thermal conductivity of the solid material.

(4) Solution. Equation (a) is solved by integrating it twice to obtain

$$T(x) = C_1 \exp(\beta x/L) + C_2, \tag{e}$$

where C_1 and C_2 are constants of integration. Using boundary conditions (c) and (d) gives C_1 and C_2

$$C_1 = \frac{Bi(T_\infty - T_o)\exp(-\beta)}{[\beta + Bi]} \quad \text{and} \quad C_2 = T_o. \tag{f}$$

Substituting into (e), rearranging and introducing the definition of the Biot number gives

$$\frac{T(x) - T_o}{T_\infty - T_o} = \frac{Bi}{Bi + \beta}\exp\beta[(x/L) - 1], \tag{g}$$

where the Biot number is defined as

$$Bi = \frac{hL}{k_s}.$$

(h)

Equation (g) gives the temperature distribution in the plate. To determine the flow rate per unit surface area needed to maintain the hot side at design temperature T_d, set $x = L$ and $T = T_d$ in (g)

$$\frac{T_d - T_o}{T_\infty - T_o} = \frac{Bi}{Bi + \beta}.$$

Solving for coolant flow parameter β and using the definitions of β and Biot number Bi, gives the required flow rate per unit surface area

$$\frac{\dot{m}}{A} = \frac{\bar{k}h}{k_s c_{pf}} \frac{(T_\infty - T_d)}{(T_d - T_o)}.$$

(i)

(5) Checking. *Dimensional check*: The Biot number Bi, coolant flow parameter β and the exponent of the exponential are dimensionless.

Limiting check: (i) If $h = 0$ ($Bi = 0$), no heat is exchanged at $x = L$ and consequently the entire plate should be at the coolant temperature T_o. Setting $Bi = 0$ in (g) gives $T(x) = T_o$.

(ii) If $h = \infty$ ($Bi = \infty$), surface temperature at $x = L$ should be the same as the ambient temperature T_∞. Setting $Bi = \infty$ and $x = L$ in (g) gives $T(L) = T_\infty$.

(iii) If the coolant flow rate is infinite ($\beta = \infty$), the entire plate should be at the coolant temperature T_o. Noting that $x/L \le 1$, setting $\beta = \infty$ in (g) gives $T(x) = T_o$.

(iv) If the coolant flow rate is zero ($\beta = 0$), condition (c) shows that the plate is insulated at $x = 0$. Thus the entire plate should be at the ambient temperature T_∞. Setting $\beta = 0$ in (g) gives $T(x) = T_\infty$.

(6) Comments. (i) As anticipated, solution (g) shows that increasing the coolant parameter β lowers the plate temperature. Note that since $(x/L) \le 1$, the exponent of the exponential in (g) is negative. (ii) The solution depends on two parameters: the Biot number Bi and coolant flow parameter β. (iii) The required flow rate per unit area can also be determined using energy balance. Conservation of energy for a control volume between the reservoir and the outlet at $x = L$, applied at design condition, gives

$$\dot{m}c_{pf}T_o = \dot{m}c_{pf}T_d + hA(1-P)(T_\infty - T_d) - k_f AP \frac{dT(L)}{dx}. \qquad \text{(j)}$$

Note that this energy balance accounts for surface convection and conduction through the fluid at $x = L$. Using (d) to eliminate $\dfrac{dT(L)}{dx}$ and solving the resulting equation for $\dfrac{\dot{m}}{A}$ gives equation (i).

Example 5.2: Transient Conduction in a Porous Plate

A porous plate of porosity P and thickness L is initially at a uniform temperature T_o. Coolant at temperature T_o flows through the plate at a rate $\dot{m}$. At time $t \geq 0$ the plate surface at the coolant inlet is maintained at constant temperature T_1 while the opposite surface is insulated. Determine the transient temperature distribution in the plate.

(1) Observations. (i) This is a one-dimensional transient conduction problem in a porous plate with coolant flow. (ii) Since one side is insulated, physical consideration requires that the entire plate be at T_1 at steady state.

(2) Origin and Coordinates. The origin is selected at the coolant side and the coordinate x points towards the insulated side.

(3) Formulation.

   (i) Assumptions. (1) Solid and fluid temperatures are identical at any given location, (2) constant porosity, (3) constant properties of both solid and fluid, (4) constant flow rate, (5) negligible changes in kinetic and potential energy and (6) no energy generation.

   (ii) Governing Equations. Based on the above assumptions, eq. (5.6) gives

$$\frac{\partial^2 T}{\partial x^2} - \frac{\dot{m}c_{pf}}{A\bar{k}}\frac{\partial T}{\partial x} = \frac{1}{\alpha}\frac{\partial T}{\partial t}. \qquad (5.6)$$

   (iii) Independent Variable with Two Homogeneous Boundary Conditions. Only the x-variable can have two homogeneous conditions.

(iv) Boundary and Initial Conditions. The two boundary conditions are

(1) $T(0,t) = T_1$,   (2) $\dfrac{\partial T(L,t)}{\partial x} = 0$

The initial condition is

(3) $T(x,0) = T_o$

(4) Solution. For convenience, the problem is first expressed in dimensionless form. Let

$$\theta = \frac{T - T_1}{T_o - T_1}, \quad \xi = \frac{x}{L}, \quad \tau = \frac{\bar{\alpha}\, t}{L^2} \text{ (Fourier number)},$$

$$\beta = \frac{1}{2} \frac{\dot{m} c_{pf}\, L}{A\bar{k}} \text{ (coolant flow rate parameter)}$$

Using, these dimensionless variables and parameters, eq. (5.6) gives

$$\frac{\partial^2 \theta}{\partial \xi^2} - 2\beta \frac{\partial \theta}{\partial \xi} = \frac{\partial \theta}{\partial \tau}. \qquad \text{(a)}$$

The boundary and initial conditions become

(1) $\theta(0,\tau) = 0$

(2) $\dfrac{\partial \theta(1,\tau)}{\partial \xi} = 0$

(3) $\theta(\xi,0) = 1$

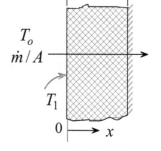

Fig. 5.6

(i) Assumed Product Solution. Let

$$\theta(\xi,\tau) = X(\xi)\,\Gamma(\tau). \qquad \text{(b)}$$

Substituting (b) into (a), separating variables and setting the separated equation equal to a $\pm \lambda_n^2$, we obtain

$$\frac{d^2 X_n}{d\xi^2} - 2\beta \frac{dX_n}{d\xi} \mp \lambda_n^2 X_n = 0, \qquad \text{(c)}$$

and

$$\frac{d\Gamma_n}{d\tau} \mp \lambda_n^2 \, \Gamma_n = 0 . \tag{d}$$

(ii) Selecting the Sign of the λ_n^2 Terms. Since there are two homogeneous conditions in the ξ-variable, the plus sign must be used in equation (c). Thus (c) and (d) become

$$\frac{d^2 X_n}{d\xi^2} - 2\beta \frac{dX_n}{d\xi} + \lambda_n^2 \, X_n = 0 , \tag{e}$$

and

$$\frac{d\Gamma_n}{d\tau} + \lambda_n^2 \, \Gamma_n = 0 . \tag{f}$$

It can be shown that the equations corresponding to $\lambda_n = 0$ yield a trivial solutions for $X_0 \Gamma_0$ and will not be detailed here.

(iii) Solutions to the Ordinary Differential Equations. The solution to (e) takes on three different forms depending on β^2 and λ_n^2 [see equations (A-5) and (A-6), Appendix A]. However, only one option yields the characteristic values λ_n. This option corresponds to $\beta^2 < \lambda_n^2$. Thus the solution is given by equation (A-6a)

$$X_n(\xi) = \exp(\beta\xi) \big[A_n \sin M_n \xi + B_n \cos M_n \xi \big], \tag{g}$$

where A_n and B_n are constants of integration and M_n is defined as

$$M_n = \sqrt{\lambda_n^2 - \beta^2} . \tag{h}$$

The solution to (f) is

$$\Gamma_n(\tau) = C_n \exp(-\lambda_n^2 \, \tau) , \tag{i}$$

where C_n is constant.

(iv) Application of Boundary and Initial Conditions. Boundary condition (1) gives

$$B_n = 0 . \tag{j}$$

Boundary condition (2) gives the characteristic equation for λ_n

$$\beta \tan M_n = -M_n . \tag{k}$$

Substituting (g) and (i) into (b)

$$\theta(\xi,\tau) = \sum_{n=1}^{\infty} a_n \exp(\beta\xi - \lambda_n^2\tau) \sin M_n\xi. \qquad (1)$$

The non-homogeneous initial condition gives

$$1 = \sum_{n=1}^{\infty} a_n \exp(\beta\xi) \sin M_n\xi. \qquad (m)$$

(v) Orthogonality. Application of orthogonality gives a_n. Comparing (c) with Sturm-Liouville equation (3.5a), shows that the weighting function $w(\xi)$ is

$$w(\xi) = \exp(-2\beta\xi). \qquad (n)$$

Multiplying both sides of equation (m) by $w(\xi)\exp(\beta\xi)\sin M_m\xi\, d\xi$, integrating from $\xi = 0$ to $\xi = 1$ and applying orthogonality, eq. (3.7), gives

$$\int_0^1 \exp(-\beta\xi)\sin M_n\xi\, d\xi = a_n \int_0^1 \sin^2 M_n\xi\, d\xi.$$

Evaluating the integrals and solving for a_n we obtain

$$a_n = \frac{4M_n^2}{(\beta^2 + M_n^2)(2M_n - \sin 2M_n)}. \qquad (o)$$

(5) Checking. *Dimensional check:* Since the problem is expressed in dimensionless form, inspection of the solution shows that all terms are dimensionless.

Limiting check: At time $t = \infty$, the entire plate should be at T_1. Setting $\tau = \infty$ in (1) gives $\theta(\xi,\infty) = 0$. Using the definition of θ gives $T(x,\infty) = T_1$.

(6) Comments. The boundary conditions used in this example are selected to simplify the solution. More realistic boundary condition at $x = 0$ and $x = L$ are given in equations (5.9) and (5.10), respectively.

REFEENCES

[1] Schneider, P.J., *Conduction Heat Transfer*, Addison-Wesley, Reading, Massachusetts, 1955.

PROBLEMS

5.1 Energy is generated in a porous spherical shell of porosity P at a volumetric rate q'''. Coolant with specific heat c_{pf} flows radially outwards at a rate $\dot{m}$. The thermal diffusivity of the solid-fluid matrix is $\overline{\alpha}$ and the conductivity is $\overline{k}$. Show that the one-dimensional transient heat conduction equation is given by

$$\frac{1}{r^2}\frac{\partial}{\partial r}\left(r^2\frac{\partial T}{\partial r}\right) - \beta\frac{r_o}{r^2}\frac{\partial T}{\partial r} + \frac{q'''}{\overline{k}} = \frac{1}{\overline{\alpha}}\frac{\partial T}{\partial t},$$

where r_o is the ouside radius of the shell and $\beta = \dfrac{\dot{m}c_{pf}}{4\pi\overline{k}r_o}$.

5.2 A simplified model for analyzing heat transfer in living tissue treats capillary blood perfusion as blood flow through porous media. Metabolic heat production can be modeled as volumetric energy generation. Consider a tissue layer of thickness δ and porosity P with metabolic heat generation q'''. Blood enters the tissue layer at temperature T_1 and leaves at temperature T_2. Blood flow rate per unit surface area is $\dot{m}/A$. Determine the one-dimensional steady state temperature distribution in the tissue.

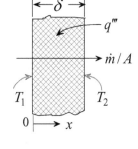

5.3 A proposed method for cooling high power density devices involves pumping coolant through micro channels etched in a sink. To explore the feasibility of this concept, model the sink as a porous disk of radius r_o and thickness δ having a volumetric energy generation q'''. Coolant enters the disk at a flow rate $\dot{m}$ and temperature T_o. Assume that the outlet surface is insulated. Determine the steady state temperature of the insulated surface.

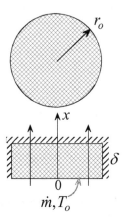

5.4   To protect the wall of a high temperature
      laboratory chamber a porous material is
      used. The wall is heated at a flux q_o'' on one
      side. On the opposite side, coolant at a mass
      flow rate $\dot{m}$ is supplied from a reservoir at
      temperature T_o. The wall area is A,
      thickness L and porosity P. Determine the
      steady state temperature of the heated
      surface.

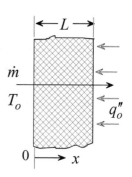

5.5   A porous shield of porosity P, surface area A
      and thickness L is heated with flux q_o''.
      Coolant at a reservoir temperature T_o is
      used to protect the heated surface. Determine
      the coolant flow rate $\dot{m}$ needed to maintain
      the heated surface at design temperature
      T_d.

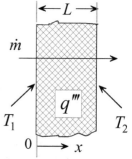

5.6   Heat is generated in a porous wall of area A
      and thickness L at a rate which depends on temperature according to

$$q''' = q_o'''(1 + \lambda T),$$

      where q_o''' and λ are constant. The porosity is P and the coolant flow
      rate is $\dot{m}$. The two surfaces are maintained at specified temperatures
      T_1 and T_2. Determine the steady state temperature distribution.
      Assume that $(-\dot{m}c_{pf}/2\bar{k}A)^2 > (q_o''\lambda/\bar{k})$.

5.7   The porosity P of a wall varies linearly along its thickness according
      to

$$P = b + cx,$$

      where b and c are constant. The wall thickness is L and its area is A.
      The coolant flow rate is $\dot{m}$. The two surfaces are maintained at
      specified temperatures T_1 and T_2. Determine the steady state
      temperature distribution.

5.8   It is proposed to design a porous suit for firemen to protect them in
      high temperature environments.   Consider a suit material with
      porosity $P = 0.02$, thermal conductivity $k_s = 0.035$ W/m–$^\circ$C and

thickness $L = 1.5\,\text{cm}$. It is estimated that an inside suit surface temperature of $55\,°\text{C}$ can be tolerated for short intervals of 30 minute duration. You are asked to estimate the amount of coolant (air) which a fireman must carry to insure a safe condition for 30 minutes when the outside suit surface temperature is $125\,°\text{C}$.

5.9   Ice at the freezing temperature T_f is stored in a cubical container of inside length L. The wall thickness is δ and its conductivity is k_s. The outside ambient temperature is T_∞ and the heat transfer coefficient is h. To prolong the time it takes to melt all the ice, it is proposed to incorporate an air reservoir inside the container and use a porous wall of porosity P. The air reservoir is designed to release air at a constant rate $\dot{m}$ and distribute it uniformly through the porous wall. The ice mass is M and its latent heat of fusion is L. The specific heat of air is c_{pf} and thermal conductivity is k_f. The air enters the porous wall at the ice temperature T_f. The bottom side of the container is insulated. Determine:

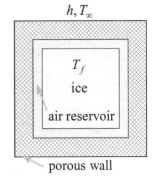

[a] The time t_m it takes for all the ice to melt.

[b] The ratio of melt time t_m using a porous wall to melt time t_o using a solid wall.

[c] The required air mass M_f.

[d] Compute t_m / t_o and M_f for the following data:

$$c_{pf} = 1005\,\text{J/kg–}°\text{C},\quad k_f = 0.025\,\text{W/m–}°\text{C},\quad M = 4\,\text{kg}$$

$$h = 26\,\text{W/m}^2\text{–}°\text{C},\quad k_s = 0.065\,\text{W/m–}°\text{C},\quad L = 20\,\text{cm},$$

$$h = 26\,\text{W/m}^2\text{–}°\text{C},\quad L = 333{,}730\,\text{J/kg},\quad T_f = 0\,°\text{C},$$

$$T_\infty = 400\,°\text{C},\quad \dot{m} = 0.00014\,\text{kg/s},\quad P = 0.3,\quad \delta = 5\,\text{cm}$$

5.10   A porous tube of porosity P, inner radius r_i, outer radius r_o and length L exchanges heat by convection along its outer surface. The heat transfer coefficient is h and the ambient temperature is T_∞.

Coolant at flow rate $\dot{m}$ is supplied to the tube from a reservoir at temperature T_o. The coolant flows uniformly outwards in the radial direction. Assume one-dimensional conduction, determine the inside and outside surface temperatures.

5.11 A two-layer composite porous wall is proposed to provide rigidity and cooling effectiveness. The thickness of layer 1 is L_1, its porosity is P_1 and conductivity is k_{s1}. The thickness of layer 2 is L_2, its porosity P_2 and conductivity is k_{s2}. Coolant at flow rate $\dot{m}$ enters layer 1 at temperature T_o. It leaves layer 2 where heat is exchanged by convection. The ambient temperature is T_∞, heat transfer coefficient is h and surface area is A.

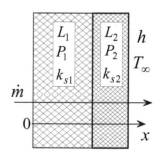

[a] Formulate the heat equations and boundary conditions for the temperature distribution in the composite wall.

[b] Solve the equations and apply the boundary conditions. (Do not determine the constants of integration).

5.12 To increase the rate of heat transfer from a fin it is proposed to use porous material with coolant flow. To evaluate this concept, consider a semi-infinite rectangular fin with a base temperature T_o. The width

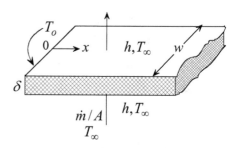

of fin is w and its thickness is δ. The surface of the fin exchanges heat with an ambient fluid by convection. The heat transfer coefficient is h and the ambient temperature is T_∞. Coolant at temperature T_∞ and flow rate per unit surface area $\dot{m}/A$ flows normal to the surface. Coolant specific heat is c_{pf} and the thermal conductivity of the solid-fluid matrix is $\bar{k}$. Assume that the coolant equilibrates locally upon entering the fin.

[a] Show that the steady state fin equation for this model is given by

$$\frac{d^2T}{dx^2} - \frac{1}{\delta^2}(4Bi + \beta)(T - T_\infty) = 0,$$

where $Bi = (1 - P)h\delta / 2\bar{k}$ and $\beta = \dot{m}c_{pf}\delta / A\bar{k}$.

[b] Determine the fin heat transfer rate.

5.13   The surface of an electronic package is cooled with fins. However, surface temperature was found to be higher than design specification. One recommended solution involves using porous material and a coolant. The coolant is to flow axially along the length of the fin. To evaluate this proposal, consider a porous rod of radius r_o and length L. One end is heated with a flux q''_o. Coolant enters the opposite end at temperature T_∞ and flow rate $\dot{m}$. The specific heat is c_{pf} and the conductivity of the fluid-solid matrix is $\bar{k}$. The cylindrical surface exchanges heat by convection. The ambient temperature is T_∞ and the heat transfer coefficient is h.

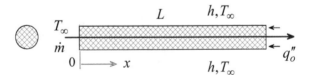

[a] Show that the steady state fin equation for this model is given by

$$\frac{d^2T}{dx^2} - \frac{\beta}{r_o}\frac{dT}{dx} - \frac{2Bi}{r_o^2}(T - T_\infty) = 0,$$

where $Bi = hr_o / \bar{k}$ and $\beta = \dot{m}c_{pf} / \pi r_o \bar{k}$.

[b] Determine the temperature of the heated surface.

5.14   A porous disk of porosity P, inner radius R_i, outer radius R_o and thickness δ is heated at its outer rim with a flux q_o''. Coolant at a mass flow rate $\dot{m}$ and temperature T_∞ flows radially through the disk. The specific heat of the coolant is c_{pf} and the conductivity of the solid-fluid matrix is $\overline{k}$. The disk exchanges heat by convection along its upper and lower surfaces. The heat transfer coefficient is h and the ambient temperature is T_∞. Assume that the Biot number is small compared to unity and the inner surface temperature is at T_∞.

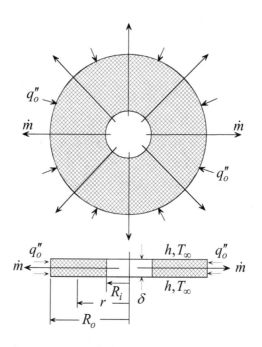

[a] Show that the steady state fin equation is given by

$$ r^2 \frac{d^2 T}{dr^2} + (1 - \beta) r \frac{dT}{dr} - \frac{4Bi}{\delta^2} r^2 (T - T_\infty) = 0, $$

where $Bi = h\delta / 2\overline{k}$ and $\beta = \dot{m} c_{pf} / 2\pi\delta\,\overline{k}$.

[b] Determine the temperature of the heated surface.

5.14 The outer surface of a porous hemispherical shell is heated with uniform flux q_o''. Coolant at flow rate $\dot{m}$ enters the inner surface and flows radially outward. The specific heat of the coolant is c_{pf} and the conductivity of the solid-fluid matrix is $\overline{k}$. The inner radius is R_i and the outer radius is R_o. The inner surface is maintained at uniform temperature T_o. Determine the steady state temperature of the heated surface.

5.15 A straight triangular fin of length L and thickness t at the base is made of porous material of porosity P. The fin exchanges heat by convection along its surface. The heat transfer coefficient is h and the ambient temperature is T_∞. The base is maintained at temperature T_o. Coolant at temperature T_∞ and flow rate per unit surface area $\dot{m}/A$ flows normal to the surface. Coolant specific heat is c_{pf} and the thermal conductivity of the solid-fluid matrix is $\overline{k}$. Determine the ratio of the heat transfer form the porous fin to that of a solid fin.

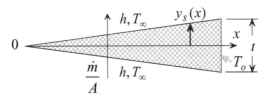

5.16 The surface temperature of a tube of radius R_i is T_o. The tube is cooled with a porous annular fin of porosity P, inner radius R_i and outer radius R_o. Coolant at a mass flow rate $\dot{m}$ and temperature T_∞ flows axially normal to the fin surface. The specific heat of the coolant is c_{pf} and the conductivity of the solid-fluid matrix is $\overline{k}$. The fin exchanges heat by convection along its upper and lower surfaces. The heat transfer coefficient

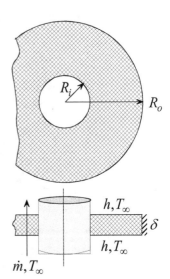

is h and the ambient temperature is T_∞. Assume that the Biot number is small compared to unity. Assume further that the coolant equilibrates locally upon entering the fin.

[a] Show that the steady state fin equation is given by

$$r^2 \frac{d^2T}{dr^2} + r\frac{dT}{dr} - \frac{(4Bi + \beta)}{\delta^2} r^2 (T - T_\infty) = 0,$$

where $Bi = (1 - P)h\delta / 2\bar{k}$ and $\beta = \dot{m}c_{pf}\delta / \pi(R_o^2 - R_i^2)\bar{k}$.

[b] Determine the heat transfer rate from the fin.

5.18 Small stainless steel ball bearings leave a furnace on a conveyor belt at temperature T_o. The balls are stacked forming a long rectangular sheet of width w and height δ. The conveyor belt, which is made of wire mesh, moves through the furnace with velocity U. The ball bearings are cooled by convection along the upper and lower surfaces. The heat transfer coefficient is h and the ambient temperature is T_∞. To accelerate the cooling process, coolant at temperature T_∞ and flow rate per unit surface area $\dot{m}/A$ flows through the balls normal to the surface. Coolant specific heat is c_{pf} and the thermal conductivity of the steel-fluid matrix is $\bar{k}$. Assume that the coolant equilibrates locally upon contacting the balls. Model the ball bearing sheet as a porous fin of porosity P.

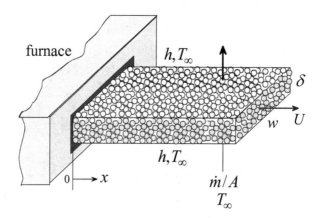

[a] Show that the steady state fin equation is given by

$$\frac{d^2T}{dx^2} - \frac{U}{\overline{\alpha}}\frac{dT}{dx} - \frac{1}{\delta^2}(4Bi + \beta)(T - T_\infty) = 0,$$

Where $Bi = (1 - P)h\delta / 2\overline{k}$, $\beta = \dot{m}c_p\delta / A\overline{k}$ and
$\overline{\alpha} = \overline{k} / \overline{\rho c}_{pf}$.

[b] Determine temperature of balls at a distance L from the furnace.

5.17 Consider transient one-dimensional cooling of a porous hollow cylinder of porosity P, inner radius R_i and outer radius R_o. Coolant at a mass flow rate per unit length $\dot{m}$ and temperature T_o is supplied from a reservoir. It flows radially outward through the cylinder. The specific heat of the coolant is c_{pf} and the conductivity and diffusivity of the solid-fluid matrix are $\overline{k}$ and $\overline{\alpha}$, respectively. Initially the cylinder is at the coolant temperature T_o. At time $t \geq 0$ the outside surface exchanges heat by convection with the ambient fluid. The heat transfer coefficient is h and the ambient temperature is T_∞. Show that the dimensionless heat equation, boundary and initial conditions for transient one-dimensional conduction are given by

$$\frac{1}{\xi}\frac{\partial}{\partial\xi}\left(\xi\frac{\partial\theta}{\partial\xi}\right) - 2\beta\frac{1}{\xi}\frac{\partial\theta}{\partial\xi} = \frac{\partial\theta}{\partial\tau},$$

$$\frac{\partial\theta(1,\tau)}{\partial\xi} = 2\beta\,\theta(1,\tau),$$

$$\xi_o\frac{\partial\theta(\xi_o,\tau)}{\partial\xi} = Bi\left[1 - \theta(\xi_o,\tau)\right],$$

$$\theta(\xi,0) = 0,$$

where

$$\theta = (T - T_o)/(T_\infty - T_o), \quad \xi = r/R_i, \quad \xi_o = R_o/R_i,$$
$$\tau = \overline{\alpha}\,t/R_i^2, \quad Bi = hR_o/k_s, \quad \beta = \dot{m}c_{pf}/4\pi\overline{k}.$$

CONDUCTION WITH PHASE CHANGE:
MOVING BOUNDARY PROBLEMS

6.1 Introduction

There are many applications in which a material undergoes phase change such as in melting, freezing, casting, ablation, cryosurgery and soldering. Conduction with phase change is characterized by a moving interface separating two phases. Such problems are usually referred to as *moving boundary* or *free boundary* problems. They are inherently transient during interface motion. The motion and location of the interface are unknown a priori and thus must be determined as part of the solution. Since material properties change following phase transformation and since a discontinuity in temperature gradient exists at the interface, it follows that each phase must be assigned its own temperature function. Furthermore, changes in density give rise to motion of the liquid phase. If the effect of motion is significant, the heat equation of the liquid phase must include a convective term. However, in most problems this effect can be neglected.

A moving front which is undergoing a phase change is governed by a boundary condition not encountered in previous chapters. This condition, which is based on conservation of energy, is non-linear. Because of this non-linearity there are few exact solutions to phase change problems.

In this chapter we will state the heat conduction equations for one-dimensional phase change problems. The interface boundary condition will be formulated and its non-linear nature identified. The governing equations will be cast in dimensionless form to reveal the important parameters governing phase change problems. A simplified model based on quasi-steady state approximation will be described. Solutions to exact problems will be presented.

6.2 The Heat Equations

Fig. 6.1 shows a region which is undergoing phase change due to some action at one of its boundaries. The temperature distribution in the two-phase region is governed by two heat equations, one for the solid phase and one for the liquid phase. We make the following assumptions:

(1) Properties of each phase are uniform and remain constant.
(2) Negligible effect of liquid phase motion due to changes in density.
(3) One-dimensional conduction.
(4) No energy generation.

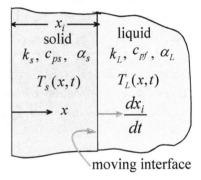

Based on these assumptions, eq. (1.8) gives

$$\frac{\partial^2 T_s}{\partial x^2} = \frac{1}{\alpha_s}\frac{\partial T_s}{\partial t} \quad 0 < x < x_i, \quad (6.1)$$

and

$$\frac{\partial^2 T_L}{\partial x^2} = \frac{1}{\alpha_L}\frac{\partial T_L}{\partial t} \quad x > x_i, \quad (6.2)$$

Fig. 6.1

where the subscripts L and s refer to liquid and solid, respectively.

6.3 Moving Interface Boundary Conditions

Continuity of temperature and conservation of energy give two boundary conditions at the solid-liquid interface. Mathematical description of these conditions follows.

(1) Continuity of temperature. We assume that the material undergoes a phase change at a fixed temperature. Continuity of temperature at the interface requires that

$$T_s(x_i,t) = T_L(x_i,t) = T_f, \quad (6.3)$$

where

$T(x,t) =$ temperature variable

T_f = fusion (melting or freezing) temperature

$x_i = x_i(t)$ = interface location at time t

(2) Energy equation. Fig. 6.2 shows a liquid element which is undergoing phase change to solid. A one-dimensional solid-liquid interface located at $x = x_i(t)$ is assumed to move in the positive x-direction. Consider an element dx_L in the liquid adjacent to the interface. The element has a fixed mass δm. During a time interval dt the element undergoes phase change to solid. Because material density changes as it undergoes phase transformation, the thickness of the

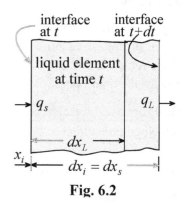

Fig. 6.2

element changes to dx_s. Consequently, the interface advances a distance $dx_i = dx_s$. Application of conservation of energy to the element as it transforms from liquid to solid during time dt, gives

$$\delta E_{in} - \delta E_{out} = \delta E, \tag{a}$$

where

δE_{in} = energy added during time dt

δE_{out} = energy removed during time dt

δE = energy change of element during time dt

The energy added by conduction through the solid phase is

$$\delta E_{in} = q_s \, dt, \tag{b}$$

where

q_s = rate of heat conducted to the element through the solid phase

Energy is removed from the element by conduction and in the form of work done by the element due to changes in volume. These two components of energy are given by

$$\delta E_{out} = q_L \, dt + pdV, \tag{c}$$

where

dV = change in volume of element from liquid to solid during dt

p = pressure

q_L = rate of heat conducted from the element through the liquid phase

The change in energy of the element is given by

$$\delta E = (\hat{u}_s - \hat{u}_L)\delta m, \tag{d}$$

δm = mass of element

$\hat{u}$ = internal energy per unit mass

Substituting (b), (c) and (d) into (a)

$$(q_s - q_L)\,dt = (\hat{u}_s - \hat{u}_L)\delta m + pdV. \tag{e}$$

Applying Fourier's law

$$q_s = -k_s A \frac{\partial T_s(x_i,t)}{\partial x}, \text{ and } q_L = -k_L A \frac{\partial T_L(x_i,t)}{\partial x}, \tag{f}$$

where

A = surface area of element normal to x

k = thermal conductivity

x = coordinate

$x_i = x_i(t)$ = interface location at time t

The mass of the element δm is given by

$$\delta m = \rho_s A dx_i, \tag{g}$$

where ρ is density. The change in volume dV of the element is related to its mass and to changes in density as

$$dV = (\frac{1}{\rho_s} - \frac{1}{\rho_L})\delta m. \tag{h}$$

Substituting (f), (g) and (h) into (e) and assuming that pressure remains constant

$$-k_s \frac{\partial T_s(x_i,t)}{\partial x} + k_L \frac{\partial T_L(x_i,t)}{\partial x} = \rho_s[(\hat{u}_s + \frac{p}{\rho_s}) - (\hat{u}_L + \frac{p}{\rho_L})]\frac{dx_i}{dt}. \tag{i}$$

However

$$\left(\hat{u}_L + \frac{p}{\rho_L}\right) - \left(\hat{u}_s + \frac{p}{\rho_s}\right) = \hat{h}_L - \hat{h}_s = \mathcal{L},\tag{j}$$

where

$\hat{h}$ = enthalpy per unit mass

$\mathcal{L}$ = latent heat of fusion

Substituting (j) into (i)

$$k_s \frac{\partial T_s(x_i,t)}{\partial x} - k_L \frac{\partial T_L(x_i,t)}{\partial x} = \rho_s \mathcal{L} \frac{dx_i}{dt}.\tag{6.4}$$

Eq. (6.4) is the interface energy equation. It is valid for both solidification and melting. However, for melting ρ_s is replaced by ρ_L for melting.

(3) Convection at the interface. There are problems where the liquid phase is not stationary. Examples include solidification associated with forced or free convection over a plate or inside a tube. For such problems equation (6.4) must be modified to include the effect of fluid motion. This introduces added mathematical complications. An alternate approach is to replace the temperature gradient in the liquid phase with Newton's law of cooling. Thus

$$k_s \frac{\partial T_s(x_i,t)}{\partial x} \pm h(T_f - T_\infty) = \rho_s \mathcal{L} \frac{dx_i}{dt},\tag{6.5}$$

where

h = heat transfer coefficient

T_∞ = temperature of the liquid phase far away from the interface

The plus sign in eq. (6.5) is for solidification and the minus sign is for melting.

6.4 Non-linearity of the Interface Energy Equation

A solution to a phase change conduction problem must satisfy the energy condition of eq. (6.4) or eq. (6.5). Careful examination of these equations shows that they are non-linear. The non-linearity is caused by the dependence of the interface velocity dx_i / dt on the temperature gradient. To reveal the non-linear nature of equations (6.4) and (6.5), we form the total derivative of T_s in eq. (6.3)

$$\frac{\partial T_s(x_i,t)}{\partial x} dx + \frac{\partial T_s(x_i,t)}{\partial t} dt = 0.$$

Dividing through by dt and noting that $dx = dx_i$, gives dx_i / dt

$$\frac{dx_i}{dt} = -\frac{\partial T_s(x_i,t)/\partial t}{\partial T_s(x_i,t)/\partial x}. \tag{6.6}$$

When this result is substituted into eq. (6.4) we obtain

$$k_s \left[\frac{\partial T_s(x_i,t)}{\partial x}\right]^2 - k_L \frac{\partial T_L(x_i,t)}{\partial x}\frac{\partial T_s(x_i,t)}{\partial x} = -\rho_s \mathcal{L}\frac{\partial T_s(x_i,t)}{\partial t}. \tag{6.7}$$

Note that both terms on the left side of eq. (6.7) are non-linear. Similarly, substituting eq. (6.6) into eq. (6.5) shows that it too is non-linear.

6.5 Non-dimensional Form of the Governing Equations: Governing Parameters

To identify the governing parameters in phase change problems, the two heat equations, (6.1) and (6.2) and the interface energy equation (6.4) are cast in non-dimensional form. The following dimensionless quantities are defined:

$$\theta_s = \frac{T_s - T_f}{T_f - T_o}, \quad \theta_L = \frac{k_L}{k_s}\frac{T_L - T_f}{T_f - T_o}, \quad \xi = \frac{x}{L}, \quad \tau = Ste\frac{\alpha_s}{L^2}t, \tag{6.8}$$

where L is a characteristic length and T_o is a reference temperature. Ste is the *Stefan number* which is defined as

$$Ste = \frac{c_{ps}(T_f - T_o)}{\mathcal{L}}, \tag{6.9}$$

where c_{ps} is the specific heat of the solid phase. Substituting eq. (6.8) into equations (6.1), (6.2) and (6.4) gives

$$\frac{\partial^2\theta_s}{\partial\xi^2} = Ste\frac{\partial\theta_s}{\partial\tau}, \tag{6.10}$$

$$\frac{\alpha_L}{\alpha_s}\frac{\partial^2\theta_L}{\partial\xi^2} = Ste\frac{\partial\theta_L}{\partial\tau}, \tag{6.11}$$

and

$$\frac{\partial \theta_s(\xi_i,t)}{\partial \xi} - \frac{\partial \theta_L(\xi_i,t)}{\partial \xi} = \frac{d\xi_i}{d\tau}. \tag{6.12}$$

Examination of the dimensionless governing equations shows that two parameters, the ratio of thermal diffusivities and the Stefan number, govern phase change problems. It is worth noting the following:

(1) By including the ratio (k_L / k_s) in the definition of θ_L, this parameter is eliminated from the interface energy equation.

(2) A phase change problem with convection at its boundary will introduce the Biot number as an additional parameter.

(3) The Stefan number, defined in eq. (6.9), represents the ratio of the sensible heat to the latent heat. Sensible heat, $c_{ps}(T_f - T_o)$, is the energy removed from a unit mass of solid at the fusion temperature T_f to lower its temperature to T_o. Latent heat, $\mathcal{L}$, is the energy per unit mass which is removed (solidification) or added (melting) during phase transformation at the fusion temperature. Note that the Stefan number in eq. (6.9) refers to the solid phase since it is defined in terms of the specific heat of the solid c_{ps}. The definition for the liquid phase is

$$Ste = \frac{c_{pL}(T_o - T_f)}{\mathcal{L}}. \tag{6.13}$$

Sensible heat, $c_{pL}(T_o - T_f)$, is the energy added to a unit mass of liquid at fusion temperature T_f to raise its temperature to T_o.

6.6 Simplified Model: Quasi-Steady Approximation

Because of the non-linearity of the interface energy equation, there are few exact solutions to conduction problems with phase change. An approximation which makes it possible to obtain solutions to a variety of problems is based on a *quasi-steady* model. In this model the Stefan number is assumed small compared to unity. A small Stefan number corresponds to sensible heat which is small compared to latent heat. To appreciate the significance of a small Stefan number, consider the limiting case of a material whose specific heat is zero, i.e. $Ste = 0$. Such a material has infinite thermal diffusivity. This means that thermal effects propagate with infinite speed and a steady state is reached instantaneously as the interface moves. Alternatively, a material with infinite latent heat

($Ste = 0$), has a stationary interface. Thus the interface moves slowly for a small Stefan number and the temperature distribution at each instant corresponds to that of steady state. In practice, quasi-steady approximation is justified for $Ste < 0.1$. Setting $Ste = 0$ in equations (6.10) and (6.11) gives

$$\frac{\partial^2 \theta_s}{\partial \xi^2} = 0, \tag{6.14}$$

$$\frac{\partial^2 \theta_L}{\partial \xi^2} = 0. \tag{6.15}$$

Note that the interface energy equation (6.12) is unchanged in this model and that temperature distribution and interface motion are time dependent. What is simplified in this approximation are the governing equations and their solutions.

Example 6.1: Solidification of a Slab at the Fusion Temperature T_f

A slab of thickness L is initially at the fusion temperature T_f. One side of the slab is suddenly maintained at constant temperature $T_o < T_f$ while the opposite side is kept at T_f. A solid-liquid interface forms and moves towards the opposite face. Use a quasi-steady state model to determine the time needed for the entire slab to solidify.

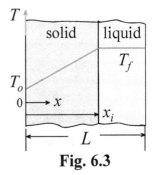

Fig. 6.3

(1) Observations. (i) Physical consideration requires that the liquid phase remains at a uniform temperature equal to the fusion temperature T_f. (ii) Solidification starts at time $t = 0$. (iii) The time it takes the slab to solidify is equal to the time needed for the interface to traverse the slab width L. Thus the problem reduces to determining the interface motion which is governed by the interface energy condition.

(2) Origin and Coordinates. Fig. 6.3 shows the origin and the coordinate axes.

(3) Formulation.

(i) Assumptions.   (1) One-dimensional conduction, (2) constant properties of the liquid and solid phases, (3) no changes in fusion temperature and (4) quasi-steady state, $Ste < 0.1$.

(ii) Governing Equations.   For quasi-steady state the conduction equations, expressed in dimensional form, are

$$\frac{\partial^2 T_s}{\partial x^2} = 0, \tag{a}$$

$$\frac{\partial^2 T_L}{\partial x^2} = 0. \tag{b}$$

(iii) Boundary and Initial Conditions.

(1) $T_s(0,t) = T_o$

(2) $T_s(x_i,t) = T_f$

(3) $T_L(x_i,t) = T_f$

(4) $T_L(L,t) = T_f$

Interface energy condition is

$$(5)\ k_s \frac{\partial T_s(x_i,t)}{\partial x} - k_L \frac{\partial T_L(x_i,t)}{\partial x} = \rho_s \mathcal{L} \frac{dx_i}{dt}.$$

The initial conditions are

(6) $T_L(x,0) = T_f$

(7) $x_i(0) = 0$

(4) Solution. Direct integration of (a) and (b) gives

$$T_s(x,t) = Ax + B, \tag{c}$$

and

$$T_L(x,t) = Cx + D, \tag{d}$$

where A, B, C, and D are constants of integration. These constants can be functions of time. Application of the first four boundary conditions gives

$$T_s(x,t) = T_o + (T_f - T_o)\frac{x}{x_i(t)}, \tag{e}$$

and

$$T_L(x,t) = T_f . \tag{f}$$

As anticipated, the liquid temperature remains constant. The interface location is determined from condition (5). Substituting (e) and (f) into condition (5) gives

$$k_s \frac{T_f - T_o}{x_i} - 0 = \rho_s \mathcal{L} \frac{dx_i}{dt} .$$

Separating variables

$$x_i dx_i = \frac{k_s (T_f - T_o)}{\rho_s \mathcal{L}} dt .$$

Integration gives

$$x_i^2 = \frac{2k_s (T_f - T_o)}{\rho_s \mathcal{L}} t + C_1 .$$

Initial condition (7) gives $C_1 = 0$. Thus

$$x_i(t) = \sqrt{\frac{2k_s (T_f - T_o)}{\rho_s \mathcal{L}} t} . \tag{6.16a}$$

In terms of the dimensionless variables of eq. (6.8), this result can be expressed as

$$\xi_i = \sqrt{2\tau} . \tag{6.16b}$$

The time t_o needed for the entire slab to solidify is obtained by setting $x_i = L$ in eq. (6.16a)

$$t_o = \frac{\rho_s \mathcal{L} L^2}{2k_s (T_f - T_o)} . \tag{6.17a}$$

Since total solidification corresponds to $\xi_i = 1$, eq. (6.16b) gives

$$\tau_o = 1/2 . \tag{6.17b}$$

(5) Checking. *Dimensional check*: eqs. (6.16a) and (6.17a) are dimensionally consistent.

Limiting check: (i) If $T_o = T_f$, no solidification takes place and the time needed to solidify the slab should be infinite. Setting $T_o = T_f$ in eq.

(6.17a) gives $t_o = \infty$. (ii) If the slab is infinitely wide, solidification time should be infinite. Setting $L = \infty$ in eq. (6.17a) gives $t_o = \infty$.

(6) Comments. (i) Initial condition (6) is not used in the solution. This is inherent in the quasi-steady state model. Since time derivatives of temperature are dropped in this model, there is no opportunity to satisfy initial conditions. Nevertheless, the solution to the temperature distribution in the liquid phase does satisfy initial condition (6). (ii) The solution confirms the observation that the liquid phase must remain at T_f. Thus analysis of the liquid phase is unnecessary. (iii) Since the liquid phase remains at the fusion temperature, fluid motion due to density changes or convection plays no role in the solution.

Example 6.2: Melting of Slab with Time Dependent Surface Temperature

The solid slab shown in Fig. 6.4 is initially at the fusion temperature T_f. The side at $x = 0$ is suddenly maintained at a time dependent temperature above T_f, given by

$$T_L(0,t) = T_o \exp \beta t ,$$

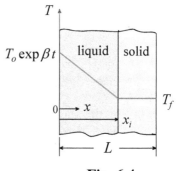

Fig. 6.4

where T_o and β are constant. A liquid-solid interface forms at $x = 0$ and moves towards the opposite side. Use a quasi-steady state model to determine the interface location and the time it takes for the slab to melt.

(1) Observations. (i) Based on physical consideration, the solid phase remains at uniform temperature equal to the fusion temperature T_f. (ii) Melting starts at time $t = 0$. (iii) The time it takes the slab to melt is equal to the time needed for the interface to traverse the slab width L.

(2) Origin and Coordinates. Fig. 6.4 shows the origin and the coordinate axes.

(3) Formulation.

  (i) Assumptions. (1) One-dimensional conduction, (2) constant properties of the liquid phase, (3) no changes in fusion temperature, (4) neglect motion of the liquid phase and (5) quasi-steady state ($Ste < 0.1$).

(ii) Governing Equations. For quasi-steady state the liquid phase conduction equation, expressed in dimensional form, is

$$\frac{\partial^2 T_L}{\partial x^2} = 0 . \tag{a}$$

Since no heat can transfer to the solid phase, its temperature remains constant. Thus

$$T_s(x,t) = T_f . \tag{b}$$

(iii) Boundary and Initial Conditions. The boundary conditions are

(1) $T_L(0,t) = T_o \exp \beta t$

(2) $T_L(x_i,t) = T_f$

Substituting (b) into eq. (6.4) gives the interface energy condition

$$(3) \; -k_L \frac{\partial T_L(x_i,t)}{\partial x} = \rho_L \mathcal{L} \frac{dx_i}{dt}$$

Note that since this is a melting problem, ρ_s in eq. (6.4) is replaced by ρ_L. The initial condition of the interface is

(4) $x_i(0) = 0$

(4) Solution. Integration of (a) gives

$$T_L(x,t) = Ax + B .$$

Application of boundary conditions (1) and (2) gives

$$T_L(x,t) = \frac{T_f - T_o \exp \beta t}{x_i} x + T_o \exp \beta t . \tag{c}$$

Substituting (c) into condition (3) and separating variables

$$-\frac{k_L}{\rho_L \mathcal{L}} (T_f - T_o \exp \beta t) dt = x_i dx_i . \tag{d}$$

Integrating and using initial condition (4)

$$-\frac{k_L}{\rho_L \mathcal{L}} \int_0^t (T_f - T_o \exp \beta t) dt = \int_0^{x_i} x_i dx_i ,$$

or

$$-\frac{k_L}{\rho_L \mathcal{L}}\left[T_f t - (T_o/\beta)\exp(\beta t) + (T_o/\beta)\right] = \frac{x_i^2}{2}.$$

Solving this result for $x_i(t)$

$$x_i(t) = \sqrt{\frac{2k_L}{\rho_L \mathcal{L}}\left[(T_o/\beta)\exp(\beta t) - T_f t - (T_o/\beta)\right]}. \qquad (6.18)$$

To determine the time t_o it takes for the slab to melt, set $x_i = L$ in eq. (6.18)

$$L = \sqrt{\frac{2k_L}{\rho_L \mathcal{L}}\left[(T_o/\beta)\exp(\beta t_o) - T_f t_o - (T_o/\beta)\right]}. \qquad (6.19)$$

This equation can not be solved for t_o explicitly. A trial and error procedure is needed to determine t_o.

(5) Checking. *Dimensional check*: Eq. (6.18) is dimensionally consistent.

Limiting check: For the special case of constant surface temperature at $x = 0$, results for this example should be identical with the solidification problem of Example 6.1 with the properties of Example 6.1 changed to those of the liquid phase. Setting $\beta = 0$ in the time dependent surface temperature gives $T(0,t) = T_o$, which is the condition for Example 6.1. However, since β appears in the denominator in eq. (6.18), direct substitution of $\beta = 0$ can not be made. Instead, the term $\exp \beta t$ in eq. (6.18) is expanded for small values of βt before setting $\beta = 0$. Thus, for small βt eq. (6.18) is written as

$$x_i(t) = \sqrt{\frac{2k_L}{\rho_L \mathcal{L}}}\sqrt{(T_o/\beta)\left[(1 + (\beta t)/1!) + \cdots\right] - T_f t - (T_o/\beta)},$$

or, for $\beta = 0$

$$x_i(t) = \sqrt{\frac{2k_L}{\rho_L \mathcal{L}}(T_o - T_f)t}. \qquad (6.20)$$

This result agrees with eq. (6.16a).

(6) **Comments.** Time dependent boundary conditions do not introduce mathematical complications in the quasi-steady state model.

6.7 Exact Solutions

6.7.1 Stefan's Solution

One of the earliest published exact solutions to phase change problems is credited to Stefan who published his work in 1891 [1]. He considered solidification of a semi-infinite liquid region shown in Fig. 6.5. The liquid is initially at the fusion temperature T_f. The surface at $x = 0$ is suddenly maintained at temperature $T_o < T_f$. Solidification begins instantaneously at $x = 0$. We wish to determine the temperature distribution and interface location. Since no heat can transfer to the liquid phase, its temperature remains constant throughout. Thus

$$T_L(x,t) = T_f . \tag{a}$$

The governing equation in the solid phase is given by eq. (6.1)

$$\frac{\partial^2 T_s}{\partial x^2} = \frac{1}{\alpha_s} \frac{\partial T_s}{\partial t} . \tag{b}$$

The boundary conditions are

(1) $T_s(0,t) = T_o$

(2) $T_s(x_i,t) = T_f$

Substituting (a) into eq. (6.4) gives

(3) $k_s \dfrac{\partial T_s(x_i,t)}{\partial x} = \rho_s \mathcal{L} \dfrac{dx_i}{dt}$

The initial condition is

(4) $x_i(0) = 0$

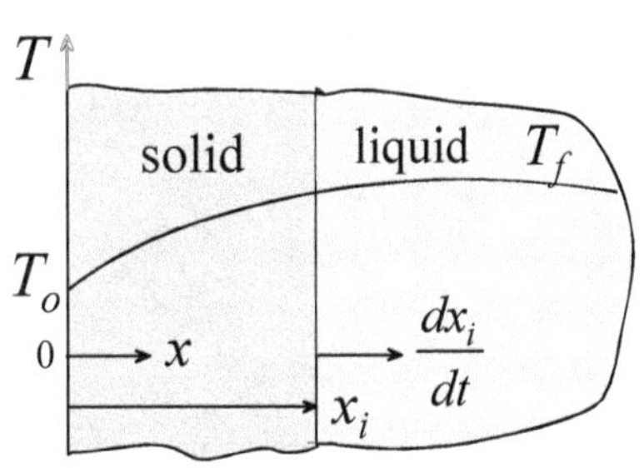

Fig. 6.5

Equation (b) is solved by similarity transformation. We assume that the two independent variables x and t can be combined into a single variable $\eta = \eta(x,t)$. Following the use of this method in solving the problem of transient conduction in a semi-infinite region, the appropriate similarity variable is

$$\eta = \frac{x}{\sqrt{4\alpha_s t}}\,.$$
(6.21)

We postulate that the solution to (b) can be expressed as

$$T_s = T_s(\eta)\,.$$
(c)

Using eq. (6.21) and (c), equation (b) transforms to

$$\frac{d^2 T_s}{d\eta^2} + 2\eta\frac{dT_s}{d\eta} = 0\,.$$
(d)

Thus the governing partial differential equation (b) is transformed into an ordinary differential equation. Note that (d) is identical to eq. (4.33) which describes transient conduction in a semi-infinite region without phase change. Following the procedure used to solve eq. (4.33), the solution to (d) is

$$T_s = A\ \mathrm{erf}\eta + B\,.$$
(6.22)

Applying boundary condition (2) and using eq. (6.21) give

$$T_f = A\ \mathrm{erf}\frac{x_i}{\sqrt{4\alpha_s t}} + B\,.$$
(e)

Since T_f is constant, it follows that the argument of the error function in (e) must also be constant. Thus we conclude that

$$x_i \propto \sqrt{t}\,.$$

Let

$$x_i = \lambda\sqrt{4\alpha_s t}\,,$$
(f)

where λ is a constant to be determined. Note that this solution to the interface location satisfies initial condition (4). Applying boundary condition (1) to eq. (6.22) and noting that $\mathrm{erf}\,0 = 0$, gives

$$B = T_o\,.$$
(g)

Boundary condition (2), (g) and eq. (6.22) give the constant A

$$A = \frac{T_f - T_o}{\mathrm{erf}\,\lambda}\,.$$
(h)

Substituting (g) and (h) into eq. (6.22) gives the temperature distribution in the solid phase

$$T_s(x,t) = T_o + \frac{T_f - T_o}{\text{erf } \lambda} \text{erf } \eta. \tag{6.23}$$

Finally, interface energy condition (3) is used to determine the constant λ. Substituting (f) and eq. (6.23) into condition (3) gives

$$k_s \frac{T_f - T_o}{\text{erf } \lambda} \left[\frac{d}{d\eta}(\text{erf } \eta) \frac{\partial \eta}{\partial x} \right]_{x_i} = \rho_s \mathcal{L} \frac{\lambda}{2} \frac{\sqrt{4\alpha_s}}{\sqrt{t}}.$$

The derivative of the error function is given by eq. (4.34). Substituting into the above and using eq. (6.21) gives

$$\lambda (\exp \lambda^2) \text{ erf } \lambda = \frac{(T_f - T_o) c_{ps}}{\sqrt{\pi} \mathcal{L}}, \tag{6.24}$$

where c_{ps} is specific heat of the solid. Equation (6.24) gives the constant λ. However, since eq. (6.24) can not be solved explicitly for λ, a trial and error procedure is required to obtain a solution. Note that λ depends on the material as well as the temperature at $x = 0$.

It is interesting to examine Stefan's solution for small values of λ. To evaluate eq. (6.24) for small λ, we note that

$$\exp \lambda^2 = 1 + \frac{\lambda^2}{1!} + \frac{\lambda^4}{2!} + \cdots \approx 1,$$

and

$$\text{erf } \lambda == \frac{2}{\sqrt{\pi}} \left(\lambda - \frac{\lambda^3}{3 \times 1!} + \frac{\lambda^5}{5 \times 2!} + \cdots \right) \approx \frac{2}{\sqrt{\pi}} \lambda.$$

Substituting into eq. (6.24) gives

$$\lambda = \sqrt{\frac{c_{ps}(T_f - T_o)}{2\mathcal{L}}}, \quad \text{for small } \lambda. \tag{6.25}$$

Thus, according to eq. (6.13), a small λ corresponds to a small Stefan number. Substituting eq. (6.25) into (f) gives

$$x_i = \sqrt{\frac{2k_s(T_f - T_o)}{\rho_s \mathcal{L}}\, t}\ , \quad \text{for small } \lambda \text{ (small } Ste\text{).} \tag{6.26}$$

This result is identical to eq. (6.16a) of the quasi-steady state model.

6.7.2 Neumann's Solution: Solidification of Semi-Infinite Region

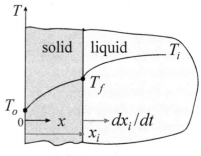

Neumann [1] solved the more general problem of phase change in a semi-infinite region which is not initially at the fusion temperature. Thus Stefan's solution is a special case of Neumann's problem. Although Neumann's solution was presented in his lectures in the 1860's, the work was not published until 1912.

Fig. 6.6

Consider the solidification of a semi-infinite region shown in Fig. 6.6. Initially the region is at a uniform temperature T_i which is above the solidification temperature T_f. The surface at $x = 0$ is suddenly maintained at a constant temperature $T_o < T_f$.   A solid-liquid interface forms instantaneously at $x = 0$ and propagates through the liquid phase. Since $T_i \neq T_f$, heat is conducted through the liquid towards the interface. Neumann's solution to this problem is based on the assumption that the temperature distribution is one-dimensional, properties of each phase are uniform and remain constant and that motion of the liquid phase is neglected. Thus the governing equations are

$$\frac{\partial^2 T_s}{\partial x^2} = \frac{1}{\alpha_s}\frac{\partial T_s}{\partial t} \qquad 0 < x < x_i, \tag{6.1}$$

and

$$\frac{\partial^2 T_L}{\partial x^2} = \frac{1}{\alpha_L}\frac{\partial T_L}{\partial t} \qquad x > x_i. \tag{6.2}$$

The boundary conditions are

(1) $T_s(0,t) = T_o$

(2) $T_s(x_i,t) = T_f$

(3) $T_L(x_i,t) = T_f$

(4) $T_L(\infty,t) = T_i$

The interface energy equation is

(5) $k_s \dfrac{\partial T_s(x_i,t)}{\partial x} - k_L \dfrac{\partial T_L(x_i,t)}{\partial x} = \rho_s \mathcal{L} \dfrac{dx_i}{dt}$

The initial conditions are

(6) $T_L(x,0) = T_i$

(7) $x_i(0) = 0$

Following the procedure used to solve Stefan's problem, the similarity method is applied to solve equations (6.1) and (6.2). The appropriate similarity variable is

$$\eta = x / \sqrt{4\alpha_s t} \ . \tag{6.21}$$

We assume that the solutions to eq. (6.1) and eq. (6.2) can be expressed as

$$T_s = T_s(\eta), \tag{a}$$

and

$$T_L = T_L(\eta). \tag{b}$$

Using eq. (6.21), (a) and (b), equations (6.1) and (6.2) transform to

$$\frac{d^2 T_s}{d\eta^2} + 2\eta \frac{dT_s}{d\eta} = 0 \qquad 0 < \eta < \eta_i, \tag{c}$$

and

$$\frac{d^2 T_L}{d\eta^2} + 2\frac{\alpha_s}{\alpha_L} \eta \frac{dT_L}{d\eta} = 0 \qquad \eta > \eta_i, \tag{d}$$

where

$$\eta_i = x_i / \sqrt{4\alpha_s t} \ . \tag{e}$$

Solutions to (c) and (d) are

$$T_s = A \ \mathrm{erf}\eta + B, \tag{f}$$

and

$$T_L = C \operatorname{erf}\sqrt{\alpha_s / \alpha_L}\, \eta + D. \tag{g}$$

Applying boundary condition (2)

$$T_f = A\operatorname{erf}\frac{x_i}{\sqrt{4\alpha_s t}} + B.$$

Thus

$$x_i \propto \sqrt{t}.$$

Let

$$x_i = \lambda\sqrt{4\alpha_s t}. \tag{h}$$

Using eq. (6.21) and (h), conditions (1) to (6) are transformed to

(1) $T_s(0) = T_o$

(2) $T_s(\lambda) = T_f$

(3) $T_L(\lambda) = T_f$

(4) $T_L(\infty) = T_i$

(5) $k_s \dfrac{dT_s(\lambda)}{d\eta} - k_L \dfrac{dT_L(\lambda)}{d\eta} = 2\rho_s \alpha_s \lambda\, \mathcal{L}$

(6) $T_L(\infty) = T_i$

Note that (h) satisfies condition (7) and that conditions (4) and (6) are identical. In the transformed problem the interface appears stationary at $\eta = \lambda$. Fig. 6.7 shows the temperature distribution of the transformed problem. Boundary conditions (1)-(4) give the four constants of integration, A, B, C and D. Solutions (f) and (g) become

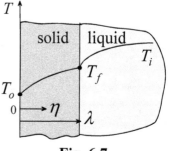

Fig. 6.7

$$T_s(x,t) = T_o + \frac{(T_f - T_o)}{\operatorname{erf}\lambda}\operatorname{erf}\frac{x}{\sqrt{4\alpha_s t}}, \tag{6.27}$$

and

$$T_L(x,t) = T_i + \frac{(T_f - T_i)}{1 - \operatorname{erf}\sqrt{(\alpha_s / \alpha_L)}\,\lambda}\left(1 - \operatorname{erf}\frac{x}{\sqrt{4\alpha_L t}}\right). \tag{6.28}$$

The constant λ appearing in the interface solution (h) and in equations (6.27) and (6.28) is still unknown. All boundary conditions are satisfied except the interface energy condition. Substituting equations (6.27) and (6.28) into condition (5) gives an equation for λ

$$\frac{\exp(-\lambda^2)}{\operatorname{erf}\lambda} - \sqrt{\frac{\alpha_s}{\alpha_L}} \frac{k_L}{k_s} \frac{T_i - T_f}{T_f - T_o} \frac{\exp(-\lambda^2 \alpha_s / \alpha_L)}{1 - \operatorname{erf}(\sqrt{\alpha_s / \alpha_L}\ \lambda)} = \frac{\sqrt{\pi}\ \mathcal{L}\lambda}{c_{ps}(T_f - T_o)}.$$

$$(6.29)$$

Note that Stefan's solution is a special case of Neumann's solution. It can be obtained from Neumann's solution by setting $T_i = T_f$ in equations (6.27)-(6.29).

6.7.3 Neumann's Solution: Melting of Semi-infinite Region

The same procedure can be followed to solve the corresponding melting problem. In this case the density ρ_s in interface condition (5) above is replaced by ρ_L. The solutions to the interface location and temperature distribution are

$$x_i = \lambda\sqrt{4\alpha_L t}\ , \tag{6.30}$$

$$T_L(x,t) = T_o + \frac{(T_f - T_o)}{\operatorname{erf}\lambda} \operatorname{erf} \frac{x}{\sqrt{4\alpha_L t}}\ , \tag{6.31}$$

and

$$T_s(x,t) = T_i + \frac{(T_f - T_i)}{1 - \operatorname{erf}\sqrt{(\alpha_L/\alpha_s)}\ \lambda}\left(1 - \operatorname{erf}\frac{x}{\sqrt{4\alpha_s t}}\right), \tag{6.32}$$

where λ is given by

$$\frac{\exp(-\lambda^2)}{\operatorname{erf}\lambda} - \sqrt{\frac{\alpha_L}{\alpha_s}} \frac{k_s}{k_L} \frac{T_f - T_i}{T_o - T_f} \frac{\exp(-\alpha_L \lambda^2 / \alpha_s)}{1 - \operatorname{erf}(\sqrt{\alpha_L / \alpha_s}\ \lambda)} = \frac{\sqrt{\pi}\ \mathcal{L}\lambda}{c_{pL}(T_o - T_f)}.$$

$$(6.33)$$

It is important to recognize that for the same material at the same $\left|T_i - T_f\right|$ and $\left|T_o - T_f\right|$, values of λ for solidification and for melting are not identical.

6.8 Effect of Density Change on the Liquid Phase

Although we have taken into consideration property change during phase transformation, we have neglected the effect of liquid motion resulting from density change. A material that expands during solidification causes the liquid phase to move in the direction of interface motion as shown in Fig. 6.8. The heat conduction equation for a fluid moving with a velocity U is obtained from eq. (1.7). Assuming constant properties and one-dimensional conduction, eq. (1.7) gives

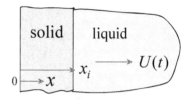

Fig. 6.8

$$\alpha_L \frac{\partial^2 T_L}{\partial x^2} = \frac{\partial T_L}{\partial t} + U \frac{\partial T_L}{\partial x}. \tag{6.34}$$

To formulate an equation for the liquid velocity U, consider phase change in which the liquid phase is not restrained from motion by an external rigid boundary. Fig. 6.9 shows a liquid element dx_L adjacent to the interface at x_i. During a time interval dt the element solidifies and expands to dx_s. The interface moves a distance dx_i given by

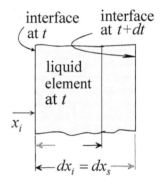

Fig. 6.9

$$dx_i = dx_s. \tag{a}$$

The expansion of the element causes the liquid phase to move with a velocity U given by

$$U = \frac{dx_i - dx_L}{dt}. \tag{b}$$

Conservation of mass for the element yields

$$\rho_L dx_L = \rho_s dx_s = \rho_s dx_i,$$

or

$$dx_L = \frac{\rho_s}{\rho_L} dx_i. \tag{c}$$

Substituting (c) into (b)

$$U = \left(1 - \frac{\rho_s}{\rho_L}\right)\frac{dx_i}{dt}. \tag{6.35}$$

Since interface velocity dx_i/dt is not constant, it follows that the velocity U is time dependent. Substituting eq. (6.35) into eq. (6.34) gives

$$\alpha_L \frac{\partial^2 T_L}{\partial x^2} - \left(1 - \frac{\rho_s}{\rho_L}\right)\frac{dx_i}{dt}\frac{\partial T_L}{\partial x} = \frac{\partial T_L}{\partial t}. \tag{6.36}$$

The solution to Neumann's problem, taking into consideration liquid phase motion, can be obtained by applying the similarity transformation method.

6.9 Radial Conduction with Phase Change

Exact analytic solutions to radial conduction with phase change can be constructed if phase transformation takes place in an infinite region. Examples include phase change around a line heat source or heat sink. Consider solidification due to a line heat sink shown in Fig. 6.10. The liquid is initially at $T_i > T_f$. Heat is suddenly removed along a line sink at a rate Q_o per unit length. Assuming constant properties in each phase and neglecting the effect of liquid motion, the heat conduction equations are

$$\frac{\partial^2 T_s}{\partial r^2} + \frac{1}{r}\frac{\partial T_s}{\partial r} = \frac{1}{\alpha_s}\frac{\partial T_s}{\partial t} \qquad 0 < r < r_i(t), \tag{6.37}$$

$$\frac{\partial^2 T_L}{\partial r^2} + \frac{1}{r}\frac{\partial T_L}{\partial r} = \frac{1}{\alpha_L}\frac{\partial T_L}{\partial t} \qquad r > r_i(t). \tag{6.38}$$

where $r_i(t)$ is the interface location. The boundary condition at the center $r = 0$ is based on the strength of the heat sink. It is expressed as

$$(1)\ \lim_{r \to 0}\left[2\pi r k_s \frac{\partial T_s}{\partial r}\right] = Q_o$$

The remaining boundary conditions are

(2) $T_s(r_i,t) = T_f$

(3) $T_L(r_i,t) = T_f$

(4) $T_L(\infty,t) = T_i$

The interface energy equation is

(5)

$$k_s \frac{\partial T_s(r_i,t)}{\partial r} - k_L \frac{\partial T_L(r_i,t)}{\partial r} = \rho_s \mathcal{L} \frac{dr_i}{dt}$$

The initial temperature is

(6) $T_L(r,0) = T_i$

Assuming that solidification begins instantaneously, the interface initial condition is

(7) $r_i(0) = 0$

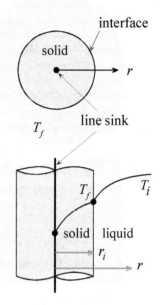

Fig. 6.10

Solution to this problem is based on the similarity method. The similarity variable is [1]

$$\eta = r^2 / 4\alpha_s t \,. \tag{6.39}$$

Assume that

$$T_s(x,t) = T_s(\eta),$$

and

$$T_L(x,t) = T_L(\eta).$$

Using eq. (6.39), equations (6.37) and (6.38) transform to

$$\frac{d^2 T_s}{d\eta^2} + \left(1 + \frac{1}{\eta}\right) \frac{dT_s}{d\eta} = 0, \tag{6.40}$$

and

$$\frac{d^2 T_L}{d\eta^2} + \left(\frac{\alpha_s}{\alpha_L} + \frac{1}{\eta}\right) \frac{dT_L}{d\eta} = 0 \,. \tag{6.41}$$

These equations can be integrated directly. Separating variables in eq. (6.40) and integrating

$$\int^{T_s} \frac{d(dT_s/d\eta)}{dT_s/d\eta} = \int^{\eta} (-\frac{d\eta}{\eta} - d\eta),$$

$$\ln \frac{dT_s}{d\eta} = -\ln \eta - \eta + \ln A,$$

$$\ln \left[\frac{1}{A} \eta \frac{dT_s}{d\eta} \right] = -\eta.$$

Rearranging, separating variables and integrating again

$$\int^{T_s} dT_s = A \int_{\eta}^{\infty} \frac{e^{-\eta}}{\eta} d\eta + B,$$

$$T_s = A \int_{\eta}^{\infty} \frac{e^{-\eta}}{\eta} d\eta + B. \tag{a}$$

where A and B are constants of integration. The choice of the upper limit in the above integral will be explained later. The same approach is used to solve eq. (6.41). The result is

$$T_L = C \int_{\eta}^{\infty} \frac{e^{-(\alpha_s/\alpha_L)\eta}}{\eta} d\eta + D, \tag{b}$$

where C and D are constants of integration. Solution (a) and boundary condition (1) give

$$\lim_{\eta \to 0} \left[2\pi r k_s \frac{dT_s}{d\eta} \frac{\partial \eta}{\partial r} \right] = \lim_{\eta \to 0} \left[-2\pi k_s r A \frac{e^{-\eta}}{\eta} \frac{2r}{4\alpha_s t} \right] =$$

$$\lim_{\eta \to 0} \left[-4\pi k_s A \frac{e^{-\eta}}{\eta} \eta \right] = -4\pi k_s A = Q_o.$$

This gives A as

$$A = -\frac{Q_o}{4\pi k_s}. \tag{c}$$

To determine $r_i(t)$ solution (a) is applied to boundary condition (2) to give

$$T_f = A \int_{\eta_i}^{\infty} \frac{e^{-\eta}}{\eta} d\eta + B, \qquad\qquad (d)$$

where η_i is the value of η at the interface. It is determined by setting $r = r_i$ in eq. (6.39)

$$\eta_i = r_i^2 / 4\alpha_s t. \qquad\qquad (6.42)$$

Since T_f remains constant at all times, this result requires that η_i be independent of time. It follows from eq. (6.42) that $r_i^2 \propto t$. Thus we let

$$r_i^2 = 4\lambda\alpha_s t. \qquad\qquad (6.43)$$

Note that this form of r_i satisfies initial condition (7). Substituting into eq. (6.42) gives

$$\eta_i = \lambda, \qquad\qquad (e)$$

where λ is a constant to be determined. Substituting (c) and (e) into (d)

$$T_f = -\frac{Q_o}{4\pi k_s} \int_{\lambda}^{\infty} \frac{e^{-\eta}}{\eta} d\eta + B.$$

Solving for B

$$B = T_f + \frac{Q_o}{4\pi k_s} \int_{\lambda}^{\infty} \frac{e^{-\eta}}{\eta} d\eta. \qquad\qquad (f)$$

Applying condition (4) to solution (b) gives

$$T_i = C \int_{\infty}^{\infty} \frac{e^{-(\alpha_s / \alpha_L)\eta}}{\eta} d\eta + D = 0 + D,$$

or

$$D = T_i. \qquad\qquad (g)$$

Boundary condition (3) and solution (b) give

$$C = \frac{T_f - T_i}{\displaystyle\int_{\lambda}^{\infty} \frac{e^{-(\alpha_s / \alpha_L)\eta}}{\eta} d\eta}. \qquad\qquad (h)$$

Interface energy equation (5) gives an equation for λ

$$\frac{Q_o}{4\pi}e^{-\lambda} - \frac{k_L(T_i - T_f)}{\displaystyle\int_\lambda^\infty \frac{e^{-(\alpha_s/\alpha_L)\eta}}{\eta}d\eta}e^{-(\alpha_s/\alpha_L)\lambda} = \lambda\,\rho_s\,\alpha_s\,\mathcal{L}. \qquad (6.44)$$

Thus all the required constants are determined. Before substituting the constants of integration into solutions (a) and (b) we will examine the integrals appearing in this solution. These integrals are encountered in other application and are tabulated in the literature [2]. The *exponential integral function Ei(-x)* is defined as

$$Ei(-x) = -\int_x^\infty \frac{e^{-v}}{v}dv . \qquad (6.45)$$

This explains why in integrating equations (6.41) and (6.42) the upper limit in the integrals in (a) and (b) is set at $\eta = \infty$. Values of exponential integral function at $x = 0$ and $x = \infty$ are

$$Ei(0) = \infty, \quad Ei(\infty) = 0. \qquad (6.46)$$

Using the definition of $Ei(-x)$ and the constants given in (c), (f), (g) and (h), equations (a), (b) and (6.44), become

$$T_s(r,t) = T_f + \frac{Q_o}{4\pi k_s}\left[Ei(-r^2/4\alpha_s t) - Ei(-\lambda)\right], \qquad (6.47)$$

$$T_L = T_i + \frac{T_f - T_i}{Ei(-\lambda\alpha_s/\alpha_L)}Ei(-r^2/4\alpha_L t), \qquad (6.48)$$

and

$$\frac{Q_o}{4\pi}e^{-\lambda} + \frac{k_L(T_i - T_f)}{Ei(-\lambda\alpha_s/\alpha_L)}e^{-(\lambda\alpha_s/\alpha_L)} = \lambda\,\rho_s\,\alpha_s\,\mathcal{L}. \qquad (6.49)$$

6.10 Phase Change in Finite Regions

Exact analytic solutions to phase change problems are limited to semi-infinite and infinite regions. Solutions to phase change in finite slabs and inside or outside cylinders and spheres are not available. Such problems are usually solved approximately or numerically.

REFERENCES

[1] Carslaw, H.S., and Jaeger, J.G., *Conduction of Heat in Solids*, 2nd edition, Oxford University Press, 1959.

[2] Selby, S.M., *Standard Mathematical Tables*, The Chemical Rubber Co., Cleveland, Ohio, 1968.

PROBLEMS

6.1   A slab of thickness L and fusion temperature T_f is initially solid at temperature $T_i < T_f$. At time $t \geq 0$ one side is heated to temperature $T_o > T_f$ while the other side is held at T_i. Assume that $Ste < 0.1$, determine the transient and steady state interface location.

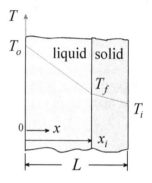

6.2   Consider freezing of a deep lake which is initially at the fusion temperature T_f. During a sudden cold wave which lasted four weeks, air temperature dropped to $-18\,^{\circ}\mathrm{C}$. Assume that lake surface temperature is approximately the same as air temperature. Justify using a quasi-steady model and determine the ice thickness at the end of four weeks. Ice properties are: $c_{ps} = 2093$ J/kg–°C, $k_s = 2.21$ W/m–°C, $\mathcal{L} = 333{,}730$ J/kg and $\rho_s = 916.8$ kg/m³.

6.3   Radiation is beamed at a semi-infinite region which is initially solid at the fusion temperature T_f. The radiation penetrates the liquid phase resulting in a uniform energy generation q'''. The surface at $x = 0$ is maintained at temperature $T_o > T_f$. Assume that $Ste < 0.1$, determine the transient and steady state interface location.

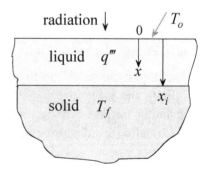

6.4 You decided to make ice during a cold day by placing water in a pan outdoors. Heat transfer from the water is by convection. Air temperature is $-10\,^\circ$C and the heat transfer coefficient is 12.5 W/m^2-$^\circ$C. Initially the water is at the fusion temperature. How thick will the ice layer be after 7 hours? Justify using a quasi-steady model to obtain an approximate answer. Properties of ice are: $c_{ps} = 2093$ J/kg-$^\circ$C, $k_s = 2.21$ W/m-$^\circ$C, $\mathcal{L} = 333{,}730$ J/kg and $\rho_s = 916.8$ kg/m^3.

6.5 An old fashioned ice cream kit consists of two concentric cylinders of radii R_a and R_b. The inner cylinder is filled with milk and ice cream ingredients while the space between the two cylinders is filled with an ice-brine mixture. To expedite the process, the inner cylinder is manually rotated. Assume that the surface of the

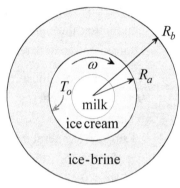

cylinder is at the brine temperature T_o and that $Ste < 0.1$. Assume further that the liquid is initially at the fusion temperature. Derive an expression for the total solidification time. Apply the result to the following case: $R_a = 10$ cm, $R_b = 25$ cm, $T_o = -20\,^\circ$C, $T_i = T_f$. Assume that ice cream has the same properties as ice, given by:

$c_{ps} = 2093$ J/kg-$^\circ$C, $k_s = 2.21$ W/m-$^\circ$C, $\mathcal{L} = 333{,}730$J/kg and $\rho_s = 916.8$ kg/m^3.

6.6 Liquid at the fusion temperature T_f is contained between two concentric cylinders of radii R_a and R_b. At time $t \geq 0$ the inner cylinder is cooled at a time dependent rate $q_o'(t)$ per unit length. The outer cylinder is insulated.

[a] Assuming that $Ste < 0.1$, derive an expression for the total solidification time.

[b] Determine the solidification time for the special case of $q_o'(t) = C/\sqrt{t}$, where C is constant.

6.7 Liquid at the fusion temperature fills a thin walled channel of length L and variable cross section area $A_c(x)$ given by

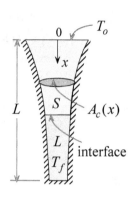

$$A_c(x) = A_o e^{-x/L},$$

where A_o is constant and x is distance along the channel. At time $t \geq 0$ the surface at $x = 0$ is maintained at $T_o < T_f$. Assume that the channel is well insulated and neglect heat conduction through the wall, use quasi-steady approximation to determine the interface location.

6.8 Consider two concentric spheres of radii R_a and R_b. The space between the spheres is filled with solid material at the fusion temperature T_f. At time $t \geq 0$ the inner sphere is heated such that its surface temperature is maintained at $T_o > T_f$. Determine the time needed for the solid to melt. Assume one-dimensional conduction and $Ste < 0.1$.

6.9 Consider a semi-infinite solid region at the fusion temperature T_f. The surface at $x = 0$ is suddenly heated with a time dependent flux given by

$$q_o''(0) = \frac{C}{\sqrt{t}},$$

where C is constant. Determine the interface location for the case of a Stefan number which is large compared to unity.

6.10 A glacier slides down on an inclined plane at a rate of 1.2 m/year. The ice front is heated by convection. The heat transfer coefficient is $h = 8 \text{ W/m}^2 - ^\circ\text{C}$. The average ambient temperature during April through September is $\overline{T}_\infty = 10^\circ\text{C}$. During the remaining months the temperature is below freezing. Melting ice at the front flows into an adjacent stream.

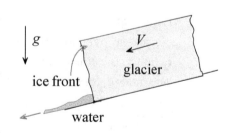

Because of continuous melting, the front recedes slowly.   Assume that conduction through the ice at the front is small compared to convection, determine the location of the front after 10 years. Ice properties are:

$$c_{ps} = 1460 \text{ J/kg–°C} \quad k_s = 3.489 \text{ W/m–°C}$$
$$\mathcal{L} = 334{,}940 \text{ J/kg}$$
$$\rho_s = 928 \text{ kg/m}^3$$

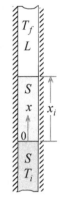

6.11 A semi-infinite liquid column at the fusion temperature T_f is suddenly brought in contact with a semi-infinite solid at a uniform temperature $T_i < T_f$. The solid does not undergo phase transformation.   The liquid column begins to solidify and grow.   Determine the interface location for the case of a Stefan number which is large compared to unity.

6.12 A slab of width L is initially solid at the fusion temperature T_f.   The slab is brought into contact with a semi-infinite solid region which is initially at uniform temperature $T_i > T_f$.   The solid region does not undergo phase change while the slab melts. Obtain an exact solution for the time needed for the entire slab to melt. Assume that the free surface of the slab is insulated.

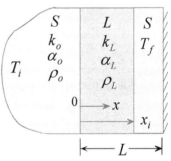

6.13   Solve Neumann's problem taking into consideration the effect of density change.

6.14 A slab of width L is initially liquid at the fusion temperature T_f.   The slab is brought into contact with a semi-infinite solid region which is initially at uniform temperature $T_i < T_f$.   The solid region does not

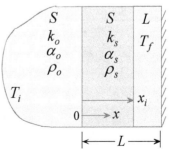

undergo phase change while the slab solidifies. Obtain an exact solution for the time needed for the entire slab to solidify. Assume that the free surface of the slab is insulated.

6.15 Under certain conditions the temperature of a liquid can be lowered below its fusion temperature without undergoing solidification. The liquid in such a state is referred to as supercooled. Consider a supercooled semi-infinite liquid region which is initially at uniform temperature $T_i < T_f$. The surface at $x = 0$ is suddenly maintained at the fusion temperature T_f. Solidification begins immediately and a solid-liquid front propagates through the liquid phase. Note that the solid phase is at uniform temperature T_f. Determine the interface location.

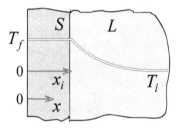

6.16 Consider melting due to a line heat source in an infinite solid region. The solid is initially at $T_i < T_f$. Heat is suddenly added along a line source at a rate Q_o per unit length. Determine the interface location. Assume constant properties in each phase and neglect fluid motion due to density change.

NON-LINEAR CONDUCTION PROBLEMS

7.1 Introduction

Non-linearity in conduction problems arises when properties are temperature dependent or when boundary conditions are non-linear. Surface radiation and free convection are typical examples of non-linear boundary conditions. In phase change problems the interface energy equation is non-linear.

Although the method of separation of variables has wide applicability, it is limited to linear problems. Various methods are used to solve non-linear problems. Some are exact and others are approximate. In this chapter we will examine the source of non-linearity and present three methods of solution. Chapters 8 and 9 deal with approximate techniques that are also applicable to non-linear problems.

7.2 Sources of Non-linearity

7.2.1 Non-linear Differential Equations

Let us examine the following heat equation for one-dimensional transient conduction

$$\frac{\partial}{\partial x}\left(k\frac{\partial T}{\partial x}\right) + q''' = \rho c_p \frac{\partial T}{\partial t}. \tag{7.1}$$

In this equation the density ρ, specific heat c_p and thermal conductivity k can be functions of temperature. Variation of ρ and/or c_p with temperature makes the transient term non-linear. Similarly, if $k = k(T)$ the first term becomes non-linear. This is evident if we rewrite eq. (7.1) as

$$k\frac{\partial^2 T}{\partial x^2} + \frac{dk}{dT}\left[\frac{\partial T}{\partial x}\right]^2 + q''' = \rho c_p \frac{\partial T}{\partial t}. \tag{7.2}$$

The term $(\partial T / \partial x)^2$ is clearly non-linear. Another example of a non-linear differential equation is encountered in fins. The governing equation for a fin exchanging heat by convection and radiation is

$$\frac{d^2 T}{dx^2} - \frac{hC}{kA}(T - T_\infty) - \frac{\varepsilon \sigma C}{kA}(T^4 - T_{sur}^4) = 0 .$$ (7.3)

The non-linearity of this equation is due to the T^4 term.

7.2.2 Non-linear Boundary Conditions

(1) Free convection. A common free convection boundary condition is expressed as

$$-k\frac{\partial T}{\partial x} = \beta(T - T_\infty)^{5/4} ,$$ (7.4)

where β is constant. Unlike forced convection where the flux is proportional to $(T - T_\infty)$, the 5/4 power associated with some free convection problems makes this boundary condition non-linear.

(2) Radiation. A typical radiation boundary condition is expressed as

$$-k\frac{\partial T}{\partial x} = \varepsilon \sigma (T^4 - T_{sut}^4) .$$ (7.5)

(3) Phase change interface. Conservation of energy at a phase change interface yields

$$k_s\frac{\partial T}{\partial x} - k_L\frac{\partial T}{\partial x} = \rho_s L \frac{dx_i}{dt} .$$ (7.6)

This condition is non-linear as shown in eq. (6.7).

7.3 Taylor Series Method

This method gives an approximate description of the temperature distribution in the vicinity of a location where the temperature and its derivatives are specified or can be determined. Taylor series expansion of the function $T(x)$ about $x = 0$ is given by

$$T(x) = T(0) + \frac{dT(0)}{dx}\frac{x}{1!} + \frac{d^2 T(0)}{dx^2}\frac{x^2}{2!} + \frac{d^3 T(0)}{dx^3}\frac{x^3}{3!} + \cdots + \frac{d^n T(0)}{dx^n}\frac{x^n}{n!} .$$ (7.7)

The first derivative can be obtained if the flux is specified. For an insulated boundary or a plane of symmetry the first derivative vanishes. Higher order derivatives can be determined from the heat conduction equation. It should be noted that the heat conduction equation is not solved in this method. The following example illustrates the use of this method.

Example 7.1: Slab with Variable Thermal Conductivity

A slab of thickness L generates heat at a volumetric rate of q'''. The slab is cooled symmetrically on both sides. The temperature at the mid-plane is T_o. The thermal conductivity of the slab depends on temperature according to

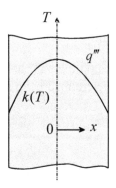

$$k = k_o(1 + \beta T + \gamma T^2),$$

where k_o, β and γ are constant. Use Taylor series expansion to determine the steady state temperature distribution.

Fig. 7.1

(1) Observations. (i) Because the conductivity depends on temperature the problem is non-linear. (ii) Since the mid-plane temperature is specified, Taylor series expansion should be about the mid-plane. (iii) Symmetry requires that the temperature gradient at the mid-plane vanish.

(2) Origin and Coordinates. Fig. 7.1 shows the origin and coordinate axis.

(3) Formulation.

(i) **Assumptions.** (1) Steady state, (2) one-dimensional conduction, (3) symmetry about the mid-plane and (4) uniform energy generation.

(ii) **Governing Equations.** Although the heat conduction equation is not solved in this method, it is needed to determine higher order temperature derivatives. For steady state, eq. (7.2) simplifies to

$$k \frac{d^2 T}{dx^2} + \frac{dk}{dT}\left[\frac{dT}{dx}\right]^2 + q''' = 0. \tag{a}$$

(iii) **Boundary Conditions.**

(1) $T(0) = T_o$

(2) $\dfrac{dT(0)}{dx} = 0$

(4) Solution.   The temperature distribution is given by Taylor series expansion, eq. (7.7)

$$T(x) = T(0) + \frac{dT(0)}{dx}\frac{x}{1!} + \frac{d^2T(0)}{dx^2}\frac{x^2}{2!} + \frac{d^3T(0)}{dx^3}\frac{x^3}{3!} + \frac{d^4T(0)}{dx^4}\frac{x^4}{4!} + \cdots$$

(b)

Conditions (1) and (2) give $T(0)$ and $dT(0)/dx$. The second derivative is determined by evaluating the heat equation (a) at $x = 0$

$$\frac{d^2T(0)}{dx^2} = -\frac{q'''}{k(0)}.$$

(c)

To determine the third derivative, equation (a) is solved for d^2T/dx^2 and differentiated with respect to x

$$\frac{d^3T}{dx^3} = \frac{q'''}{k^2}\frac{dk}{dT}\frac{dT}{dx} + \frac{1}{k^2}\left[\frac{dk}{dT}\right]^2\left[\frac{dT}{dx}\right]^3 - \frac{1}{k}\frac{d^2k}{dT^2}\left[\frac{dT}{dx}\right]^3 - \frac{2}{k}\frac{dk}{dT}\frac{dT}{dx}\frac{d^2T}{dx^2}.$$

(d)

Evaluating (d) at $x = 0$ and using boundary condition (2) gives

$$\frac{d^3T(0)}{dx^3} = 0.$$

(e)

The fourth derivative is obtained by differentiating (d) with respect to x, setting $x = 0$ and using boundary condition (2) and equation (c). The result is

$$\frac{d^4T(0)}{dx^4} = -\frac{3(q''')^2}{[k(0)]^3}\frac{dk(0)}{dT}.$$

(f)

Substituting (c), (e) and (f) into (b) and using boundary condition (1) gives

$$T(x) = T_o - \frac{q'''}{k(0)}\frac{x^2}{2!} - \frac{3(q''')^2}{[k(0)]^3}\frac{dk(0)}{dT}\frac{x^4}{4!},$$

(g)

where $k(T)$ is given by

$$k = k_o(1 + \beta T + \gamma T^2). \tag{h}$$

Using boundary condition (1), equation (h) yields

$$k(0) = k(T_o) = k_o(1 + \beta T_o + \gamma T_o^2), \tag{i}$$

and

$$\frac{dk(0)}{dT} = k_o(\beta + 2\gamma T_o). \tag{j}$$

Substituting (i) and (j) into (g)

$$T(x) = T_o - \frac{q'''}{k_o(1 + \beta T_o + \gamma T_o^2)}\frac{x^2}{2!} - \frac{3(q''')^2(\beta + 2\gamma T_o)}{(k_o)^2(1 + \beta T_o + \gamma T_o^2)^3}\frac{x^4}{4!}. \tag{k}$$

(5) Checking. *Dimensional check*: Equation (k) is dimensionally correct.

Limiting check: The solution corresponding to constant thermal conductivity is obtained by setting $\beta = \gamma = 0$ in (k)

$$T(x) = T_o - \frac{q'''}{2k_o}x^2. \tag{l}$$

This is the exact solution for constant k. It satisfies the two boundary conditions and the corresponding heat equation.

Symmetry: The temperature solution is expressed in terms of even powers of x. Thus it is symmetrical with respect to $x = 0$.

(6) Comments. (i) The slab thickness does not enter into the solution since two boundary conditions are given at $x = 0$. (ii) Solution (k) is approximate. Its accuracy deteriorates as the distance x is increased. (iii) This problem can be solved exactly by direct integration of eq. (7.1) which simplifies to

$$\frac{d}{dx}\left[k\frac{dT}{dx}\right] + q''' = 0. \tag{m}$$

Substituting (h) into (m), separating variables, integrating and applying the two boundary conditions gives

$$T + \frac{\beta}{2}T^2 + \frac{\gamma}{3}T^3 = T_o + \frac{\beta}{2}T_o^2 + \frac{\gamma}{3}T_o^3 - \frac{q'''}{2k_o}x^2. \tag{n}$$

Although this is an exact solution, it is implicit.

7.4 Kirchhoff Transformation

7.4.1 Transformation of Differential Equations

This method deals with the non-linearity associated with temperature dependent thermal conductivity in equation (7.1). We introduce a new temperature variable $\theta(T)$ defined as [1]

$$\theta(T) = \frac{1}{k_o} \int_0^T k(T)dT .$$
(7.8)

Note that if $k(T)$ is specified, eq. (7.8) gives a relationship between θ and T. To transform eq. (7.1) in terms of the variable θ, eq. (7.8) is first differentiated to obtain $d\theta / dT$

$$\frac{d\theta}{dT} = \frac{k}{k_o} .$$
(a)

Using this result we construct $\partial T / \partial t$ and $\partial T / \partial x$

$$\frac{\partial T}{\partial t} = \frac{dT}{d\theta} \frac{\partial \theta}{\partial t} = \frac{k_o}{k} \frac{\partial \theta}{\partial t} ,$$
(b)

and

$$\frac{\partial T}{\partial x} = \frac{dT}{d\theta} \frac{\partial \theta}{\partial x} = \frac{k_o}{k} \frac{\partial \theta}{\partial x} .$$
(c)

Substituting (b) and (c) into eq. (7.1)

$$\frac{\partial^2 \theta}{\partial x^2} + \frac{q'''}{k_o} = \frac{1}{\alpha} \frac{\partial \theta}{\partial t} ,$$
(7.9)

where α is the thermal diffusivity, defined as

$$\alpha = \alpha(T) = \frac{k}{\rho c_p} .$$

The following observations are made regarding the above transformation:

(1) The non-linear conduction term in eq. (7.1) is transformed into a linear form.

(2) The diffusivity α is a function of temperature since ρ and c_p are temperature dependent. Thus, the transient term in eq. (7.9) is non-linear.

(3) Since thermal diffusivity does not play a role in steady state stationary problems, the transformed equation in this special case is linear.

(4) In problems where $\alpha = k/\rho c_p$ can be assumed constant, equation (7.9) becomes linear.

7.4.2 Transformation of Boundary Conditions

To complete the transformation, boundary conditions must also be expressed in terms of the variable θ. Successful transformation is limited to the following two conditions:

(1) *Specified Temperature.* Let the temperature at a boundary be

$$T = T_o. \tag{7.10}$$

The definition of θ in eq. (7.8) gives

$$\theta = F(T). \tag{a}$$

Substituting eq. (7.10) into (a)

$$\theta = F(T_o) \equiv \theta_o. \tag{7.11}$$

Thus specified temperature at a boundary is transformed into specified θ. Note that T_o can be a function of time and location.

(2) *Specified Heat Flux.* This boundary condition is expressed as

$$-k\frac{\partial T}{\partial x} = q_o''. \tag{7.12}$$

However

$$\frac{\partial T}{\partial x} = \frac{dT}{d\theta}\frac{\partial \theta}{\partial x} = \frac{k_o}{k}\frac{\partial \theta}{\partial x},$$

or

$$\frac{\partial T}{\partial x} = \frac{k_o}{k}\frac{\partial \theta}{\partial x}. \tag{b}$$

Substituting (b) into eq. (7.12)

$$-k_o \frac{\partial \theta}{\partial x} = q_o''. \tag{7.13}$$

Thus, the transformed condition is a specified flux. Furthermore, since k_o is constant, eq. (7.13) is linear.

Example 7.2: Two-dimensional Conduction in a Cylinder with Variable Conductivity

A cylinder of radius r_o and length L is insulated at one end and heated with uniform flux q_o'' at the other end. The cylindrical surface is maintained at a uniform temperature T_o. The thermal conductivity varies with temperature according to

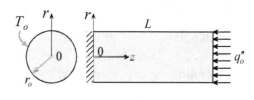

Fig. 7.2

$$k = k_o(1 + \beta T).$$

Use Kirchhoff method to determine the steady state temperature distribution in the cylinder.

(1) Observations. (i) Because the conductivity depends on temperature the problem is non-linear. (ii) The boundary conditions consist of specified temperature and specified heat flux. (iii) Kirchhoff transformation can be used to solve this problem.

(2) Origin and Coordinates. Fig. 7.2 shows the origin and coordinate axes.

(3) Formulation.

(i) **Assumptions.** (1) Steady state, (2) two-dimensional conduction and (3) axisymmetric temperature distribution.

(ii) **Governing Equations.** Based on the above assumptions, eq. (1.11), modified for variable k, becomes

$$\frac{1}{r}\frac{\partial}{\partial r}\left[kr\frac{\partial T}{\partial r}\right] + \frac{\partial}{\partial z}\left[k\frac{\partial T}{\partial z}\right] = 0. \tag{a}$$

(iii) **Boundary Conditions.** The four boundary conditions are

(1) $\dfrac{\partial T(0,z)}{\partial r} = 0,$ (2) $T(r_o,z) = T_o,$

(3) $\dfrac{\partial T(r,0)}{\partial z} = 0,$ (4) $k\dfrac{\partial T(r,L)}{\partial z} = q_o''.$

(4) Solution. This non-linear problem lends itself to a solution by Kirchhoff transformation. Introducing the transformation variable θ

$$\theta(T) = \frac{1}{k_o} \int_0^T k(T)\, dT, \tag{b}$$

where $k(T)$ is given as

$$k(T) = k_o(1 + \beta T). \tag{c}$$

Substituting (c) into (b)

$$\theta(T) = \frac{1}{k_o} \int_0^T k_o(1 + \beta T)\, dT = T + \frac{\beta}{2}T^2. \tag{d}$$

This gives a relationship between θ and T. To determine θ, the governing equation and boundary conditions are transformed using (b). Thus

$$\frac{\partial T}{\partial r} = \frac{dT}{d\theta}\frac{\partial\theta}{\partial r} = \frac{k_o}{k}\frac{\partial\theta}{\partial r} \quad \text{and} \quad \frac{\partial T}{\partial z} = \frac{dT}{d\theta}\frac{\partial\theta}{\partial z} = \frac{k_o}{k}\frac{\partial\theta}{\partial z}. \tag{e}$$

Substituting into (a)

$$\frac{\partial^2\theta}{\partial r^2} + \frac{1}{r}\frac{\partial\theta}{\partial r} + \frac{\partial^2\theta}{\partial z^2} = 0. \tag{f}$$

Using (d) and (e), the boundary conditions transform to

(1) $\dfrac{\partial\theta(0,z)}{\partial r} = 0$

(2) $\theta(r_o,z) = T_o + \beta T_o^2/2 \equiv \theta_o$

(3) $\dfrac{\partial\theta(r,0)}{\partial z} = 0$

(5) $k_o \dfrac{\partial \theta(r,L)}{\partial z} = q''_o$

Thus the governing equation and boundary conditions are transformed to a linear problem which can be solved by the method of separation of variables. Details of the solution to (f) will not be presented here. Once $\theta(r,z)$ is determined the solution to quadratic equation (d) gives $T(r,z)$

$$T(r,z) = \sqrt{\frac{1}{\beta^2} + \frac{2}{\beta} \theta(r,z)} - \frac{1}{\beta}.$$

(5) Comments. The same approach can be used to solve this problem if in addition the cylinder generates heat volumetrically.

7.5 Boltzmann Transformation

The Boltzmann transformation approach is based on the similarity method which we encountered in the solution of transient conduction in semi-infinite regions and in phase change problems. It is limited to semi-infinite domains and applies to restricted initial and boundary conditions. The idea is to introduce a *similarity variable* which combines two independent variables and transforms the differential equation from partial to ordinary. However, when applying this method to variable thermal conductivity, the non-linear partial differential equation is transformed to a non-linear ordinary differential equation whose solution may still present difficulties.

Example 7.3: Transient Conduction in a Semi-infinite Region with Variable Conductivity

A semi-infinite region is initially at a uniform temperature T_i. The surface at $x = 0$ is suddenly maintained at a constant temperature T_o. The thermal conductivity is temperature dependent. Use Boltzmann transformation to determine the transient temperature distribution.

(1) Observations. (i) Since the region is semi-infinite and the initial temperature is uniform, the problem lends itself to solution by the similarity method. (ii) The problem is non-linear because the thermal conductivity depends on temperature.

(2) Origin and Coordinates. Fig. 7.3 shows the origin and coordinate axis.

(3) Formulation.

(i) Assumptions. (1) One-dimensional transient conduction and (2) uniform initial temperature.

(ii) Governing Equations. Equation (7.1) simplifies to

$$\frac{\partial}{\partial x}\left(k\frac{\partial T}{\partial x}\right) = \rho c_p \frac{\partial T}{\partial t}.$$ (a)

(iii) Boundary and Initial Conditions.

(1) $T(0,t) = T_o$

(2) $T(\infty,t) = T_i$

(3) $T(x,0) = T_i$

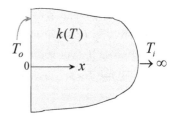

Fig. 7.3

(4) Solution. Following the method used to solve transient conduction in a semi-infinite region with constant properties, introduce a similarity variable $\eta(x,t)$, defined as

$$\eta(x,t) = \frac{x}{\sqrt{t}}.$$ (b)

Applying this variable to (a) gives

$$\frac{d}{d\eta}\left(k\frac{dT}{d\eta}\right) + \frac{\rho c_p}{2}\eta\frac{dT}{d\eta} = 0.$$ (7.14)

The three conditions transform to

(1) $T(0) = T_o$

(2) $T(\infty) = T_i$

(3) $T(\infty) = T_i$

We make the following observations regarding this result:

(1) The problem is successfully transformed into an ordinary differential equation. The variables x and t are replaced by η.

(2) Equation (b) is second order which can satisfy two boundary conditions. Since two of the three conditions become identical in the transformed problem, equation (b) has the required number of conditions.

(3) Equation (7.14) is non-linear. Since k, ρ and c_p depend on temperature, it follows that both terms in eq.(7.14) are non-linear.

(4) For the special case of constant ρ and c_p, eq.(7.14) is solved by *successive approximation* [1].

(5) Comments. Several restrictions are imposed on this problem to obtain an approximate solution. They are: semi-infinite region, uniform initial temperature, constant boundary temperature and constant ρ and c_p.

7.6 Combining Boltzmann and Kirchhoff Transformations

Some of the shortcomings of the Boltzmann transformation can be eliminated by combining it with the Kirchhoff transformation. To illustrate this approach we return to Example 7.3. We introduce the Kirchhoff transformation

$$\theta(T) = \frac{1}{k_o} \int_0^T k(T)dT .$$

(7.8)

Applying eq. (7.8) to eq. (7.14) transforms it to

$$\frac{d^2\theta}{d\eta^2} + \frac{1}{2\alpha}\eta\frac{d\theta}{d\eta} = 0 ,$$

(7.15)

where α is the thermal diffusivity, defined as

$$\alpha = \frac{k}{\rho c_p} .$$

(a)

The two boundary conditions transform to

(1) $\theta(0) = \dfrac{1}{k_o} \displaystyle\int_0^{T_o} k(T)dT \equiv \theta_o(T_o)$

(2) $\theta(\infty) = \dfrac{1}{k_o} \displaystyle\int_0^{T_i} k(T)dT \equiv \theta_i(T_i)$

The second term in eq.(7.15) is non-linear because α is temperature dependent. However, although the three properties k, ρ and c_p are temperature dependent, their ratio in the definition of α may be a weak function temperature for some materials. In such a case α can be assumed constant and eq.(7.15) becomes linear. Based on this assumption eq.(7.15) can be separated and integrated directly.   The solution, expressed in terms of the original variables, is given by

$$\theta(x,t) = A \operatorname{erf} \frac{x}{\sqrt{4\alpha t}} + B, \tag{b}$$

where A and B are constant. Applying the two boundary conditions, the solution becomes

$$\theta(x,t) = \theta_o + (\theta_i - \theta_o) \operatorname{erf} \frac{x}{\sqrt{4\alpha t}}. \tag{c}$$

Eq. (7.8) is used to convert (c) to the variable T. Once $k(T)$ is specified, this equation gives θ in terms of T. For example, consider the special case where the conductivity is given by

$$k(T) = k_o(1 + \beta T). \tag{d}$$

Substituting into eq. (7.8) gives

$$\theta(T) = k_o(T + \beta T^2/2).$$

The constants θ_o and θ_i become

$$\theta_o = k_o(T_o + \beta T_o^2/2),$$

$$\theta_i = k_o(T_i + \beta T_i^2/2).$$

7.7 Exact Solutions

In the previous sections we have presented various techniques for dealing with certain non-linear problems in conduction. However, none of these methods are suitable for solving radiation problems. In this section we illustrate how a simple transformation is used to analyze fins with surface radiation. Consider the semi-infinite constant area fin shown in Fig. 7.4. The fin exchanges heat with the surroundings by both convection and radiation. We assume that the ambient fluid and the radiating surroundings are at the same temperature T_∞. The base of the fin is maintained at T_o. The heat equation for this fin is given by

$$\frac{d^2 T}{dx^2} - m_c^2(T - T_\infty) - m_r^2(T^4 - T_\infty^4) = 0, \tag{7.16}$$

where

$$m_c^2 = \frac{hC}{kA_c}, \text{ and } m_r^2 = \frac{\varepsilon\sigma C}{kA_c}. \tag{a}$$

Equation (7.16) is based on conservation of energy for an element dx using Fourier's law of conduction, Newton's law of cooling and Stefan-Boltzmann radiation law. Equation (7.16) is a special case of eq. (7.3) with $T_{sur} = T_\infty$. The two boundary conditions are

(1) $T(0) = T_o$

(2) $T(\infty) = T_\infty$

We introduce the following transformation [2]

$$\frac{dT}{dx} = \psi. \tag{b}$$

Differentiating (b)

$$\frac{d^2T}{dx^2} = \frac{d\psi}{dx}. \tag{c}$$

Solving (b) for dx

$$dx = \frac{dT}{\psi}. \tag{d}$$

Substituting (d) into (c)

$$\frac{d^2T}{dx^2} = \psi\frac{d\psi}{dT}. \tag{e}$$

Using (e) to eliminate $\dfrac{d^2T}{dx^2}$ in eq. (7.16) gives

$$\psi\frac{d\psi}{dT} = m_c^2(T - T_\infty) + m_r^2(T^4 - T_\infty^4). \tag{7.17}$$

This equation is separable and thus can be integrated directly to give

$$\frac{\psi^2}{2} = m_c^2\left[(T^2/2) - (T_\infty T)\right] + m_r^2\left[(T^5/5) - (T_\infty^4 T)\right] + C, \tag{f}$$

where C is constant of integration. Solving (f) for ψ and using (b) gives the temperature gradient at any location x

$$\psi = \frac{dT}{dx} = \mp\sqrt{2}\left\{m_c^2\left[(T^2/2)-T_\infty T\right]+m_r^2\left[(T^5/5)-T_\infty^4 T\right]+C\right\}^{1/2}. \quad (g)$$

If $T_o > T_\infty$, the fin loses heat to the surroundings. That is, $dT/dx < 0$ and thus the negative sign in (g) must be used. On the other hand the positive sign is used if $T_o < T_\infty$. To determine the constant C, equation (g) is applied at $x = \infty$ where

$$T(\infty) = T_\infty, \text{ and } \frac{dT(\infty)}{dx} = 0. \quad (h)$$

Substituting (h) into (g) and solving for C gives

$$C = \left[(1/2)m_c^2 T_\infty^2 + (5/4)m_r^2 T_\infty^5\right]. \quad (i)$$

With the temperature gradient given in (g), the heat transfer rate can be determined by applying Fourier's law at the base $x = 0$ and using boundary condition (1) and equation (i). The result is

$$q = -kA_c \frac{dT(0)}{dx} =$$
$$\pm\sqrt{2}kA_c\left\{(m_c^2/2)(T_o^2 - 2T_\infty T_o + T_\infty^2) + (m_r^2/5)(T_o^5 - 5T_\infty^4 T_o + 4T_\infty^5)\right\}^{1/2}.$$
$$(7.18)$$

where the positive sign is for $T_o > T_\infty$ and the negative for $T_o < T_\infty$. The following observations are made regarding this solution:

(1) Although an exact solution is obtained for the heat transfer rate from the fin, a solution for the temperature distribution can not be determined.

(2) The more general case of $T_{sur} \neq T_\infty$ can be solved following the same approach. In this case $T(\infty)$, which is needed to determine C, is not equal to T_∞. However, it can be determined by applying eq.(7.16) at $x = \infty$ where

$$\frac{d^2 T(\infty)}{dx^2} = 0. \quad (j)$$

Substituting (j) into eq.(7.16) gives

$$m_c^2 \left[T(\infty) - T_\infty \right] + m_r^2 \left[T^4(\infty) - T_{sur}^4 \right] = 0 . \tag{k}$$

This equation is solved for $T(\infty)$ by a trial and error procedure.

REFERENCES

[1] Ozisik, M.N., *Heat Conduction*, Wiley, New York, 1980.

[2] Thomas, L.C. *Heat Transfer*, Prentice Hall, Englewood, New Jersey, 1992.

PROBLEMS

7.1   A solid cylinder of radius R_o generates heat at a volumetric rate q'''. The temperature at the center is T_o.   The thermal conductivity depends on temperature according to

$$k(T) = k_o (1 - \beta T),$$

where k_o and β are constant.   Use Taylor series method to determine the steady state one-dimensional temperature distribution in the cylinder.

7.2   A slab of thickness L generates heat at a volumetric rate q'''. One side is insulated while the other side is maintained at a uniform temperature T_o. The thermal conductivity depends on temperature according to

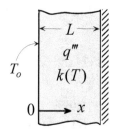

$$k(T) = k_o (1 + \beta T),$$

where k_o and β are constant.

[a] Use Taylor series method to determine the one-dimensional steady state temperature distribution in the slab.

[b] Calculate the temperature at the mid-section using the following data: $k_o = 120 \ \text{W/m}-^\circ\text{C}$, $L = 4 \ \text{cm}$, $q''' = 5 \times 10^6 \ \text{W/m}^3$, $T_o = 50\,^\circ\text{C}$, $\beta = 0.005 \ 1/^\circ\text{C}$

7.3   Heat is generated at a volumetric rate q''' in a slab of thickness L. One side of the slab is heated with uniform flux q''_o while the other side is cooled by convection. The heat transfer coefficient is h and the ambient temperature is T_∞. The thermal conductivity depends on temperature according to

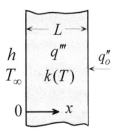

$$k(T) = k_o(1 + \beta T),$$

where k_o and β are constant. Determine the steady state one-dimensional temperature distribution using Taylor series method.

7.4   A long hollow cylinder of inner radius R_i and outer radius R_o is cooled by convection at its inner surface. The heat transfer coefficient is h and the ambient temperature is T_∞. Heat is added at the outer surface with a uniform flux q''_o. The thermal conductivity depends on temperature according to

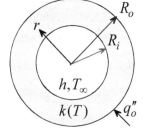

$$k(T) = k_o(1 + \beta T),$$

where k_o and β are constant. Use Taylor series method to determine the steady state one-dimensional temperature distribution.

7.5   A long solid cylinder of outer radius R_o generates heat volumetrically at a rate q'''. Heat is removed from the outer surface by convection. The ambient temperature is T_∞ and the heat transfer coefficient is h. The thermal conductivity depends on temperature according to

$$k(T) = k_o(1 - \beta T),$$

where k_o and β are constant. Use Taylor series method to determine the steady state one-dimensional temperature distribution.

7.6   A fin with a constant cross section area A_c exchanges heat by convection along its surface. The heat transfer coefficient is h and the ambient temperature is T_∞. The heat transfer rate at the base is q_f and the temperature is T_o. The thermal conductivity varies with temperature according to

$$k(T) = k_o(1 + \beta T),$$

where k_o and β are constant. Use Taylor series method to determine the steady state temperature distribution.

7.7   Consider conduction in a region with surface convection at one of its boundaries.   Discuss the application of Kirchhoff transformation to such a boundary condition for a temperature dependent thermal conductivity

7.8   Heat is generated volumetrically at a rate q''' in a long hollow cylinder of inner radius R_i and outer radius R_o. The inner surface is insulated and the outer surface is maintained at a uniform temperature T_o. The thermal conductivity depends on temperature according to

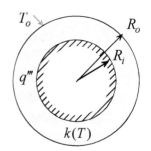

$$k(T) = k_o(1 + \beta T),$$

where k_o and β are constant. Use Kirchhoff transformation to determine the one-dimensional steady state temperature distribution. Calculate the temperature of the insulated surface for the following case: $k_o = 73.6$ W/m$-^\circ$C, $q''' = 1.45 \times 10^8$ W/m^3, $R_i = 1.3$cm, $R_o = 3.9$cm, $T_o = 100^\circ$C, $\beta = 6.2 \times 10^{-4}$ 1/$^\circ$C.

What will the surface temperature be if the conductivity is assumed constant?

7.9   An electric cable of radius R_i generates heat volumetrically at a rate q'''. The cable is covered with an insulation of outer radius R_o. The outside insulation surface is maintained at uniform temperature T_o. The conductivity of the cable is k_1 and that of the insulation is k_2. Both depend

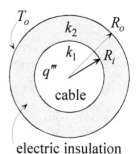

electric insulation

on temperature according to

$$k_1 = k_{01}(1 + \beta T),$$

and

$$k_2 = k_{02}(1 + \beta T),$$

where k_{01}, k_{02} and β are constant.   Determine the steady state temperature at the center of the cable using Kirchhoff transformation.

7.10 Consider two-dimensional steady state conduction in a rectangular plate. One side is at a uniform temperature T_o while the other sides are maintained at zero. The thermal conductivity depends on temperature according to

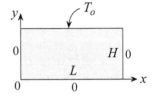

$$k(T) = k_o(1 + \beta T + \gamma T^2),$$

where k_o, β and γ are constant. Use Kirchhoff transformation to determine the heat flux along the boundary $(x,0)$.

7.11 A long solid cylinder of radius r_o is initially at a uniform temperature T_i. At time $t \geq 0$ the surface is maintained at temperature T_o. The thermal conductivity depends on temperature according to

$$k(T) = k_o(1 - \beta T),$$

where k_o and β are constant.   Use Kirchhoff transformation to determine the one-dimensional transient temperature distribution. Assume constant thermal diffusivity.

7.12 A slab of thickness L is initially at zero temperature. The slab generates heat at a volumetric rate q'''. At time $t \geq 0$ one side is maintained at a uniform temperature T_o while the other side is kept at zero. The thermal conductivity

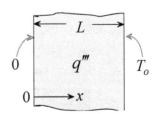

depends on temperature according to

$$k(T) = k_o(1 - \beta T),$$

where k_o and β are constant. Use Kirchhoff transformation to determine the one-dimensional transient temperature. Assume constant thermal diffusivity.

7.13 An infinite region is initially at uniform temperature T_i. Heat is suddenly added along a line source at a rate Q_o per unit length. The thermal conductivity depends on temperature according to

$$k(T) = k_o(1 - \beta T),$$

where k_o and β are constant. Use Kirchhoff transformation to determine the transient temperature distribution in the region. Assume constant thermal diffusivity.

7.14 The surface of a semi-infinite region which is initially at a uniform temperature T_i is suddenly heated with a time dependent flux given by

$$q_o = \frac{C}{\sqrt{t}}.$$

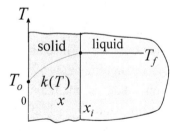

The thermal conductivity varies with temperature as

$$k(T) = k_o(1 + \beta T),$$

where k_o and β are constant. Use Kirchhoff transformation to determine the one-dimensional transient temperature distribution in the region. Assume that the thermal diffusivity is constant.

7.15 Consider solidification of a semi-infinite liquid region which is initially at the fusion temperature T_f. At $t \geq 0$ the surface at $x = 0$ is suddenly maintained at a temperature $T_o < T_f$. The conductivity of the solid depends

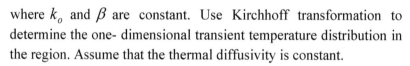

on temperature according to

$$k(T) = k_o(1 + \beta T),$$

where k_o and β are constant. Use Kirchhoff and Boltzmann transformations to determine the interface location $x_i(t)$. Assume constant thermal diffusivity.

7.16 Consider melting of a semi-infinite solid region which is initially at the fusion temperature T_f. The boundary at $x = 0$ is suddenly heated with a time dependent flux given by

$$q_o''(t) = \frac{C}{\sqrt{t}},$$

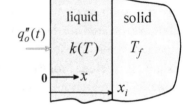

where C is constant. The thermal conductivity of the liquid depends on temperature according to

$$k(T) = k_o(1 + \beta T),$$

where k_o and β are constant. Use Kirchhoff and Boltzmann transformations to determine the interface location $x_i(t)$. Assume constant thermal diffusivity.

7.17 Consider a constant area semi-infinite fin which generates heat at a volumetric rate q'''. The fin exchanges heat with the surroundings by convection and radiation. The ambient fluid and the radiating surroundings are at T_∞. The base of the fin is maintained at T_o. Assume constant properties. Determine the steady state heat transfer rate from the fin.

APPROXIMATE SOLUTIONS:
THE INTEGRAL METHOD

There are various situations where it is desirable to obtain approximate analytic solutions. An obvious case is when an exact solution is not available or can not be easily obtained. Approximate solutions are also obtained when the form of the exact solution is not convenient to use. Examples include solutions that are too complex, implicit or require numerical integration. The *integral method* is used extensively in fluid flow, heat transfer and mass transfer. Because of the mathematical simplifications associated with this method, it can deal with such complicating factors as phase change, temperature dependent properties and non-linearity.

8.1 Integral Method Approximation: Mathematical Simplification

In *differential formulation* the basic laws are satisfied exactly at every point. On the other hand, in the integral method the basic laws are satisfied in an average sense. The mathematical consequence of this compromise is a reduction of the number of independent variables and/or a reduction of the order of the governing differential equation. Thus major mathematical simplifications are associated with this approach. This explains why it has been extensively used to solve a wide range of problems.

8.2 Procedure

The following procedure is used in the application of the integral method to the solution of conduction problems:

(1) Integral formulation. The first step is the *integral formulation* of the principle of conservation of energy. This formulation results in the *heat-balance integral* which is the governing equation for the problem. Note that each problem has its own heat-balance integral. Each of the following two approaches leads to the formulation of the heat-balance integral:

(a) *Control volume formulation.* In this approach the principle of conservation of energy is applied to a *finite* control volume encompassing the entire region in which temperature variation takes place. Thus, energy exchange at a boundary is accounted for in the energy balance for the control volume.

(b) *Integration of the governing differential equation.* This approach can be used if the differential equation governing the problem is available. Recall that differential formulation of conservation of energy is applied to an *infinitesimal* element or control volume. This leads to a differential equation governing the temperature distribution. This differential equation is integrated term by term over the entire region in which temperature variation takes place. In transient conduction problems, integration of the time derivative term is facilitated by applying the following *Leibnitz's rule* [1]:

$$\int_{a(t)}^{b(t)} \frac{\partial T}{\partial t} dx = T(a,t)\frac{da}{dt} - T(b,t)\frac{db}{dt} + \frac{d}{dt}\int_{a(t)}^{b(t)} T(x,t)dx, \quad (8.1)$$

where $a(t)$ and $b(t)$ are the limits of the spatial variable x.

(2) Assumed temperature profile. An appropriate temperature profile is assumed which satisfies known boundary conditions. The assumed profile can take on different forms. However, a polynomial is usually used in Cartesian coordinates. The assumed profile is expressed in terms of a single unknown parameter or variable which must be determined.

(3) Determination of the unknown parameter or variable. Application of the assumed temperature profile to the heat-balance integral results in an equation whose solution gives the unknown variable.

8.3 Accuracy of the Integral Method

Since basic laws are satisfied in an average sense, integral solutions are inherently approximate. The following observations are made regarding the accuracy of this method:

(1) Since the assumed profile is not unique (several forms are possible), the accuracy of integral solutions depends on the form of the assumed profile. In general, errors involved in this method are acceptable in typical engineering applications.

(2) The accuracy is not very sensitive to the form of the assumed profile.

(3) While there are general guidelines for improving the accuracy, no procedure is available for identifying assumed profiles that will result in the most accurate solutions.

(4) An assumed profile which satisfies conditions at a boundary yields more accurate information at that boundary than elsewhere.

8.4 Application to Cartesian Coordinates

The integral method will be applied to two problems to illustrate its use in constructing approximate solutions. The procedure outlined in Section 8.2 will be followed. In each case the heat-balance integral is obtained using both control volume formulation and integration of the governing differential equation.

Example 8.1: Constant Area Fin

The base of the fin shown in Fig. 8.1 is maintained at a specified temperature T_o while the tip is insulated. Heat exchange with the ambient fluid at T_∞ is by convection. The heat transfer coefficient is h. The cross-sectional area is A_c and the length is L. Use the integral method to determine the fin heat transfer rate.

(1) Observations. (i) Constant area fin. (ii) Specified temperature at the base and insulated tip. (iii) Surface heat exchange is by convection.

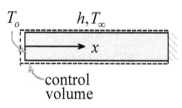

control volume

Fig. 8.1

(2) Origin and Coordinates. Fig. 8.1 shows the origin and coordinate axis.

(3) Formulation and Solution.

   (i) Assumptions. (1) Steady state, (2) fin approximations are valid $(Bi < 0.1)$, (3) no energy generation and (4) uniform h and T_∞.

(ii) Integral Formulation. The heat-balance integral for this problem will be formulated using two methods.

(a) *Control volume formulation.* A finite control volume is selected which encompasses the entire fin as shown in Fig. 8.1. Application of conservation of energy, eq. (1.6), to the control volume gives

$$\dot{E}_{in} - \dot{E}_{out} = 0. \tag{a}$$

Pretending that heat is added at the base and removed from the surface, Fourier's law of conduction gives

$$\dot{E}_{in} = -kA_c \frac{dT(0)}{dx}, \tag{b}$$

where A_c is the cross-sectional area. Since surface temperature varies along the fin, Newton's law must be integrated along the surface to determine the total heat convected to the ambient fluid. Thus, assuming constant h

$$\dot{E}_{out} = hC \int_0^L (T - T_\infty)dx, \tag{c}$$

where C is the perimeter. Substituting (b) and (c) into (a)

$$\frac{dT(0)}{dx} + m^2 \int_0^L (T - T_\infty)dx = 0, \tag{8.2}$$

where

$$m^2 = \frac{hC}{kA_c}. \tag{d}$$

Equation (8.2) is the heat-balance integral for this fin.

(b) *Integration of the governing differential equation.* Equation (8.2) can be formulated by integrating the fin equation term by term. Differential formulation of conservation of energy for a fin is given by eq. (2.9)

$$\frac{d^2T}{dx^2} - m^2(T - T_\infty) = 0. \tag{2.9}$$

Multiplying eq. (2.9) by dx, assuming constant h and integrating over the length of the fin from $x = 0$ to $x = L$, gives

$$\int_0^L \frac{d^2T}{dx^2}dx - m^2 \int_0^L (T - T_\infty)dx = 0.$$

The integral of the second derivative is the first derivative. Thus the above becomes

$$\frac{dT(L)}{dx} - \frac{dT(0)}{dx} - m^2 \int_0^L (T - T_\infty)dx = 0. \qquad (e)$$

Since the tip is insulated, it follows that

$$\frac{dT(L)}{dx} = 0. \qquad (f)$$

Thus (e) becomes

$$\frac{dT(0)}{dx} + m^2 \int_0^L (T - T_\infty)dx = 0. \qquad (8.2)$$

This is identical to the heat-balance integral obtained using control volume formulation.

(iii) Assumed Temperature Profile. Assume a polynomial of the second degree

$$T(x) = a_0 + a_1 x + a_2 x^2. \qquad (g)$$

The coefficients a_0, a_1 and a_2 must be such that (g) satisfies the boundary conditions on the temperature. In addition to condition (f), specified temperature at the base gives

$$T(0) = T_o. \qquad (h)$$

Applying (f) and (h) to (g) gives

$$a_0 = T_o \text{ and } a_2 = -a_1/2L.$$

Thus, (g) becomes

$$T = T_o + a_1(x - x^2/2L). \qquad (i)$$

(iv) Determination of the Unknown Coefficient. The coefficient a_1 is the only remaining unknown. It is determined by satisfying the heat-balance integral. Substituting (i) into eq. (8.2)

$$a_1 + m^2 \int_0^L \left[(T_o - T_\infty) + a_1(x - x^2/2L)\right] dx = 0.$$

Performing the integration and solving the resulting equation for a_1 yields

$$a_1 = -\frac{m^2 L(T_o - T_\infty)}{1 + m^2 L^2 /3}. \tag{j}$$

Substituting (j) into (i) and rearranging gives the dimensionless temperature distribution $T^*(x)$

$$T^*(x) = \frac{T(x) - T_\infty}{T_o - T_\infty} = 1 - \frac{m^2 L^2}{1 + m^2 L^2 /3}(1 - x/2L)(x/L). \tag{8.3}$$

The tip temperature $T^*(L)$ is obtained by setting $x = L$

$$T^*(L) = \frac{T(L) - T_\infty}{T_o - T_\infty} = 1 - (1/2)\frac{m^2 L^2}{1 + m^2 L^2 /3}. \tag{8.4}$$

The fin heat transfer rate q_f is obtained by substituting eq. (8.3) into (b) and using (d)

$$q^* = \frac{q_f}{\sqrt{hCkA_c}\,(T_o - T_\infty)} = \frac{mL}{1 + m^2 L^2 /3}, \tag{8.5}$$

where q^* is the dimensionless heat transfer rate.

(4) Checking. *Dimensional check*: Each term in eq. (8.3) and eq. (8.5) is dimensionless.

Limiting check: If $h = 0$, no heat leaves the fin. Thus the entire fin should be at the base temperature. Setting $h = m = 0$ in eqs. (8.3) and (8.5) gives $T(x) = T_o$ and $q_f = 0$.

(5) Comments. The exact solution to this problem is given by equations (2.14) and (2.15). Using these equations the exact solutions to the tip temperature and fin heat transfer rate are expressed as

$$T_e^*(L) = \frac{T_e(L) - T_\infty}{T_o - T_\infty} = \frac{1}{\cosh mL}, \tag{8.6}$$

and

$$q_e^* = \tanh mL ,\tag{8.7}$$

where the subscript e denotes exact. Note that the integral and exact solutions depend on the parameter mL. Table 8.1 gives the percent error as a function of this parameter. Two observations are made regarding this result. (1) The heat transfer error is smaller than tip temperature error. (2) The error increases with increasing values of the parameter mL. This is due to the fact that for large values of mL the form of the assumed profile becomes increasingly inappropriate. This is evident when a large mL is associated with a large fin length L. As $L \to \infty$ the integral solution gives $T^* = -1/2$ while the exact solution gives $T^* = 0$. Thus the percent error approaches infinity as $L \to \infty$. On the other hand, as $L \to \infty$ the integral solution for heat transfer rate, eq. (8.5), gives $q^* = 0$ while the exact solution, eq. (8.7), gives $q^* = 1$. Thus the maximum error in determining the heat transfer rate is 100%. More details regarding the solution to this problem are found in Reference [2].

Table 8.1

	Percent error								
mL	0	0.5	1.0	2.0	3.0	4.0	5.0	10	∞
$T^*(L)$	0	0.248	3.56	48.2	225.8	819	2619	502700	∞
q^*	0	0.125	1.52	11.1	24.63	36.8	46.4	70.9	100

Example 8.2: Semi-infinite Region with Time-Dependent Surface Flux

A semi-infinite region which is initially at uniform temperature T_i is suddenly heated at its surface with a time-dependent flux $q_o''(t)$. Determine the transient temperature in the region and the surface temperature.

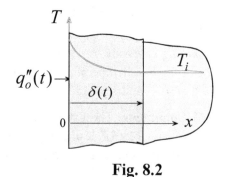

Fig. 8.2

(1) Observations. (i) Semi-infinite region. (ii) Time dependent flux at the surface.

(2) Origin and Coordinates. Fig. 8.2 shows the origin and coordinate axes.

(3) Formulation and Solution.

(i) **Assumptions.** (1) One-dimensional conduction, (2) uniform initial temperature and (3) constant properties.

(ii) **Integral Formulation.** The heat-balance integral for this problem will be formulated using two methods.

(a) *Control volume formulation.* The effect of surface heating is confined to a layer of thickness $\delta(t)$ known as the *penetration depth* or the *thermal layer.* At the edge of this layer the temperature is T_i. Temperature variation within this layer is shown in Fig. 8.2. A finite control volume is selected which encompasses the entire thermal layer. Application of conservation of energy to the control volume gives

$$\dot{E}_{in} - \dot{E}_{out} = \dot{E},\qquad\text{(a)}$$

where $\dot{E}$ is the time rate of change of energy within the control volume. Energy added at the surface $x = 0$ is specified as

$$\dot{E}_{in} = q_o''(t).\qquad\text{(b)}$$

Since at $x = \delta(t)$ the temperature gradient is approximately zero, it follows that no heat is conducted at this boundary. Thus

$$\dot{E}_{out} \cong 0.\qquad\text{(c)}$$

For constant density and specific heat, the rate of energy change within the element is

$$\dot{E} = \rho c_p \frac{d}{dt} \int_0^{\delta(t)} \left[T(x,t) - T_i\right] dx.\qquad\text{(d)}$$

Substituting (b)-(d) into (a)

$$q_o''(t) = \rho c_p \frac{d}{dt} \int_0^{\delta(t)} \left[T(x,t) - T_i\right] dx.\qquad\text{(8.8)}$$

Equation (8.8) is the heat-balance integral for this problem.

(b) *Integration of the governing differential equation.* The governing differential equation for one-dimensional transient conduction is obtained from eq. (1.8)

$$k\frac{\partial^2 T}{\partial x^2} = \rho c_p \frac{\partial T}{\partial t}. \tag{e}$$

Multiplying both sides of (e) by dx and integrating from $x = 0$ to $x = \delta(t)$

$$k\int_0^{\delta(t)} \frac{\partial^2 T}{\partial x^2}\,dx = \rho c_p \int_0^{\delta(t)} \frac{\partial T}{\partial t}\,dx. \tag{f}$$

Noting that the integral of the second derivative is the first derivative and using Leibnitz's rule, (f) becomes

$$k\frac{\partial T(\delta,t)}{\partial x} - k\frac{\partial T(0,t)}{\partial x} = \rho c_p\left[T(0,t)\frac{d0}{dt} - T(\delta,t)\frac{d\delta}{dt} + \frac{d}{dt}\int_0^{\delta(t)} T(x,t)\,dx\right]. \tag{g}$$

This result is simplified using the following exact and approximate boundary conditions

(1) $-k\dfrac{\partial T(0,t)}{\partial x} = q_o''(t)$

(2) $\dfrac{\partial T(\delta,t)}{\partial x} \cong 0$

(3) $T(\delta,t) \cong T_i$

Substituting into (g)

$$q_o''(t) = \rho c_p\left[-T_i\frac{d\delta}{dt} + \frac{d}{dt}\int_0^{\delta(t)} T(x,t)\,dx\right]. \tag{h}$$

However

$$\frac{d\delta}{dt} = \frac{d}{dt}\int_0^{\delta(t)} dx.$$

Substituting into (h) gives the heat-balance integral

$$q_o''(t) = \rho c_p\frac{d}{dt}\int_0^{\delta(t)} \left[T(x,t) - T_i\right]dx. \tag{8.8}$$

(iii) Assumed Temperature Profile.   Assume a polynomial of the second degree

$$T(x) = a_0 + a_1 x + a_2 x^2 . \tag{8.9}$$

Boundary conditions (1), (2) and (3) give the coefficients a_0, a_1, and a_2

$$a_0 = T_i + q_o''(t)\delta(t)/2k , \quad a_1 = -q_o''(t)/k , \quad a_2 = q_o''(t)/2k\delta(t) .$$

Equation (8.9) becomes

$$T(x,t) = T_i + \frac{q_o''(t)}{2k\delta(t)}[\delta(t) - x]^2 . \tag{8.10}$$

(iv) Determination of the Unknown Variable $\delta(t)$. The penetration depth $\delta(t)$ is determined by satisfying the heat-balance integral for the problem. Substituting eq. (8.10) into eq. (8.8)

$$\delta^2 - 2\delta x + x^2$$

$$q_o''(t) = \rho c_p \frac{d}{dt} \int_0^{\delta(t)} \frac{q_o''(t)}{2k\delta(t)}[\delta(t) - x]^2 \, dx .$$

Integrating with respect to x

$$q_o''(t) = \frac{\rho c_p}{6k} \frac{d}{dt}\left[q_o''(t)\delta^2(t)\right]. \tag{i}$$

This is a first order ordinary differential equation for $\delta(t)$. The initial condition on $\delta(t)$ is

$$\delta(0) = 0 . \tag{j}$$

Multiplying (i) by dt, integrating and using (j) gives

$$\frac{6k}{\rho c_p} \int_0^t q_o''(t)\,dt = \int_0^t d\left[q_o''(t)\delta^2(t)\right] =$$

$$q_o''(t)\delta^2(t) - q_o''(0)\delta^2(0) = q_o''(t)\delta^2(t) .$$

Solving for $\delta(t)$ and noting that $\alpha = k/\rho c_p$

$$\delta(t) = \left[\frac{6\alpha}{q_o''(t)} \int_0^t q_o''(t)\,dt\right]^{1/2} . \tag{8.11}$$

When eq. (8.11) is substituted into eq. (8.10) we obtain the transient temperature.

(4) Checking. *Dimensional check*: Eqs. (8.10), (i) and (8.11) are dimensionally consistent.

Limiting check: (i) If $q_o''(t) \to 0$, the temperature remains at T_i. Setting $q_o''(t) = 0$ in eq. (8.10) gives $T(x,t) = T_i$. (ii) If $\alpha \to \infty$, the penetration depth $\delta \to \infty$. Setting $\alpha = \infty$ in eq. (8.11) gives $\delta = \infty$.

(5) Comments. The accuracy of the integral solution can be examined by comparison with the exact solution. For the special case of constant heat flux, $q_o''(t) = q_o''$, equation (8.11) becomes

$$\delta(t) = \sqrt{6\alpha t} . \tag{k}$$

Substituting (k) into eq. (8.10)

$$T(x,t) = T_i + \frac{q_o''}{2k\sqrt{6\alpha t}}\left[\sqrt{6\alpha t} - x\right]^2 . \tag{l}$$

Specifically, the temperature at the surface, $T(0,t)$ is obtained by setting $x = 0$ in (l)

$$T(0,t) = T_i + \frac{q_o''}{2k}\sqrt{6\alpha t} .$$

This result can be rewritten in dimensionless form

$$\frac{T(0,t) - T_i}{q_o''\sqrt{\alpha t}/k} = \sqrt{\frac{3}{2}} = 1.225 .$$

The exact solution gives

$$\left[\frac{T(0,t) - T_i}{q_o''\sqrt{\alpha t}/k}\right]_{exact} = \sqrt{\frac{4}{\pi}} = 1.128 .$$

Thus the error in predicting surface temperature is 8.6%.

8.5 Application to Cylindrical Coordinates

The procedure outlined in Section 8.2 will be followed to illustrate how the integral method is used to obtain solutions to problems where the temperature varies radially.

Example 8.3: Cylindrical Fin

The cylindrical fin shown in Fig. 8.3 is maintained at a specified temperature T_o at the base and is insulated at the tip. The heat transfer coefficient is h and the ambient temperature is T_∞. The fin has an inner radius r_i, outer radius r_o and thickness w. Of interest is the determination of the temperature distribution and the heat transfer rate using the integral method.

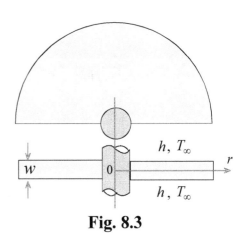

h, T_∞

h, T_∞

Fig. 8.3

(1) Observations. (i) Fin thickness is constant. (ii) Specified temperature at the base and insulated tip. (iii) Surface heat exchange is by convection.

(2) Origin and Coordinates. Fig. 8.3 shows the origin and coordinate axis.

(3) Formulation and Solution.

    (i) Assumptions. (1) Steady state, (2) fin approximations are valid ($Bi < 0.1$), and (3) uniform h and T_∞.

    (ii) Integral Formulation. The heat-balance integral will be formulated using two methods.

(a) *Control volume formulation.* A control volume is selected which encompasses the entire fin. Application of conservation of energy to the control volume gives

$$\dot{E}_{in} - \dot{E}_{out} = 0. \tag{a}$$

Assuming that the base is at a higher temperature than the ambient, Fourier's law of conduction gives

$$\dot{E}_{in} = -2\pi r_i wk \frac{dT(r_i)}{dr}. \tag{b}$$

Integration of Newton's law of cooling along the surface gives the heat removed by convection. Thus, assuming constant h

$$\dot{E}_{out} = 4\pi h \int_{r_i}^{r_o} (T - T_\infty) r \, dr \, . \tag{c}$$

Substituting (b) and (c) into (a)

$$\frac{dT(r_i)}{dr} + \frac{2h}{kwr_i} \int_{r_i}^{r_o} (T - T_\infty) r \, dr = 0 \, . \tag{8.12}$$

This is the heat-balance integral for the fin.

(b) *Integration of the governing differential equation.* Differential formulation of energy conservation for this fin is given by equation (2.24)

$$. \frac{d^2 T}{dr^2} + \frac{1}{r} \frac{dT}{dr} - \frac{2h}{kw} (T - T_\infty) = 0 \, . \tag{2.24}$$

This equation can be rewritten as

$$\frac{1}{r} \frac{d}{dr} \left(r \frac{dT}{dr} \right) - \frac{2h}{kw} (T - T_\infty) = 0 \, . \tag{d}$$

Multiplying (d) by $2\pi r dr$ and integrating over the surface from $r = r_i$ to $r = r_o$, gives

$$\int_{r_i}^{r_o} \frac{1}{r} \frac{d}{dr} \left(r \frac{dT}{dr} \right) 2\pi r \, dr - \frac{2h}{kw} \int_{r_i}^{r_o} (T - T_\infty) 2\pi r \, dr = 0 \, ,$$

or

$$\int_{r_i}^{r_o} d \left(r \frac{dT}{dr} \right) - \frac{2h}{kw} \int_{r_i}^{r_o} (T - T_\infty) r \, dr = 0 \, . \tag{e}$$

The first term can be integrated directly. Thus (e) becomes

$$r_o \frac{dT(r_o)}{dr} - r_i \frac{dT(r_i)}{dr} - \frac{2h}{kw} \int_{r_i}^{r_o} (T - T_\infty) r \, dr = 0 \, .$$

Since the tip is insulated it follows that

$$\frac{dT(r_o)}{dr} = 0 \, . \tag{f}$$

Thus (e) becomes

$$\frac{dT(r_i)}{dr} + \frac{2h}{kwr_i} \int_{r_i}^{r_o} (T - T_\infty) r\, dr = 0. \qquad (8.12)$$

This is identical to the heat-integral equation obtained using control volume formulation.

(iii) Assumed Temperature Profile. Assume a polynomial of the second degree

$$T(r) = a_0 + a_1 r + a_2 r^2. \qquad (g)$$

The coefficients a_0, a_1 and a_2 must be such that (g) satisfies the boundary conditions on the temperature. In addition to condition (f), specified temperature at the base gives

$$T(r_i) = T_o. \qquad (h)$$

Using conditions (f) and (h), the assumed profile becomes

$$T(r) = T_o + r_i [(r_i / 2r_o) + (r / r_i) - (r^2 / 2r_i r_o) - 1] a_1. \qquad (i)$$

Thus the only unknown is the coefficient a_1.

(iv) Determination of the Unknown Coefficient. The coefficient a_1 is determined using the heat-balance integral, eq. (8.12). Substituting (i) into eq. (8.12)

$$\left(1 - \frac{r_i}{r_o}\right) a_1 + \frac{2h}{kwr_i} \int_{r_i}^{r_o} \{(T_o - T_\infty) +$$

$$r_i [(r_i / 2r_o) + (r / r_i) - (r^2 / 2r_i r_o) - 1] a_1\} r\, dr = 0. \qquad (j)$$

Performing the integration and solving for the coefficient a_1 gives

$$a_1 = \frac{\dfrac{m^2 r_o^2}{r_i} (T_\infty - T_o)(1 - R^2)}{2(1 - R) + m^2 r_o^2 \left[\dfrac{(R-2)(1-R^2)}{2} + \dfrac{2(1-R^3)}{3R} - \dfrac{R(1-R^4)}{4R}\right]}, \qquad (k)$$

where

$$m^2 = \frac{2h}{kw},$$
(l)

and

$$R = \frac{r_i}{r_o}.$$
(m)

Substituting (k) into (i) gives the temperature distribution

$$\frac{T(r) - T_o}{T_\infty - T_o} = \frac{m^2 r_o^2 \left[\dfrac{R-2}{2} + \dfrac{r}{r_i} - \dfrac{Rr^2}{2r_i^2} \right](1 - R^2)}{2(1-R) + m^2 r_o^2 \left[\dfrac{(R-2)(1-R^2)}{2} + \dfrac{2(1-R^3)}{3R} - \dfrac{(1-R^4)}{4R} \right]}.$$
(8.13)

Heat transfer from the fin is determined by applying Fourier's law at $r = r_i$

$$q = -k2\pi r_i w \frac{dT(r_i)}{dr}.$$
(n)

Substituting (8.13) into (n) and using the dimensionless fin heat transfer rate q^* introduced in the definition of fin efficiency, eq. (2.18), gives

$$q^* = \eta_f = \frac{q}{2\pi h(r_o^2 - r_i^2)(T_o - T_\infty)}$$

$$= \frac{2(1-R)}{2(1-R) + m^2 r_o^2 \left[\dfrac{(R-2)(1-R^2)}{2} + \dfrac{2(1-R^3)}{3R} - \dfrac{(1-R^4)}{4R} \right]}.$$
(8.14)

(4) Checking. *Dimensional check*: Equations (8.13) and (8.14) are dimensionally correct.

Limiting check: (i) For $r_o = \infty$, the dimensionless heat transfer rate q^* should vanish. Setting $r_o = \infty$ in eq. (8.14) gives $q^* = 0$. (ii) For $h = 0$, no heat leaves the fin and thus the entire fin should be at T_o. Setting $h = m = 0$ in eq. (8.13) gives $T(r) = T_o$.

(5) Comments. Fin equation (2.24) can be solved exactly [3]. Thus, the accuracy of the integral solution can be evaluated. Examination of eq. (8.14) shows that q^* depends on two parameters: R and mr_o. Table 8.2 compares the integral solution with the exact solution for $R = 0.2$ and various values of the parameter mr_o. At large values of mr_o the assumed polynomial profile departs further from the actual temperature distribution and consequently the integral solution becomes increasingly less accurate as mr_o is increased. Alternate profiles involving $\log r$ have been shown to significantly improve the accuracy of problems in cylindrical coordinates [4].

Table 8.2

	Percent error in q^* for $r_i / r_o = 0.2$					
mr_o	0.2	0.5	1	2	3	4
% Error	1.7	3.5	16.8	33.0	43.4	51.2

8.6 Non-linear Problems [5]

The integral method can be used to solve non-linear problems. Two examples will be presented. The first involves temperature dependent properties and the second deals with phase change.

Example 8.4: Semi-infinite Region with Temperature Dependent Properties

A semi-infinite region is initially at uniform temperature T_i. The region is suddenly heated at its surface with uniform flux q_o''. The conductivity, density and specific heat depend on temperature according to the following:

$$k(T) = k_o(1 + \beta_1 T), \qquad \text{(a)}$$

$$c_p(T) = c_{po}(1 + \beta_2 T), \qquad \text{(b)}$$

and

$$\rho(T) = \rho_o(1 + \beta_3 T), \qquad \text{(c)}$$

where the coefficients β_n are constant and the subscript o

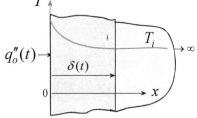

Fig. 8.4

indicates a reference state. Determine the temperature distribution in the region.

(1) Observations. (i) Semi-infinite region. (ii) Temperature dependent properties.

(2) Origin and Coordinates. Fig. 8.4 shows the origin and coordinate axes.

(3) Formulation and Solution.

(i) **Assumptions.** (1) One-dimensional transient conduction, (2) no energy generation and (3) uniform initial temperature.

(ii) **Integral Formulation.** The heat-balance integral will be formulated by integrating the heat conduction equation for variable properties. Equation (1.7) simplifies to

$$\frac{\partial}{\partial x}\left(k\frac{\partial T}{\partial x}\right) = \rho c_p \frac{\partial T}{\partial t}, \tag{8.15}$$

where k, ρ and c_p are functions of temperature. Note that both terms in eq. (8.15) are non-linear. The boundary conditions are

$$(1) \quad -k\frac{\partial T(0,t)}{\partial x} = q_o''$$

$$(2) \quad T(\delta,t) \cong T_i$$

$$(3) \quad \frac{\partial T(\delta,t)}{\partial x} \cong 0$$

where $\delta(t)$ is the penetration depth. Equation (8.15) will be simplified by introducing the following Kirchhoff transformation

$$\theta(T) = \frac{1}{\rho_o c_{po}} \int_0^T \rho(T)\, c_p(T)\, dT . \tag{8.16}$$

Differentiating eq. (8.16) gives

$$\frac{d\theta}{dT} = \frac{\rho c_p}{\rho_o c_{po}} . \tag{d}$$

Thus

$$\frac{\partial \theta}{\partial x} = \frac{d\theta}{dT}\frac{\partial T}{\partial x}, \quad \text{or} \quad \frac{\partial T}{\partial x} = \frac{\rho_o c_{po}}{\rho c_p}\frac{\partial \theta}{\partial x},$$

$$\frac{\partial \theta}{\partial t} = \frac{d\theta}{dT}\frac{\partial T}{\partial t}, \quad \text{or} \quad \frac{\partial T}{\partial t} = \frac{\rho_o c_{po}}{\rho c_p}\frac{\partial \theta}{\partial t}.$$

Substituting into eq. (8.15)

$$\frac{\partial}{\partial x}\left(\alpha(T)\frac{\partial \theta}{\partial x}\right) = \frac{\partial \theta}{\partial t}, \tag{8.17}$$

where $\alpha(T)$ is the thermal diffusivity, defined as

$$\alpha(T) = \frac{k(T)}{\rho(T)c_p(T)}. \tag{8.18}$$

In equation (8.17) only the left side term is non-linear. The three boundary conditions must be transformed in terms of the new variable $\theta(x,t)$. Using the definitions α and θ, boundary condition (1) becomes

$$-\alpha_s\frac{\partial \theta(0,t)}{\partial x} = \frac{q_o''}{\rho_o c_{po}}, \tag{e}$$

where α_s is the thermal diffusivity at surface temperature $\theta(0,t)$. Boundary condition (2) transforms to

$$\theta(\delta,t) = \frac{1}{\rho_o c_{po}}\int_0^{T_i}\rho(T)\,c_p(T)\,dT \equiv \theta_i. \tag{f}$$

Boundary condition (3) gives

$$\frac{\partial \theta(\delta,t)}{\partial x} \cong 0. \tag{g}$$

The objective now is to solve eq. (8.17) for $\theta(x,t)$ using the integral method. Once $\theta(x,t)$ is determined, temperature distribution $T(x,t)$ can be obtained from eq.(8.16) using (b) and (c)

$$\theta(T) = \frac{1}{\rho_o c_{po}}\int_0^T\rho_o(1+\beta_3 T)\,c_{po}(1+\beta_2 T)\,dT.$$

Carrying out the integration gives

$$\theta(T) = T + \frac{\beta_2 + \beta_3}{2}T^2 + \frac{\beta_2\beta_3}{3}T^3. \tag{8.19}$$

Equation (8.19) relates the transformation variable θ to the temperature variable T.

The heat-balance integral for this problem will be formulated by integrating eq. (8.17) term by term. Multiplying eq. (8.17) by dx and integrating from $x = 0$ to $x = \delta(t)$

$$\int_0^{\delta(t)} \frac{\partial}{\partial x}\left(\alpha \frac{\partial\theta}{\partial x}\right) dx = \int_0^{\delta(t)} \frac{\partial\theta}{\partial t} dx . \tag{h}$$

Using Leibnitz's rule to rewrite the integral on the right side, (h) gives

$$\alpha_i \frac{\partial\theta(\delta,t)}{\partial x} - \alpha_s \frac{\partial\theta(0,t)}{\partial x} = \theta(0,t)\frac{d0}{dt} - \theta(\delta,t)\frac{d\delta}{dt} + \frac{d}{dt}\int_0^{\delta(t)} \theta(x,t)\, dx, \tag{i}$$

where α_i is thermal diffusivity at $\theta(\delta,t)$. This result is simplified using conditions (e), (f) and (g)

$$\frac{q_o''}{\rho_o c_{po}} = -\theta_i \frac{d\delta}{dt} + \frac{d}{dt}\int_0^{\delta(t)} \theta(x,t)\, dx . \tag{j}$$

However

$$\frac{d\delta}{dt} = \frac{d}{dt}\int_0^{\delta(t)} dx .$$

Substituting into (j) gives the heat-balance integral

$$\frac{q_o''}{\rho_o c_{po}} = \frac{d}{dt}\int_0^{\delta(t)} [\theta(x,t) - \theta_i]\, dx . \tag{20}$$

(iii) Assumed Temperature Profile. Assume a polynomial of the second degree

$$\theta(x) = a_0 + a_1 x + a_2 x^2 . \tag{k}$$

Boundary conditions (e), (f) and (g) give the coefficients a_0, a_1, and a_2. The result is

$$a_0 = \theta_i + q_o'' \delta / 2\rho_o c_{po} \alpha_s$$

$$a_1 = -q_o'' / \rho_o c_{po} \alpha_s$$

$$a_2 = q_o'' / 2\delta \rho_o c_{po} \alpha_s$$

Equation (k) becomes

$$\theta(x,t) = \theta_i + \frac{q_o''}{\rho_o c_{po} \alpha_s} \left[\delta(t)/2 - x + x^2 / 2\delta(t) \right]. \qquad (8.21)$$

(iv) Determination of the Unknown Variable $\delta(t)$. The thermal layer $\delta(t)$ is obtained by satisfying the heat-balance integral for the problem. Substituting eq. (8.21) into eq. (8.20)

$$\frac{q_o''}{\rho_o c_{po}} = \frac{d}{dt} \int_0^{\delta(t)} \left[\frac{q_o''}{\rho_o c_{po} \alpha_s} \left[\delta(t)/2 - x + x^2 / 2\delta(t) \right] \right] dx,$$

or

$$1 = \frac{1}{6} \frac{d}{dt} \left[(1/\alpha_s)\delta^2(t) \right]. \qquad (l)$$

This is a first order ordinary differential equation for $\delta(t)$. The initial condition on $\delta(t)$ is

$$\delta(0) = 0. \qquad (m)$$

Integrating (l) and using initial condition (m) gives $\delta(t)$

$$\delta(t) = \sqrt{6\alpha_s t}. \qquad (8.22)$$

Since α_s depends on surface temperature $\theta(0,t)$, eq. (8.22) does not give $\delta(t)$ directly. Setting $x = 0$ in eq. (8.21) gives $\theta(0,t)$

$$\theta(0,t) = \theta_i + \frac{q_o''}{2\rho_o c_{po} \alpha_s} \delta(t).$$

Using (8.22) to eliminate $\delta(t)$

$$\theta(0,t) = \theta_i + \frac{q_o''}{\rho_o c_{po}} \sqrt{\frac{3}{2} \frac{t}{\alpha_s}}. \qquad (8.23)$$

Using eqs. (8.18) and (8.19) we obtain α_s in terms of $\theta(0,t)$. Using this result with eq. (8.23) gives α_s as a function of time. Eq. (8.22) is then used to obtain $\delta(t)$. Substituting $\delta(t)$ into eq. (8.21) gives $\theta(x,t)$. The transformation equation (8.19) gives the temperature distribution $T(x,t)$.

(4) Checking. *Dimensional check*: Equations (8.21) and (8.22) are dimensionally correct.

Limiting check: For the special case of constant properties the solution agrees with the result of Example 8.2.

(5) Comments. (i) The more general case of time dependent heat flux at the surface can be easily solved following the same procedure. (ii) The accuracy of the solution can be improved by using a cubic polynomial

$$\theta(x,t) = a_0 + a_1 x + a_2 x^2 + a_3 x^3 .$$
(n)

A fourth boundary condition is needed to determine the four coefficients. Since the temperature gradient vanishes at $x \geq \delta$ it follows that

$$\frac{\partial^2 \theta(\delta,t)}{\partial x^2} \cong 0 .$$
(o)

Using this condition, the assumed profile becomes

$$\theta(x,t) = \theta_i + \frac{q_o'' \delta}{3\alpha_s \rho_o c_{po}}(1 - x/\delta)^3 .$$
(8.24)

Based on this profile the solution to $\delta(t)$ is

$$\delta(t) = \sqrt{12\alpha_s t} .$$
(8.25)

Example 8.5: Conduction with Phase Change

A semi-infinite region is initially solid at the fusion temperature T_f. At $t > 0$ the surface at $x = 0$ is suddenly maintained at $T_o > T_f$. Determine the transient interface location $x_i(t)$.

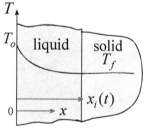

(1) Observations. (i) Semi-infinite region. (ii) Solid phase remains at T_f.

(2) Origin and Coordinates. Fig. 8.5 shows the origin and coordinate axes.

Fig. 8.5

(3) Formulation and Solution.

(i) **Assumptions.** (1) One-dimensional transient conduction, (2) constant properties and (3) uniform initial temperature.

(ii) **Integral Formulation.** The heat-balance integral for the liquid phase will be formulated by integrating the differential equation. The heat equation for constant properties is obtained from eq. (1.8)

$$\alpha \frac{\partial^2 T}{\partial x^2} = \frac{\partial T}{\partial t}. \tag{a}$$

Multiplying by dx and integrating from $x = 0$ to $x = x_i(t)$

$$\alpha \int_0^{x_i(t)} \frac{\partial^2 T}{\partial x^2} dx = \int_0^{x_i(t)} \frac{\partial T}{\partial t} dx. \tag{b}$$

Evaluating the integrals and using Leibnitz's rule, (b) becomes

$$\alpha \frac{\partial T(x_i,t)}{\partial x} - \alpha \frac{\partial T(0,t)}{\partial x} = T(0,t)\frac{d0}{dt} - T(x_i,t)\frac{dx_i}{dt} + \frac{d}{dt}\int_0^{x_i(t)} T(x,t)\,dx. \tag{c}$$

This result is simplified using the following conditions

(1) $T(0,t) = T_o$

(2) $T(x_i,t) = T_f$

(3) $-k\frac{\partial T(x_i,t)}{\partial x} = \rho \mathcal{L}\frac{dx_i}{dt}$

Using boundary conditions (2) and (3), equation (c) becomes

$$-\alpha\rho\frac{\mathcal{L}}{k}\frac{dx_i}{dt} - \alpha\frac{\partial T(0,t)}{\partial x} = -T_f\frac{dx_i}{dt} + \frac{d}{dt}\int_0^{x_i(t)} T(x,t)\,dx. \tag{d}$$

Equation (d) is the heat-balance integral for this problem. Note that properties in (d) refer to the liquid phase.

(iii) **Assumed Temperature Profile.** Assume a polynomial of the second degree

$$T(x,t) = a_0 + a_1(x - x_i) + a_2(x - x_i)^2.$$
(e)

The coefficients in (c) are determined using the three boundary conditions. However, condition (3) in its present form leads to a second order differential equation for $x_i(t)$. An alternate approach is to combine conditions (2) and (3). The total derivative of $T(x_i,t)$ in condition (2) is

$$dT_f = \frac{\partial T(x_i,t)}{\partial x} dx + \frac{\partial T(x_i,t)}{\partial t} dt = 0 \qquad \text{at } x = x_i.$$
(f)

Setting $dx = dx_i$ in (f) gives

$$\frac{\partial T(x_i,t)}{\partial x}\frac{dx_i}{dt} + \frac{\partial T(x_i,t)}{\partial t} = 0.$$
(g)

Substituting condition (3) into (g) gives

$$\left[\frac{\partial T(x_i,t)}{\partial x}\right]^2 = \frac{\rho \mathcal{L}}{k}\frac{\partial T}{\partial t}.$$

Using (a) to eliminate $\partial T/\partial t$ in the above and recalling that $\alpha = k/\rho c_p$, we obtain

$$\left[\frac{\partial T(x_i,t)}{\partial x}\right]^2 = \frac{\mathcal{L}}{c_p}\frac{\partial T^2(x_i,t)}{\partial x^2}.$$
(h)

Using boundary conditions (1), (2) and equation (h) gives the coefficients a_0, a_1 and a_2

$$a_0 = T_f,$$
(i)

$$a_1 = \frac{\alpha \rho \mathcal{L}}{kx_i}\left(1 - (1+\mu)^{1/2}\right),$$
(j)

$$a_2 = \frac{a_2 x_i + (T_o - T_f)}{x_i^2},$$
(k)

where

$$\mu = \frac{2c_p}{\mathcal{L}}(T_o - T_f).$$
(l)

The only remaining unknown is $x_i(t)$. It is obtained using the heat-balance integral (d). Substituting (e) into (d) gives

$$x_i \frac{dx_i}{dt} = 6\alpha \frac{1-(1+\mu)^{1/2}+\mu}{5+(1+\mu)^{1/2}+\mu}. \tag{m}$$

Solving this equation for $x_i(t)$ and using the initial condition on the interface location, $x_i(0) = 0$, gives

$$x_i = 2\lambda_a \sqrt{\alpha t}, \tag{n}$$

where

$$\lambda_a = \left[3\frac{1-(1+\mu)^{1/2}+\mu}{5+(1+\mu)^{1/2}+\mu} \right]^{1/2}. \tag{o}$$

(4) Checking. *Dimensional check:*   Equations (h), (m) and (n) are dimensionally consistent.

Limiting Check: (i) If $\mathcal{L} = \infty$, the interface remains stationary ($x_i = 0$). Setting $\mathcal{L} = \infty$ in (l) gives $\mu = 0$.   When this is substituted into (o) we obtain $\lambda_a = 0$. According to (n) $\lambda_a = 0$ corresponds to $x_i = 0$. (ii) If $T_o = T_f$, the interface remains stationary. Setting $T_o = T_f$ into (l) gives $\mu = 0$. This in turn gives $\lambda_a = 0$ and $x_i = 0$.

(5) Comments. (i) comparing the definition of the parameter μ in (l) with the definition of the Stefan number *Ste* in eq. (6.13) gives

$$\mu = 2\,Ste.$$

Thus a large μ corresponds to a large Stefan number. As the Stefan number is increased the interface moves more rapidly.

(ii) This problem is identical to Stefan's problem. The exact solution to the interface location, x_{ie}, is given by

$$x_{ie} = 2\lambda\sqrt{\alpha t}, \tag{p}$$

where λ is obtained from Stefan's solution

$$\lambda e^{\lambda^2} \operatorname{erf} \lambda = \frac{c_p(T_o - T_f)}{\sqrt{\pi}\,\mathcal{L}} = \frac{\mu}{2\sqrt{\pi}}. \tag{6.24}$$

The two solutions are compared by taking the ratio of equations (n) and (p) to obtain

$$\frac{x_i}{x_{ie}} = \frac{\lambda_a}{\lambda}.$$

The result is shown in Table 8.3. The error is seen to increase as μ (or the Stefan number) is increased. This is due to the fact that the assumed temperature profile becomes progressively inappropriate as μ increases. Nevertheless, at $\mu = 4$ the error is 7.3%. It should be pointed out that the Stefan number for many engineering applications ranges from 0 to 3. This corresponds to a range of μ from 0 to 6.

Table 8.3

μ	x_i / x_{ie}
0	1.000
0.4	1.026
0.8	1.042
1.2	1.052
1.6	1.059
2.0	1.064
2.4	1.068
2.8	1.070
3.2	1.072
3.4	1.073
4.0	1.073

8.7 Energy Generation

The following example illustrates the application of the integral method to the solution of systems with energy generation.

Example 8.6: Semi-infinite Region with Energy Generation [5]

A semi-infinite region is initially at uniform temperature $T_i = 0$. At time $t > 0$ energy is generated at a variable rate $q'''(t)$. The surface at $x = 0$ is maintained at $T(0,t) = 0$. Use the integral method to determine the temperature distribution in the region.

(1) Observations. (i) Semi-infinite region. (ii) Transient conduction. (iii) Time dependent energy generation.

(2) Origin and Coordinates. Fig. 8.6 shows the origin and coordinate axes.

(3) Formulation and Solution.

(i) **Assumptions.** (1) Constant properties, (2) one-dimensional transient conduction, and (3) uniform initial temperature and energy generation.

(ii) **Integral Formulation.** The heat-

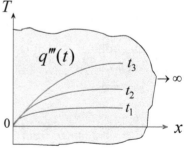

Fig. 8.6

balance integral will be formulated by integrating the differential equation for this problem. The heat equation for constant properties with energy generation is obtained from equation (1.8)

$$\alpha \frac{\partial^2 T}{\partial x^2} + \frac{q'''(t)}{\rho c_p} = \frac{\partial T}{\partial t}. \tag{a}$$

Multiplying by dx and integrating from $x = 0$ to $x = \delta(t)$

$$\alpha \int_0^{\delta(t)} \frac{\partial^2 T}{\partial x^2} dx + \int_0^{\delta(t)} \frac{q'''(t)}{\rho c_p} dx = \int_0^{\delta(t)} \frac{\partial T}{\partial t} dx. \tag{b}$$

Evaluating the first two integrals and using Leibnitz's rule to rewrite the third integral, (b) becomes

$$\alpha \frac{\partial T(\delta,t)}{\partial x} - \alpha \frac{\partial T(0,t)}{\partial x} + \frac{q'''}{\rho c_p} \delta(t) = T(0,t)\frac{d0}{dt} - T(\delta,t)\frac{d\delta}{dt} + \frac{d}{dt}\int_0^{\delta(t)} T(x,t)\,dx. \tag{c}$$

This result is simplified using the following exact and approximate boundary and initial conditions

(1) $T(0,t) = 0$

(2) $\dfrac{\partial T(\delta,t)}{\partial x} \cong 0$

Substituting these conditions into (c) gives

$$-\alpha \frac{\partial T(0,t)}{\partial x} + \frac{q'''}{\rho c_p} \delta(t) = -T(\delta,t)\frac{d\delta}{dt} + \frac{d}{dt}\int_0^{\delta(t)} T(x,t)\,dx. \tag{d}$$

Equation (d) is the heat-balance integral for this problem.

(iii) Assumed Temperature Profile. Assume a third degree polynomial of the form

$$T(x) = a_0 + a_1 x + a_2 x^2 + a_3 x^3. \tag{e}$$

Two additional boundary conditions are needed to determine the four coefficients in the assumed profile. The third condition is

$$(3) \quad \frac{\partial^2 T(\delta,t)}{\partial x^2} \cong 0$$

A fourth condition is obtained by evaluating (a) at $x = \delta$ and using boundary condition (3)

$$\frac{q'''(t)}{\rho c_p} = \frac{\partial T(\delta,t)}{\partial t}.$$

This equation can be integrated directly to give the temperature at $x = \delta$

$$T(\delta,t) = \frac{1}{\rho c_p} \int_0^t q'''(t)dt . \qquad (f)$$

The integral in (f) can be evaluated for a given energy generation rate $q'''(t)$. Let

$$Q(t) = \int_0^t q'''(t)dt . \qquad (g)$$

Substituting (g) into (f) gives the fourth boundary condition as

$$(4) \ T(\delta,t) = \frac{Q(t)}{\rho c_p}$$

Using the four boundary conditions, the coefficients in (c) can be determined. The assumed profile becomes

$$T(x,t) = \frac{Q(t)}{\rho c_p}\left[1 - (1 - x/\delta)^3\right]. \qquad (h)$$

Thus the only unknown is the penetration depth $\delta(t)$. It is determined using the heat-balance integral equation. Substituting boundary condition (4) and the assumed profile (h) into (d)

$$-12\alpha Q(t) + 4q'''(t)\delta^2 = 3\delta^2\frac{dQ}{dt} - \delta Q(t)\frac{d\delta}{dt}. \qquad (i)$$

Differentiating (g) gives

$$q'''(t) = \frac{dQ}{dt}. \qquad (j)$$

Substituting into (i) gives

$$\delta^2 Q(t)\frac{dQ}{dt} + \delta Q^2(t)\frac{d\delta}{dt} = 12\alpha Q^2(t),$$

or

$$\frac{d}{dt}\left[\delta^2 Q^2(t)\right] = 24\alpha Q^2(t). \tag{k}$$

Integration of (k) gives

$$\delta^2 Q^2(t)\Big|_0^t = 24\alpha \int_0^t Q^2(t)dt. \tag{l}$$

However, the initial condition on $\delta(t)$ is

$$\delta(0) = 0$$

Substituting into (l)

$$\delta(t) = \frac{\left[24\alpha \int_0^t Q^2(t)\,dt\right]^{1/2}}{Q(t)}. \tag{m}$$

When (m) is substituted into (h) we obtain the transient temperature distribution.

(4) Checking. *Dimensional check*: Equations (h) and (m) are dimensionally consistent.

Limiting check: (i) If $q''' = 0$, the temperature remains at the initial value. Setting $q'''(t) = 0$ in (g) gives $Q = 0$. When this is substituted into (h) gives $T(x,t) = 0$.

(ii) If $q''' = \infty$, the temperature of the entire region should be infinite. Setting $q''' = \infty$ in (g) gives $Q(t) = \infty$. When this is substituted into (h) we obtain $T(x,t) = \infty$.

(5) Comments. For the special case of constant energy generation rate, (g) gives

$$Q(t) = q'''t.$$

Substituting this result into (k) gives

$$\delta(t) = 2\sqrt{2\alpha t}.$$

Equation (h) becomes

$$T(x,t) = \frac{q'''t}{\rho c_p}\left[1 - (1 - x/2\sqrt{2\alpha t})^3\right].$$

REFERENCES

[1] Hildebrand, F.B., *Advance Calculus for Applications*, 2nd edition, Prentice-Hall, Englewood Cliffs, New Jersey, 1976.

[2] Arpaci, V.S., *Conduction Heat Transfer*, Addison-Wesley, Reading, Massachusetts, 1966.

[3] Incropera, F.P., and DeWitt, D.P. *Introduction to Heat Transfer*, 3rd edition, Wiley, New York, 1996.

[4] Lardner, T.J., and Pohle, F.P., "Application of Heat-Balance Integral to the Problems of Cylindrical Geometry," *ASME Transactions. J. Applied Mechanics*, pp. 310-312, 1961.

[5] Goodman, T.R., "Application of Integral Methods to Transient Non-linear Heat Transfer". In *Advances in Heat Transfer*, volume 1, 1964, Academic Press, New York.

PROBLEMS

8.1 The base of a fin of length L is maintained at constant temperature T_o while heat is added at uniform flux q_o'' at the tip. The heat transfer coefficient is h and the ambient temperature is T_∞. Assume a second degree polynomial temperature profile and obtain an integral solution for the steady state temperature distribution.

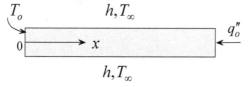

8.2 A hollow sphere of inner radius R_i and outer radius R_o is initially at uniform temperature T_i. At time $t > 0$ the outer surface is insulated while the inner surface is heated with uniform flux q_o''. Using (i) control volume formulation, and (ii) integration of the governing heat equation, show that the heat-balance integral is given by

$$\frac{R_i^2}{\rho c_p} q_o'' = \frac{d}{dt} \int_{R_i}^{\delta(t)} (T - T_i) r^2 dr .$$

8.3   A circular electronic device of
      radius R_i and thickness w
      dissipates energy at a rate P.
      The   device   is   press-fitted
      inside a cylindrical fin of inner

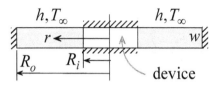

      radius R_i, outer radius R_o and thickness w. Heat is removed from
      the fin's surface by convection. The heat transfer coefficient is h and
      the ambient temperature is T_∞. The plane surfaces of the device
      and the fin tip are insulated. Assume a second degree polynomial
      temperature profile, determine the temperature at the interface R_i.

8.4   The surface temperature of a
      tube of outer radius R_i
      is T_i. The tube is cooled
      with a porous annular fin of
      porosity P, inner radius R_i
      and outer radius R_o. Coolant
      at   mass   flow   rate $\dot{m}$ and

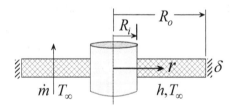

      temperature T_∞ flows axially normal to the fin's surface. The specific
      heat of the coolant is c_{pf} and the conductivity of the fluid-solid
      matrix is $\bar{k}$. The fin exchanges heat by convection along its upper
      and lower surfaces. The heat transfer coefficient is h and the ambient
      temperature is T_∞. The fin equation is

$$r^2 \frac{d^2 T}{dr^2} + r\frac{dT}{dr} - \frac{(4Bi + \beta)}{\delta^2} r^2 (T - T_\infty) = 0,$$

      where $Bi = (1 - P)h\delta / 2\bar{k}$ and $\beta = \dot{m} c_{pf} \delta / \pi (R_o^2 - R_i^2) \bar{k}$.

      Use the integral method to determine the heat transfer rate from the
      fin. Assume a second-degree polynomial temperature profile.

8.5   At time $t > 0$ the surface at $x = 0$ of a semi-infinite region is heated
      with a constant heat flux q_o''. The region is initially at uniform

temperature T_i. Assume a third degree polynomial temperature profile and determine the surface temperature $T(0,t)$. Compare your answer with the result of Example 8.2.

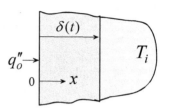

8.6   At time $t > 0$ the surface at $x = 0$ of a semi-infinite region is maintained at constant temperature T_o. Initially the region is at uniform temperature T_i. Use a second degree polynomial temperature profile to determine the surface heat flux. Compare your answer with the exact solution.

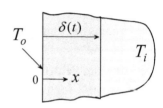

8.7   A long hollow cylinder of inner radius R_i and outer radius R_o is initially at zero temperature. At time $t > 0$ the inside surface is heated with constant flux q_o''. The outside surface is insulated. Use a second degree polynomial temperature profile to determine the time required for the temperature of the insulated surface to begin to change.

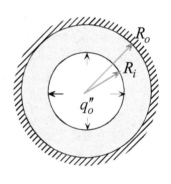

8.8   A semi-infinite constant area fin is initially at uniform temperature T_∞. At time $t > 0$ the base is maintained at temperature T_o. The fin begins to exchange heat by convection with the surroundings. The heat transfer coefficient is h and the ambient temperature is T_∞. Use a second degree polynomial temperature profile to determine the fin heat transfer rate.

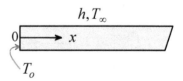

8.9   A long constant area fin is initially at a uniform temperature T_∞. At time $t \geq 0$ the temperature at the base is heated with uniform

flux q''_o. The fin begins to
exchange heat by convection
with the surroundings. The heat
transfer coefficient is h and the
ambient temperature is T_∞.   Use a second degree polynomial
temperature profile to determine the base temperature.

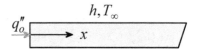

8.10 A semi-infinite solid cylinder is
initially at zero temperature.
The cylinder is insulated along
its surface.   At time $t > 0$
current is passed through the cylinder. Simultaneously heat is
removed from its base with uniform flux q''_o. Use a third degree
polynomial temperature profile to determine the base temperature.

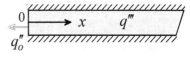

8.11 A semi-infinite region is initially at temperature T_i. At time $t > 0$
the surface is allowed to exchange heat by convection with an
ambient fluid at temperature
$T_\infty = 0$.   The   heat   transfer
coefficient is h. Determine the
surface temperature. Assume a
second   degree   polynomial
temperature profile.

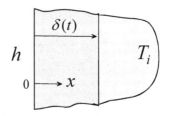

8.12 Two plates are attached as shown. The thickness of plate 1 is L_1 and
its conductivity is k_1. The thickness of plate 2 is L_2 and its
conductivity is k_2. The two plates are initially at uniform
temperature T_i. Plate 1 is suddenly heated with uniform flux q''_o
while the exposed surface of
plate 2 is insulated. Determine
the temperature of the heated
surface   at   the   time   the
temperature of plate 2 begins to
change. Use a second degree
polynomial profile.

8.13 A semi-infinite region is initially
solid at the fusion temperature T_f.
At time $t > 0$ the surface at $x = 0$
is heated with uniform flux q_o''.
Determine the interface location
and surface temperature. Assume a
second degree polynomial temperature profile.

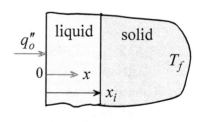

8.14 Consider a large plate of thickness L which is initially at uniform
temperature T_i. At time $t > 0$ one surface begins to exchange heat
with the surroundings by radiation while the other surface is
insulated. The temperature of the surroundings is 0 K.

[a] Show that the heat-balance integral for this problem is given by

$$-\frac{\varepsilon\sigma}{\rho c_p}T_s^4 = \frac{d}{dt}\int_0^{\delta(t)}(T - T_i)\,dx,$$

where $T_s = T(0,t)$ is surface temperature.

[b] Use a third degree polynomial temperature
profile to determine the surface temperature
T_s.

[c] Is the above heat-balance integral limited to
$\delta(t) \le L$?

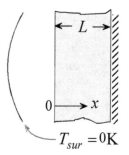

9

PERTURBATION SOLUTIONS

9.1 Introduction

Perturbation methods are analytical techniques used to obtain approximate solutions to problems in mathematics as well as in various engineering disciplines. The basic idea in this approach is to construct an approximate solution to a problem using an exact solution to a slightly different problem. The mathematical simplification associated with this method is linearization and/or replacement of a variable or its derivative with a specified function.

Before the methodology of perturbation solutions is presented, the following terms are defined.

Perturbation quantity ε : A quantity which appears as a parameter or variable in the dimensionless formulation of a problem. It can appear in the differential equation and/or boundary condition. Its magnitude is assumed small compared to unity.

Basic Problem: This is the problem corresponding to $\varepsilon = 0$. It is also known as the zero-order problem.

Basic Solution: This is the solution to the basic or zero-order problem. It represents the lowest approximate solution to the problem.

Regular Perturbation Problem: A problem whose perturbation solution is uniformly valid everywhere in the domains of the independent variables.

Singular Perturbation Problem: A problem whose perturbation solution breaks down in some regions of an independent variable.

Asymptotic Expansion: A representation of a function in a series form in terms of the perturbation quantity ε. As an example, the asymptotic expansion of the temperature distribution θ may take the form

$$\theta = \sum_{n=0}^{\infty} \varepsilon^n \theta_n ,$$

(9.1)

where n can be an integer or a fraction. The first term in the series, $n = 0$, corresponds to the *zero-order solution*. The second term represents the *first-order solution*, etc.

Parameter Perturbation. An expansion in which the perturbation quantity is a parameter (constant).

Coordinate Perturbation. An expansion in which the perturbation quantity is a coordinate (variable). The coordinate can be either spatial or temporal.

This chapter presents an abridged treatment of perturbation solutions. It is limited to parameter perturbation of regular perturbation problems. More details on perturbation techniques are found in References 1 and 2.

9.2 Solution Procedure

The following procedure is used to construct perturbation solutions:

(1) Identification of the Perturbation Quantity. Non-dimensional formulation of a problem can reveal the relevant perturbation parameter. Physical understanding of a problem is important in determining how a problem is non-dimensionalized.

(2) Introduction of an Asymptotic Solution. Assuming that the expansion proceeds in integer powers of ε, eq. (9.1) gives

$$\theta = \theta_0 + \varepsilon \theta_1 + \varepsilon^2 \theta_2 + ...$$

(9.2)

(3) Formulation of the θ_n Problems.   Substitution of the assumed solution, eq. (9.2), into the governing equations and boundary conditions and equating terms of identical powers of ε, gives the formulation of the n-order problems. This step breaks up the original problem into n problems.

(4) Solutions. The *n*-order problems are solved consecutively and their respective boundary conditions are satisfied.

Before illustrating how this procedure is applied, examples of perturbation problems in conduction will be described.

9.3 Examples of Perturbation Problems in Conduction

Perturbation problems in conduction may be associated with the governing heat equation, boundary conditions or both. The following describes various applications.

(1) Cylinder with Non-circular Inner Radius. Fig. 9.1 shows a cylinder with a distorted inner radius $R_i(\varphi)$. As an example, consider an inner radius which varies with the angle φ according to

$$R_i(\varphi) = a\left[1 - \varepsilon \sin^2 \varphi\right] \qquad (9.3)$$

For steady state and uniform conditions at the inner and outer radii the temperature distribution is two-dimensional.   For $\varepsilon \ll 1$, the departure of R_i from a circle of radius a is small. The basic problem corresponding to $\varepsilon = 0$ is a cylinder of constant inner radius $R_i = a$. Note that the basic problem is one-dimensional.

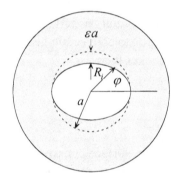

Fig. 9.1

(2) Tapered Plate. The plate shown in Fig. 9.2 is slightly tapered having a half width W which varies with axial distance x. An example is a linearly tapered plate described by

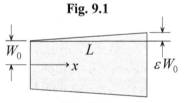

$$W / W_0 = 1 + \varepsilon\,(x/L), \qquad (9.4)$$

Fig. 9.2

where L is length and W_0 is the half width at $x = 0$. The basic problem in this example is a plate of uniform width $W = W_0$.

(3) Cylinder with Eccentric Inner Radius. Fig. 9.3 shows a hollow cylinder with an inner radius whose center is eccentrically located. The inner and outer radii are R_i and R_o and the eccentricity is e. The perturbation parameter for this example is defined as

$$\varepsilon = \frac{e}{R_o - R_i}. \qquad (9.5)$$

For uniform conditions along the inner and outer radii the temperature distribution is two-dimensional. However, the basic problem corresponding to $e = \varepsilon = 0$ represents a hollow cylinder with a concentric inner radius. This basic problem is one-dimensional.

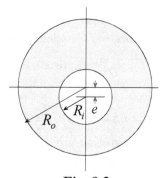

Fig. 9.3

(4) Surface Radiation.   Since radiation heat transfer rate is proportional to the fourth power of the absolute temperature, problems involving radiation are non-linear. Perturbation solutions can be obtained to such problems provided that radiation is small compared to another mode of heat transfer. For example, if radiation is small compared to convection, a perturbation parameter can be defined as the ratio of radiation to convection heat transfer rate. The perturbation parameter in this case takes the form

$$\varepsilon = \frac{e\,\sigma T_o^4}{hT_o} = \frac{e\,\sigma T_o^3}{h}, \qquad (9.6)$$

where

$e = $ surface emissivity
$h = $ heat transfer coefficient
$T_o = $ characteristic surface temperature
$\sigma = $ Stefan-Boltzmann radiation constant

In eq. (9.6) the numerator, $e\sigma T_o^4$, is the radiation heat transfer rate. The denominator, hT_o, represents the convection rate.   Thus in the basic problem corresponding to $\varepsilon = 0$, radiation is neglected.

(5) Conduction with Phase Change. The Stefan number, which was introduced in Chapter 6, represents the ratio of the sensible heat to latent heat. Thus problems involving small Stefan numbers lend themselves to perturbation solutions using the Stefan number as the perturbation parameter. The definition of the Stefan number in eq. (6.13) becomes

$$\varepsilon = Ste = \frac{c_{pf}(T_o - T_f)}{\mathcal{L}}. \qquad (9.7)$$

The basic problem corresponding to $\varepsilon = 0$ represents a quasi-steady state model in which temperature distribution corresponds to a stationary interface.

(6) Temperature Dependent Properties. Conduction problems in which properties are temperature dependent are non-linear. If the variation in properties is small, a perturbation parameter can be defined and used to constructed approximate solutions. For example, the perturbation parameter for a temperature dependent thermal conductivity may be defined as

$$\varepsilon = \frac{k_2 - k_1}{k_1}, \tag{9.8}$$

where k_1 and k_2 correspond to thermal conductivities at two characteristic temperatures of the problem.

9.4 Perturbation Solutions: Examples

The procedure outlined in Section 9.2 will be followed in obtaining several perturbation solutions. The selected examples are simplified conduction problems for which exact analytic solutions are available. Thus, the accuracy of the perturbation solutions can be evaluated.

Example 9.1 Transient Conduction with Surface Radiation and Convection

A thin foil of surface area A_s is initially at a uniform temperature T_i. The foil is cooled by radiation and convection. The ambient fluid and surroundings are at zero kelvin. The heat transfer coefficient is h and surface emissivity is e. Assume that the Biot number is small compared to unity. Identify an appropriate perturbation parameter for the case of small radiation compared to convection and obtain a second-order perturbation solution for the transient temperature of the foil.

(1) Observations. (i) The lumped-capacity method can be used if the Biot number is small compared to unity. (ii) A perturbation solution can be constructed for the case of small radiation compared to convection. (iii) A perturbation parameter which is a measure of the ratio of radiation to convection is appropriate for this problem.

(2) Formulation

(i) Assumptions. (1) The Biot number is small compared to unity, (2) the simplified radiation model of eq. (1.19) is applicable, (3) constant properties and (4) no energy generation.

(ii) Governing Equations. Application of conservation of energy, eq. (1.6), to the foil gives

$$\dot{E}_{in} + \dot{E}_g - \dot{E}_{out} = \dot{E}. \tag{1.6}$$

However, since no energy is added or generated, the above simplifies to

$$\dot{E}_{out} = -\dot{E}. \tag{a}$$

Energy leaving the foil is formulated using Newton's law of cooling and Stefan-Boltzmann radiation law

$$\dot{E}_{out} = hA_sT + e\sigma A_s T^4, \tag{b}$$

where σ is Stefan-Boltzmann constant. Equation (b) is based on the assumption that the ambient fluid temperature and the surroundings temperature are at zero degree absolute. The time rate of change of energy within the foil, $\dot{E}$, is

$$\dot{E} = \rho c_p V \frac{dT}{dt}, \tag{c}$$

where c_p is specific heat, V is volume and ρ is density. Substituting (b) and (c) into (a)

$$hA_sT + e\sigma A_s T^4 = -\rho c_p V \frac{dT}{dt}. \tag{d}$$

The initial condition on (d) is

$$T(0) = T_i. \tag{e}$$

(iii) Identification of the Perturbation Parameter ε. The radiation term in (d) is non-linear. To identify an appropriate perturbation parameter for this problem, equation (d) is expressed in dimensionless form using the following dimensionless temperature and time variables

$$\theta = \frac{T}{T_i}, \quad \tau = \frac{hA_s}{\rho c_p V}t. \tag{f}$$

Substituting into (d) and (e) and rearranging gives

$$\frac{e\sigma T_i^4}{hT_i}\theta^4 + \theta = -\frac{d\theta}{d\tau}, \tag{g}$$

and

$$\theta(0) = 1. \tag{h}$$

The dimensionless coefficient of the first term in (g) represents the ratio of radiation to convection heat transfer. Defining this ratio as the perturbation parameter ε gives

$$\varepsilon = \frac{e\sigma T_i^4}{hT_i} = \frac{e\sigma T_i^3}{h}. \tag{9.9}$$

Substituting eq. (9.9) into (g)

$$\varepsilon\,\theta^4 + \theta = -\frac{d\theta}{d\tau}. \tag{9.10}$$

(iv) Asymptotic Solution. Assume a perturbation solution in the form of the asymptotic expansion given in eq. (9.2)

$$\theta = \theta_0 + \varepsilon\,\theta_1 + \varepsilon^2\theta_2 + \dots \tag{9.2}$$

(v) Formulation of the θ_n Problems. Substituting eq. (9.2) into eq. (9.10) gives

$$\varepsilon\left[\theta_0 + \varepsilon\theta_1 + \varepsilon^2\theta_2 + \dots\right]^4 + \theta_0 + \varepsilon\theta_1 + \varepsilon^2\theta_2 + \dots = -\frac{d\theta_0}{d\tau} - \varepsilon\frac{d\theta_1}{d\tau} - \varepsilon^2\frac{d\theta_2}{d\tau} + \dots$$

Expanding the first term and retaining terms of order ε^2, the above becomes

$$\varepsilon\left[\theta_0^4 + 4\varepsilon\theta_0^3\theta_1 + \varepsilon^2(4\theta_0^3\theta_2 + 6\theta_0^2\theta_1^2) + \dots\right] + \theta_0 + \varepsilon\theta_1 + \varepsilon^2\theta_2^2 + \dots =$$

$$-\frac{d\theta_0}{d\tau} - \varepsilon\frac{d\theta_1}{d\tau} - \varepsilon^2\frac{d\theta_2}{d\tau} + \dots \tag{i}$$

Equating terms of identical powers of ε gives the governing equations for θ_n

$$\varepsilon^0: \qquad\qquad \theta_0 = -\frac{d\theta_0}{d\tau}. \tag{j-0}$$

ε^1:
$$\theta_0^4 + \theta_1 = -\frac{d\theta_1}{d\tau}.$$
(j-1)

ε^2
$$4\theta_0^3\theta_1 + \theta_2 = -\frac{d\theta_2}{d\tau}.$$
(j-2)

Initial conditions on equations (j) are obtained by substituting eq. (9.2) into (h)

$$\theta_0(0) + \varepsilon\,\theta_1(0) + \varepsilon^2\,\theta_2(0) + \ldots = 1.$$

Equating terms of identical powers of ε yields the initial conditions on θ_n

ε^0:
$$\theta_0(0) = 1.$$
(k-0)

ε^1:
$$\theta_1(0) = 0.$$
(k-1)

ε^2:
$$\theta_2(0) = 0.$$
(k-2)

The following observations are made regarding equations (j) and (k):

(i) The zero-order problem described by (j-0) represents a foil which is exchanging heat by convection only. This equation can also be obtained by setting $\varepsilon = 0$ in eq. (9.10).

(ii) In the first-order problem, (j-1), radiation is approximated by the solution to the zero-order problem. That is, the radiation term is set equal to θ_0^4. In the second-order problem, (j-2), the radiation approximation is improved by setting it equal to $4\theta_0^3\theta_1$. The problem is linearized by replacing the radiation term with a specified function.

(iii) Solutions to the n-order problems must proceed consecutively, beginning with the zero-order problem, $n = 0$, and progressing to $n = 1$, $n = 2$, ...etc.

(iv) Each problem is described by a first order differential equation requiring one initial condition. Equations (k) give the respective initial condition of each problem.

(3) Solutions

Zero-order solution: The solution to (j-0) subject to initial condition $(k - 0)$ is

$$\theta_0(\tau) = e^{-\tau}.$$
(l-0)

First-order solution: Substituting (l-0) into (j-1)

$$e^{-4\tau} + \theta_1 = -\frac{d\theta_1}{d\tau}.$$

Using initial condition (k-1), the solution to this equation gives $\theta_1(\tau)$

$$\theta_1(\tau) = \frac{e^{-\tau}}{3}(e^{-3\tau} - 1). \tag{1-1}$$

Second-order solution: With θ_0 and θ_1 known, substitution of (l-0) and
(1−1) into (j-2) gives the governing equation for the second order problem

$$(4/3)e^{-7\tau} - (4/3)e^{-4\tau} + \theta_2 = -\frac{d\theta_2}{d\tau}.$$

Solving this equation for $\theta_2(\tau)$ and using initial condition (k-2) gives

$$\theta_2(\tau) = (4/9)e^{-\tau}(1 - e^{-3\tau}) - (4/18)e^{-\tau}(1 - e^{-6\tau}). \tag{1-2}$$

Substituting (l-0), (l-1) and (l-2) into eq. (9.2) gives the perturbation solution

$$\theta(\tau) = e^{-\tau} + (\varepsilon/3)\, e^{-\tau}(e^{-3\tau} - 1) +$$
$$\varepsilon^2 \left[(4/9)e^{-\tau}(1 - e^{-3\tau}) - (4/18)e^{-\tau}(1 - e^{-6\tau})\right] + \dots \tag{m}$$

Rearranging (m)

$$\theta(\tau) = e^{-\tau}\left[1 + (\varepsilon/3)\,(e^{-3\tau} - 1) + (2/9)\varepsilon^2\,(1 - 2e^{-3\tau} + e^{-6\tau})\right] + \dots \tag{9.11}$$

(4) Checking. *Limiting checks:* (i) Setting $\varepsilon = 0$ in eq. (9.11) gives the correct solution corresponding to cooling with no radiation. (ii) At $\tau = \infty$ the foil should be at equilibrium with the ambient temperature. That is $T(\infty) = 0$ or $\theta(\infty) = 0$. Setting $\tau = \infty$ in eq. (9.11) gives $\theta(\infty) = 0$.

Initial condition check: At $\tau = 0$ eq. (9.11) reduces to $\theta(0) = 1$. Thus the perturbation solution satisfies initial condition (h).

(5) Comments. The exact solution to eq. (9.10) can be obtained by separating the variables θ and τ and integrating directly using initial condition (h)

$$\int_0^\tau d\tau = \int_1^\theta \frac{d\theta}{\varepsilon\,\theta^4 + \theta}.$$

Performing the integration, rearranging the results and setting $\theta = \theta_e$ gives

$$\theta_e(\tau) = \left[(1+\varepsilon)e^{3\tau} - \varepsilon\right]^{-(1/3)}, \qquad (9.12)$$

where θ_e is the exact solution. This result is used to examine the accuracy of the perturbation solution. Fig. 9.4 gives the ratio of the perturbation to exact solution for various values of the perturbation parameter ε. As anticipated, the error associated with the perturbation solution increases as ε is increased. Nevertheless, agreement between

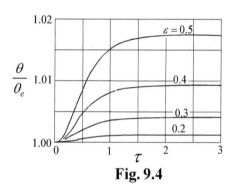

Fig. 9.4

the two solutions is surprisingly good. This suggests that the perturbation solution may converge to the exact solution. To explore this possibility, eq. (9.12) is rewritten in a form resembling eq. (9.11). Binomial expansion of eq. (9.12) gives

$$\theta_e(\tau) = e^{-\tau}\left[1 + (\varepsilon/3)\,(e^{-3\tau} - 1) + (2/9)\,\varepsilon^2\,(1 - 2e^{-3\tau} + e^{-6\tau})\right] -$$

$$(10/81)\,\varepsilon^3\left[1 - (42/10)e^{-3\tau} + (24/10)e^{-6\tau}\right] + ... \qquad (9.12a)$$

Comparing eq. (9.11) with eq. (9.12a) shows that the exact solution contains the first three terms of the perturbation solution. This trend points to convergence of the perturbation solution.

Example 9.2: Variable Thermal Conductivity

Consider steady state one-dimensional conduction in a wall of thickness L. One side is maintained at temperature T_i while the opposite side is at T_o. The thermal conductivity varies with temperature according to

$$k(T) = k_o\left[1 + \beta(T - T_o)\right], \qquad (9.13)$$

where T is temperature, β is a constant and k_o is the thermal conductivity at T_o. Identify an appropriate perturbation parameter for the case of small variation of thermal conductivity with temperature and obtain a second-order perturbation solution for the temperature distribution in the wall.

(1) Observations. (i) The term $\beta(T - T_o)$ is the source of non-linearity of the problem. This term gives a measure of the variation in thermal conductivity. (ii) A perturbation solution can be constructed for $\beta(T - T_o) \ll 1$.

(2) Formulation

(i) Assumptions. (1) Steady state, (2) one-dimensional conduction and (3) no energy generation.

(ii) Governing Equations. Based on the above assumptions, the heat equation is

$$\frac{d}{dx}\left(k\frac{dT}{dx}\right) = 0. \qquad (9.14)$$

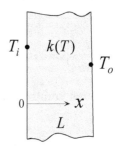

(iii) Boundary Conditions. The boundary conditions for the coordinate axis shown in Fig. 9.5 are

(1) $T(0) = T_i$

(2) $T(L) = T_o$

Fig. 9.5

(iv) Identification of the Perturbation Parameter ε. To identify an appropriate perturbation parameter, eq. (9.14) is cast in dimensionless form using the following dimensionless variables

$$\theta = \frac{T - T_o}{T_i - T_o}, \quad \xi = \frac{x}{L}.$$

Substituting into eq. (9.14)

$$\frac{d}{d\xi}\left[[1 + \beta(T_i - T_o)\theta]\frac{d\theta}{d\xi}\right] = 0.$$

This result suggests that the perturbation parameter ε can be defined as

$$\varepsilon = \beta(T_i - T_o). \qquad (a)$$

Equation (9.14) becomes

$$\frac{d}{d\xi}\left[(1+\varepsilon\theta)\frac{d\theta}{d\xi}\right]=0.$$
(9.15)

Note that eq. (9.15) is non-linear. The two boundary conditions become

(1) $\theta(0)=1$

(2) $\theta(1)=0$

(v) Asymptotic Solution. Assume a perturbation solution in the form of the asymptotic expansion given in eq. (9.2)

$$\theta=\theta_0+\varepsilon\theta_1+\varepsilon^2\theta_2+...$$
(9.2)

(vi) Formulation of the θ_n Problems. Substituting eq. (9.2) into eq. (9.15)

$$\frac{d}{d\xi}\left\{\left[1+\varepsilon(\theta_0+\varepsilon\theta_1+\varepsilon^2\theta_2+...)\right]\frac{d}{d\xi}(\theta_0+\varepsilon\theta_1+\varepsilon^2\theta_2+...)\right\}=0$$

Differentiating and expanding the above gives

$$\frac{d^2\theta_0}{d\xi^2}+\varepsilon\frac{d^2\theta_1}{d\xi^2}+\varepsilon^2\frac{d^2\theta_2}{d\xi^2}+\varepsilon\theta_0\frac{d^2\theta_0}{d\xi^2}+\varepsilon\left[\frac{d\theta_0}{d\xi}\right]^2+$$

$$\varepsilon^2\theta_0\frac{d^2\theta_1}{d\xi^2}+2\varepsilon^2\frac{d\theta_0}{d\xi}\frac{d\theta_1}{d\xi}+\varepsilon^2\theta_1\frac{d^2\theta_0}{d\xi^2}=0.$$

Equating terms of identical powers of ε yields the governing equations for θ_n

$\varepsilon^0:$ $$\frac{d^2\theta_0}{d\xi^2}=0,$$ (b-0)

$\varepsilon^1:$ $$\frac{d^2\theta_1}{d\xi^2}+\theta_0\frac{d^2\theta_0}{d\xi^2}+\left[\frac{d\theta_0}{d\xi}\right]^2=0,$$ (b-1)

$\varepsilon^2:$ $$\frac{d^2\theta_2}{d\xi^2}+\theta_0\frac{d^2\theta_1}{d\xi^2}+2\frac{d\theta_0}{d\xi}\frac{d\theta_1}{d\xi}+\theta_1\frac{d^2\theta_0}{d\xi^2}=0.$$ (b-2)

Substituting eq. (9.2) into the first boundary condition

$$\theta_0(0) + \varepsilon\, \theta_1(0) + \varepsilon^2\, \theta_2(0) + ... = 1.$$

Equating terms of identical powers of ε yields

$\varepsilon^0:$ $\qquad\qquad\qquad\qquad$ $\theta_0(0) = 1,$ $\qquad\qquad\qquad$ (c-0)

$\varepsilon^1:$ $\qquad\qquad\qquad\qquad$ $\theta_1(0) = 0,$ $\qquad\qquad\qquad$ (c-1)

$\varepsilon^2:$ $\qquad\qquad\qquad\qquad$ $\theta_2(0) = 0.$ $\qquad\qquad\qquad$ (c-2)

Similarly, the second boundary condition gives

$$\theta_0(1) + \varepsilon\, \theta_1(1) + \varepsilon^2\, \theta_2(1) + ... = 0.$$

Equating terms of identical powers of ε

$\varepsilon^0:$ $\qquad\qquad\qquad\qquad$ $\theta_0(1) = 0,$ $\qquad\qquad\qquad$ (d-0)

$\varepsilon^1:$ $\qquad\qquad\qquad\qquad$ $\theta_1(1) = 0,$ $\qquad\qquad\qquad$ (d-1)

$\varepsilon^2:$ $\qquad\qquad\qquad\qquad$ $\theta_2(1) = 0.$ $\qquad\qquad\qquad$ (d-2)

The following observations are made regarding equations (b):

(i) The zero-order problem described by (b-0) represents one-dimensional conduction in a wall having constant thermal conductivity. This equation can also be obtained by setting $\varepsilon = 0$ in eq. (9.15).

(ii) In the first-order problem, (b-1), the effect of conductivity variation is approximated by the solution to the zero-order problem. In the second order problem, (b-2), conductivity effect is approximated by the zero and first-order solutions. Note that (b-1) and (b-2) are linear.

(iii) Solutions to the n-order problems must proceed consecutively, beginning with the zero-order problem, $n = 0$, and progressing to $n = 1, n = 2,$ ....etc.

(3) Solutions

Zero-order solution: The solution to (b-0) subject to boundary conditions (c-0) and (d-0) is

$$\theta_0 = 1 - \xi.$$ $\qquad\qquad\qquad$ (e-0)

First-order solution: Substituting (e-0) into (b-1) gives

$$\frac{d^2\theta_1}{d\xi^2} = -1.$$

Integrating twice and using boundary conditions (c-1) and (d-1) `give the first-order solution

$$\theta_1 = (\xi/2)(1-\xi). \tag{e-1}$$

Second-order solution: Substituting (e-0) and (e-1) into (b-2)

$$\frac{d^2\theta_2}{d\xi^2} = 2 - 3\xi.$$

Integrating and applying boundary conditions (c-2) and (d-2) gives the second order solution

$$\theta_2 = (\xi/2)(2\xi - \xi^2 - 1). \tag{e-2}$$

Substituting (e-0), (e-1) and (e-2) into eq. (9.2) gives the perturbation solution

$$\theta = (1-\xi) + \varepsilon(\xi/2)(1-\xi) + \varepsilon^2(\xi/2)(2\xi - \xi^2 - 1) + ... \tag{9.16}$$

(4) Checking. *Limiting check*: Setting $\varepsilon = 0$ in eq. (9.16) gives the correct solution corresponding to constant conductivity.

Boundary conditions check: Evaluating eq. (9.16) at $\xi = 0$ and $\xi = 1$ gives $\theta(0) = 1$ and $\theta(1) = 0$. Thus the perturbation solution satisfies the two boundary conditions.

(5) Comments. The exact solution to the problem can be obtained by direct integration of eq. (9.15) and using the two boundary conditions. The result is

$$\theta_e = \frac{1}{\varepsilon}\sqrt{1 + \varepsilon(2+\varepsilon)(1-\xi)} - \frac{1}{\varepsilon}.$$

Table 9.1

θ and θ_e at $\xi = 0.5$

ε	θ	θ_e	% Error
0	0.5	0.5	0.0
0.1	0.512	0.512	0.002
0.2	0.523	0.523	0.038
0.3	0.532	0.533	0.113
0.4	0.54	0.541	0.258
0.5	0.547	0.55	0.473

The accuracy of the perturbation solution can be evaluated by comparing it with the exact solution. Table 9.1 examines the temperature at the mid-plane, $\xi = 0.5$, using both solutions for various values of the perturbation parameter ε. Note that the error increases as ε is increased. Nevertheless, even at the relatively large value of $\varepsilon = 0.5$ the error is less than 0.5%.

Example 9.3: Fin with Convection and Radiation

A constant area semi-infinite fin of radius r_o exchanges heat by convection and radiation. The heat transfer coefficient is h, thermal conductivity is k and surface emissivity is e. For simplicity, the ambient fluid and the surroundings are assumed to be at $T_\infty = T_{sur} = 0\,K$. The base is maintained at constant temperature T_o. Identify an appropriate perturbation parameter for the case of small radiation compared to convection and obtain a second-order perturbation solution for steady state temperature distribution in the fin and for the heat transfer rate.

(1) Observations. (i) A perturbation solution can be constructed for the case of small radiation compared to convection. An appropriate perturbation parameter is the ratio of radiation to convection heat transfer rates. (ii) This problem is non-linear due to radiation. (iii) An exact solution to the heat transfer rate is presented in Section 7.7.

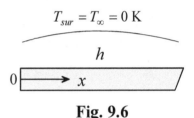

Fig. 9.6

(2) Formulation

(i) Assumptions. (1) Steady state, (2) fin approximation is valid (the Biot number is small compared to unity), (3) the simplified radiation model of eq. (1.19) is applicable, (4) constant properties and (5) ambient fluid and surroundings are at zero degree kelvin.

(ii) Governing Equations. The heat equation for this fin is formulated in Section 7.7. Setting $T_\infty = 0$ in eq. (7.16) gives

$$\frac{d^2T}{dx^2} - \frac{hC}{kA_c}T - \frac{e\sigma C}{kA_c}T^4 = 0, \qquad (9.17)$$

where

$$A_c = \pi r_o^2 = \text{cross section area}$$

$$C = 2\pi r_o = \text{perimeter}$$

(iii) Boundary Conditions. The boundary conditions are

(1) $T(0) = T_o$

(2) $T(\infty) = 0$

(iv) Identification of the Perturbation Parameter ε. To express eq. (9.17) and the boundary conditions in dimensionless form, the following dimensionless variables are defined

$$\theta = \frac{T}{T_o}, \quad \xi = \frac{x}{r_o} . \tag{a}$$

Substituting (a) into eq. (9.17)

$$\frac{k}{2hr_o} \frac{d^2\theta}{d\xi^2} - \theta - \frac{e\sigma T_o^4}{hT_o}\theta^4 = 0 . \tag{b}$$

The coefficient of the first term in (b) is the reciprocal of the Biot number, defined as

$$Bi = \frac{2hr_o}{k} . \tag{c}$$

The coefficient of the non-linear radiation term is the ratio of radiation to convection heat transfer at the base. Defining this ratio as the perturbation parameter ε, we obtain

$$\varepsilon = \frac{e\sigma T_o^3}{h} . \tag{9.18}$$

Substituting (c) and eq. (9.18) into (b) gives the heat equation for the fin

$$\frac{1}{Bi} \frac{d^2\theta}{d\xi^2} - \theta - \varepsilon\theta^4 = 0 . \tag{9.19}$$

The two boundary conditions become

(1) $\theta(0) = 1$

(2) $\theta(\infty) = 0$

(v) Asymptotic Solution. Assume a perturbation solution in the form of the asymptotic expansion given in eq. (9.2)

$$\theta = \theta_0 + \varepsilon\, \theta_1 + \varepsilon^2 \theta_2 + ... \tag{9.2}$$

(vi) Formulation of the θ_n Problems. Use eq. (9.2) into eq. (9.19)

$$\frac{1}{Bi}\left[\frac{d^2\theta_0}{d\xi^2} + \varepsilon\frac{d^2\theta_1}{d\xi^2} + \varepsilon^2\frac{d^2\theta_2}{d\xi^2} + ...\right] - \left[\theta_0 + \varepsilon\,\theta_1 + \varepsilon^2\theta_2 + ...\right] -$$
$$\varepsilon\left[\theta_0 + \varepsilon\,\theta_1 + \varepsilon^2\theta_2 + ...\right]^4 = 0. \tag{d}$$

Expanding the fourth power term, equation (d) becomes

$$\frac{1}{Bi}\left[\frac{d^2\theta_0}{d\xi^2} + \varepsilon\frac{d^2\theta_1}{d\xi^2} + \varepsilon^2\frac{d^2\theta_2}{d\xi^2} + ...\right] - \left[\theta_0 + \varepsilon\,\theta_1 + \varepsilon^2\theta_2 + ...\right] -$$
$$\varepsilon\left[\theta_0^4 + 4\varepsilon\,\theta_0^3\theta_1 + ...\right] + ... = 0$$

Equating terms of identical powers of ε yields the governing equations for θ_n

$$\varepsilon^0: \qquad\qquad \frac{1}{Bi}\frac{d^2\theta_0}{d\xi^2} - \theta_0 = 0, \tag{e-0}$$

$$\varepsilon^1: \qquad\qquad \frac{1}{Bi}\frac{d^2\theta_1}{d\xi^2} - \theta_1 - \theta_0^4 = 0, \tag{e-1}$$

$$\varepsilon^2: \qquad\qquad \frac{1}{Bi}\frac{d^2\theta_2}{d\xi^2} - \theta_2 - 4\theta_0^3\theta_1 = 0. \tag{e-2}$$

Substituting eq. (9.2) into the first boundary condition gives

$$\theta_0(0) + \varepsilon\,\theta_1(0) + \varepsilon^2\,\theta_2(0) + ... = 1.$$

Equating terms of identical powers of ε yields

$$\varepsilon^0: \qquad\qquad \theta_0(0) = 1, \tag{f-0}$$
$$\varepsilon^1: \qquad\qquad \theta_1(0) = 0, \tag{f-1}$$
$$\varepsilon^2: \qquad\qquad \theta_2(0) = 0. \tag{f-2}$$

Similarly, the second boundary condition gives

$$\theta_0(\infty) + \varepsilon\theta_1(\infty) + \varepsilon^2\theta_2(\infty) + \ldots = 0.$$

Equating terms of identical powers of ε

ε^0: $\qquad\qquad\qquad\qquad \theta_0(\infty) = 0,$ $\qquad\qquad\qquad\qquad$ (g-0)

ε^1: $\qquad\qquad\qquad\qquad \theta_1(\infty) = 0,$ $\qquad\qquad\qquad\qquad$ (g-1)

ε^2: $\qquad\qquad\qquad\qquad \theta_2(\infty) = 0.$ $\qquad\qquad\qquad\qquad$ (g-2)

The following observations are made regarding equations (e):

(i) The zero-order problem described by (e-0) represents a fin which is exchanging heat by convection only. This equation can also be obtained by setting $\varepsilon = 0$ in eq. (9.19).

(ii) In the first-order problem, (e-1), radiation is approximated by the solution to the zero-order problem. That is, the radiation term is set equal to θ_0^4. In the second-order problem, (e-2), the radiation approximation term is improved by setting it equal to $4\theta_0^3\theta_1$.

(iii) Although eq. (9.19) is non-linear, the perturbation problems described by equations (e) are linear.

(3) Solutions

Zero-order solution: The solution to (e-0) subject to boundary conditions (f-0) and (g-0) is

$$\theta_0 = \exp(-\sqrt{Bi}\ \xi).$$ (h-0)

First-order solution: Substituting (h-0) into (e-1)

$$\frac{d^2\theta_1}{d\xi^2} - Bi\,\theta_1 = Bi\exp(-4\sqrt{Bi}\ \xi).$$

This equation is of the type described in Appendix A. Its solution, based on boundary conditions (f-1) and (g-1), is

$$\theta_1 = (1/15)\left[\exp(-4\sqrt{Bi}\ \xi) - \exp(-\sqrt{Bi}\ \xi)\right].$$ (h-1)

Second-order solution: Substituting (h-0) and (h-1) into (e-2) gives the governing equation for the second-order problem

$$\frac{d^2\theta_2}{d\xi^2} - Bi\theta_2 = (4/15)\, Bi\left[\exp(-7\sqrt{Bi}\,\xi) - \exp(-4\sqrt{Bi}\,\xi)\right].$$

Using Appendix A, the solution to this equation subject to conditions (f-2) and (g-2) is

$$\theta_2 = (1/900)\left[11\exp(-\sqrt{Bi}\,\xi) + 5\exp(-7\sqrt{Bi}\,\xi) - 16\exp(-4\sqrt{Bi}\,\xi)\right].$$

(h-2)

Substituting equations (h-0), (h-1) and (h-2) into eq. (9.2) gives the perturbation solution

$$\theta = \exp(-\sqrt{Bi}\,\xi) + (1/15)\,\varepsilon\left[\exp(-4\sqrt{Bi}\,\xi) - \exp(-\sqrt{Bi}\,\xi)\right] +$$

$$\frac{\varepsilon^2}{900}\left[11\exp(-\sqrt{Bi}\,\xi) + 5\exp(-7\sqrt{Bi}\,\xi) - 16\exp(-4\sqrt{Bi}\,\xi)\right] + \ldots$$

(9.20)

The heat transfer rate from the fin is determined by applying Fourier's law at $x = 0$

$$q = -kA_c\, \frac{dT(0)}{dx}.$$

(i)

Expressed in terms of θ and ξ, equation (i) becomes

$$q = -\pi r_o k T_o\, \frac{d\theta(0)}{d\xi}.$$

(j)

Substituting eq. (9.20) into (j) and rearranging the result gives the dimensionless fin heat transfer rate

$$q^* = \frac{q}{\pi k r_o T_o \sqrt{Bi}} = 1 + \frac{\varepsilon}{5} - \frac{\varepsilon^2}{50} + \ldots$$

(9.21)

(4) Checking. *Limiting check*: (i) Setting $\varepsilon = 0$ in eq. (9.20) gives the correct temperature distribution for a fin with no radiation. Similarly, setting $\varepsilon = 0$ in eq. (9.21) gives the heat transfer rate for a fin with no radiation. (ii) If any of the quantities A_c, k, and T_o vanishes, the heat transfer rate q must also vanish. Eq. (9.21) satisfies this condition.

Boundary conditions check: Evaluating eq. (9.20) at $\xi = 0$ and $\xi = \infty$ gives $\theta(0) = 1$ and $\theta(\infty) = 0$. Thus the perturbation solution satisfies the two boundary conditions.

(5) Comments. The exact solution to the heat transfer rate, q_e, from a semi-infinite fin with surface convection and radiation is obtained in Section 7.7. Setting $T_\infty = 0$ in eq. (7.18) gives

$$q_e = \sqrt{2\,kA} \left[(hc/2kA)T_o^2 + (e\sigma c/5kA)T_o^5 \right]^{1/2}.$$

Introducing the definition ε and expressing the result in dimensionless form gives

$$q_e^* = \left[1 + 2\varepsilon/5 \right]^{1/2}. \qquad (9.22)$$

With the exact solution known, the accuracy of the perturbation solution can be evaluated. Table 9.2 shows that the two solutions are remarkably close

Table 9.2	
ε	q^*/q_e^*
0	1.0
0.2	0.99997
0.4	0.9998
0.6	0.9993
0.8	0.9985
1.0	0.9973

even for $\varepsilon = 1.0$. This leads one to suspect that the perturbation solution converges to the exact solution. Binomial expansion of eq. (9.22) gives

$$q_e^* = 1 + \frac{\varepsilon}{5} - \frac{\varepsilon^2}{50} + \frac{\varepsilon^3}{250} + ... \qquad (9.22a)$$

Comparing eq. (9.21) with eq. (9.22a) suggests that the perturbation solution appears to converge to the exact solution.

Example 9.4: Conduction with Phase Change: Stefan's Problem

A semi-infinite liquid region is initially at the fusion temperature T_f. The boundary at $x = 0$ is suddenly maintained at constant temperature $T_o < T_f$. Assume that the Stefan number is small compared to unity, obtain a second-order perturbation solution for the interface position $x_i(t)$.

(1) Observations. (i) A perturbation solution can be constructed for the case of a small Stefan number (small sensible heat compared to latent heat). (ii) The Stefan number can be used as the perturbation parameter for this problem. (iii) This problem is non-linear due to the interface boundary condition. (iv) The liquid phase remains at the fusion temperature. (v) An exact solution to this problem is given in Section 6.7.

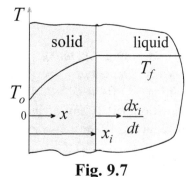

Fig. 9.7

(2) Formulation

(i) Assumptions. (1) Constant properties, (2) semi-infinite region and (3) one-dimensional conduction.

(ii) Governing Equations. Formulation of this problem is found in Section 6.7. The heat equation for the solid phase is

$$\frac{\partial^2 T}{\partial x^2} = \frac{1}{\alpha}\frac{\partial T}{\partial t}.$$

(a)

(iii) Boundary Conditions. The boundary conditions are

(1) $T(0,t) = T_o$

(2) $T(x_i,t) = T_f$

(3) $k\dfrac{\partial T(x_i,t)}{\partial x} = \rho\,\mathcal{L}\dfrac{dx_i}{dt}$

where k is thermal conductivity of the solid phase, $\mathcal{L}$ is latent heat of fusion and ρ is density. The initial condition is

(4) $x_i(0) = 0$

(iv) Identification of the Perturbation Parameter ε. Following the procedure used in Section 6.5, the following dimensionless variables are introduced

$$\theta = \frac{T - T_f}{T_f - T_o}, \quad \xi = \frac{x}{L}, \quad \tau = \varepsilon\frac{\alpha t}{L^2}, \quad \varepsilon = \frac{c_{ps}(T_f - T_o)}{\mathcal{L}}.$$

(b)

where c_{ps} is specific heat of solid and L is characteristic length. Substituting (b) into the heat equation and boundary conditions gives

$$\frac{\partial^2 \theta}{\partial \xi^2} = \varepsilon\frac{\partial \theta}{\partial \tau}.$$

(c)

(1) $\theta(0,\tau) = -1$

(2) $\theta(\xi_i,\tau) = 0$

(3) $\dfrac{\partial\theta(\xi_i,\tau)}{\partial\xi} = \dfrac{d\xi_i}{d\tau}$

The interface initial condition is

(4) $\xi_i(0) = 0$

where ξ_i is the dimensionless interface location.

(v) Asymptotic Solution. An asymptotic expansion is assumed for each of the two variables θ and ξ_i. Let

$$\theta = \theta_0 + \varepsilon\,\theta_1 + \varepsilon^2\theta_2 + \ldots \tag{d}$$

$$\xi_i = \xi_{i0} + \varepsilon\,\xi_{i1} + \varepsilon^2\,\xi_{i2} + \ldots \tag{e}$$

(vi) Formulation of the θ_n Problems. Substituting (d) into (c)

$$\frac{\partial^2\theta_0}{\partial\xi^2} + \varepsilon\frac{\partial^2\theta_1}{\partial\xi^2} + \varepsilon^2\frac{\partial^2\theta_2}{\partial\xi^2} + \ldots = \varepsilon\frac{\partial\theta_0}{\partial\tau} + \varepsilon^2\frac{\partial\theta_1}{\partial\tau} + \ldots \tag{f}$$

Equating terms of identical powers of ε

$$\varepsilon^0: \qquad\qquad \frac{\partial^2\theta_0}{\partial\xi^2} = 0, \tag{g-0}$$

$$\varepsilon^1: \qquad\qquad \frac{\partial^2\theta_1}{\partial\xi^2} = \frac{\partial\theta_0}{\partial\tau}, \tag{g-1}$$

$$\varepsilon^2: \qquad\qquad \frac{\partial^2\theta_2}{\partial\xi^2} = \frac{\partial\theta_1}{\partial\tau}. \tag{g-2}$$

The first boundary condition gives

$$\theta_0(0,\tau) + \varepsilon\,\theta_1(0,\tau) + \varepsilon^2\,\theta_2(0,\tau) + \ldots = -1.$$

Equating terms of identical powers of ε

$$\theta_0(0,\tau) = -1, \tag{h-0}$$

$$\theta_1(0,\tau) = 0, \tag{h-1}$$

$$\theta_2(0,\tau) = 0. \tag{h-2}$$

Boundary conditions (2) and (3) require special attention. Both conditions are evaluated at the interface location ξ_i. However, according to (e) the variable ξ_i depends on the perturbation parameter ε. Thus ε appears implicitly in these conditions. This presents a problem since equating terms of identical powers of ε can not be done in an equation in which ε appears implicitly. To circumvent this difficulty we use Taylor series expansion about $\xi = 0$ to approximate θ and $\partial\theta/\partial\xi$ at $\xi = \xi_i$. Thus boundary condition (2) becomes

$$\theta(\xi_i, \tau) = \theta(0, \tau) + \frac{\xi_i}{1!}\frac{\partial\theta(0,\tau)}{\partial\xi} + \frac{\xi_i^2}{2!}\frac{\partial\theta^2(0,\tau)}{\partial\xi^2} + \frac{\xi_i^3}{3!}\frac{\partial\theta^3(0,\tau)}{\partial\xi^3} + ... = 0$$

(i)

Substituting (d) and (e) into (i) gives

$$\theta(\xi_i, \tau) = \theta_0(0,\tau) + \varepsilon\,\theta_1(0,\tau) + \varepsilon^2\,\theta_2(0,\tau) + ...$$

$$+ \left[\xi_{i0} + \varepsilon\,\xi_{i1} + \varepsilon^2\,\xi_{i2} + ...\right]\left[\frac{\partial\theta_0(0,\tau)}{\partial\xi} + \varepsilon\frac{\partial\theta_1(0,\tau)}{\partial\xi} + \varepsilon^2\frac{\partial\theta_2(0.\tau)}{\partial\xi} + ...\right]$$

$$+ \frac{1}{2}\left[\xi_{i0} + \varepsilon\,\xi_{i1} + \varepsilon^2\,\xi_{i2} + ...\right]^2\left[\frac{\partial^2\theta_0(0,\tau)}{\partial\xi^2} + \varepsilon\frac{\partial^2\theta_1(0,\tau)}{\partial\xi^2} + \varepsilon^2\frac{\partial^2\theta_2(0,\tau)}{\partial\xi^2} + ...\right]$$

$$+ \frac{1}{6}\left[\xi_{i0} + \varepsilon\,\xi_{i1} + \varepsilon^2\,\xi_{i2} + ...\right]^3\left[\frac{\partial^3\theta_0(0,\tau)}{\partial\xi^3} + \varepsilon\frac{\partial^3\theta_1(0,\tau)}{\partial\xi^3} + \varepsilon^2\frac{\partial^3\theta_2(0,\tau)}{\partial\xi^3} + ...\right] = 0$$

Note that ε appears explicitly in this equation. Expanding, equating terms of identical powers of ε and using (g-0) and equations (h), the above gives

$$\varepsilon^0: \qquad\qquad \xi_{i0}\frac{\partial\theta_0(0,\tau)}{\partial\xi} = 1, \qquad\qquad (\text{j-0})$$

$$\varepsilon^1:$$

$$\xi_{i0}\frac{\partial\theta_1(0,\tau)}{\partial\xi} + \xi_{i1}\frac{\partial\theta_0(0,\tau)}{\partial\xi} + \frac{\xi_{i0}^2}{2}\frac{\partial^2\theta_1(0,\tau)}{\partial\xi^2} + \frac{\xi_{i0}^3}{6}\frac{\partial^3\theta_1(0,\tau)}{\partial\xi^3} = 0, \quad (\text{j-1})$$

$$\varepsilon^2:$$

$$\xi_{i0}\frac{\partial\theta_2(0,\tau)}{\partial\xi} + \xi_{i1}\frac{\partial\theta_1(0,\tau)}{\partial\xi} + \frac{\xi_{i0}^2}{2}\frac{\partial^2\theta_2(0,\tau)}{\partial\xi^2} + \xi_{i2}\frac{\partial\theta_0(0,\tau)}{\partial\xi} +$$

$$\qquad\qquad\qquad\qquad\qquad\qquad\qquad\qquad\qquad\qquad (\text{j-2})$$

$$\xi_{i0}\xi_{i1}\frac{\partial^2\theta_1(0,\tau)}{\partial\xi^2} + \frac{\xi_{i0}^3}{6}\frac{\partial^3\theta_2(0,\tau)}{\partial\xi^3} + \frac{\xi_{i0}^2\xi_{i1}}{6}\frac{\partial^3\theta_1(0,\tau)}{\partial\xi^3} = 0.$$

Using Taylor series expansion of $\partial\theta/\partial\xi$ about $\xi = 0$, boundary condition (3) becomes

$$\frac{\partial \theta(0,\tau)}{\partial \xi} + \xi_i \frac{\partial \theta^2(0,\tau)}{\partial \xi^2} + \frac{\xi_i^2}{2!} \frac{\partial \theta^3(0,\tau)}{\partial \xi^3} + \frac{\xi_i^3}{3!} \frac{\partial \theta^4(0,\tau)}{\partial \xi^4} + ... = \frac{d\xi_i}{d\tau}. \quad (k)$$

Substituting (d) and (e) into (k) and making use of (g-0)

$$\frac{\partial \theta_0(0,\tau)}{\partial \xi} + \varepsilon \frac{\partial \theta_1(0,\tau)}{\partial \xi} + \varepsilon^2 \frac{\partial \theta_2(0,\tau)}{\partial \xi} + ...$$

$$\left[\xi_{i0} + \varepsilon \xi_{i1} + \varepsilon^2 \xi_{i2} + ... \right] \left[\varepsilon \frac{\partial^2 \theta_1(0,\tau)}{\partial \xi^2} + \varepsilon^2 \frac{\partial^2 \theta_2(0,\tau)}{\partial \xi^2} + ... \right] +$$

$$\frac{\left[\xi_{i0} + \varepsilon \xi_{i1} + \varepsilon^2 \xi_{i2} + ... \right]^2}{2} \left[\varepsilon \frac{\partial^3 \theta_1(0,\tau)}{\partial \xi^3} + \varepsilon^2 \frac{\partial^3 \theta_2(0,\tau)}{\partial \xi^3} + ... \right] +$$

$$\frac{\left[\xi_{i0} + \varepsilon \xi_{i1} + \varepsilon^2 \xi_{i2} + ... \right]^3}{6} \left[\varepsilon \frac{\partial^4 \theta_1(0,\tau)}{\partial \xi^4} + \varepsilon^2 \frac{\partial^4 \theta_2(0,\tau)}{\partial \xi^4} + ... \right] =$$

$$\frac{d\xi_{i0}}{d\tau} + \varepsilon \frac{d\xi_{i1}}{d\tau} + \varepsilon^2 \frac{d\xi_{i2}}{d\tau} + ...$$

Expanding and equating terms of identical powers of ε

$$\varepsilon^0 : \qquad \frac{\partial \theta_0(0,\tau)}{\partial \xi} = \frac{d\xi_{i0}}{d\tau}, \qquad (l\text{-}0)$$

$$\varepsilon^1 : \frac{\partial \theta_1(0,\tau)}{\partial \xi} + \xi_{i0} \frac{\partial^2 \theta_1(0,\tau)}{\partial \xi^2} + \frac{\xi_{i0}^2}{2} \frac{\partial^3 \theta_1(0,\tau)}{\partial \xi^3} + \frac{\xi_{i0}^3}{6} \frac{\partial^4 \theta_1(0,\tau)}{\partial \xi^4} = \frac{d\xi_{i1}}{d\tau}$$

$$(l\text{-}1)$$

$$\varepsilon^2 : \begin{aligned} &\frac{\partial \theta_2(0,\tau)}{\partial \xi} + \xi_{i0} \frac{\partial^2 \theta_2(0,\tau)}{\partial \xi^2} + \xi_{i1} \frac{\partial^2 \theta_1(0,\tau)}{\partial \xi^2} + \frac{\xi_{i0}^2}{2} \frac{\partial^3 \theta_2(0,\tau)}{\partial \xi^3} + \\ &\xi_{i0} \xi_{i1} \frac{\partial^3 \theta_1(0,\tau)}{\partial \xi^3} + \frac{\xi_{i0}^3}{6} \frac{\partial^4 \theta_2(0,\tau)}{\partial \xi^4} + \frac{\xi_{01}^2 \xi_{i1}}{2} \frac{\partial^4 \theta_1(0,\tau)}{\partial \xi^4} = \frac{d\xi_{i2}}{d\tau} \end{aligned}$$

$$(l\text{-}2)$$

Substituting (e) into initial condition (4)

$$\xi_{i0}(0) + \varepsilon \xi_{i1}(0) + \varepsilon^2 \xi_{i2}(0) = 0.$$

Equating terms of identical powers of ε

$$\varepsilon^0: \qquad\qquad \xi_{i0}(0) = 0, \qquad\qquad\qquad \text{(m-0)}$$

$$\varepsilon^1: \qquad\qquad \xi_{i1}(0) = 0, \qquad\qquad\qquad \text{(m-1)}$$

$$\varepsilon^2: \qquad\qquad \xi_{i2}(0) = 0. \qquad\qquad\qquad \text{(m-2)}$$

Having completed the formulation of the zero, first and second order problems, the following observations are made:

(i) The zero-order problem described by equations (g-0), (h-0), (j-0), (l-0), and (m-0) is identical with the quasi-steady approximation of the Stefan problem detailed in Example 6.1. Note that in the quasi-steady approximation the transient term in the heat equation is neglected.

(ii) In the first-order problem the transient term is approximated by the solution to the zero order problem, as indicated in equation (g-1).

(3) Solutions

Zero-order solution: The solution to (g-0) which satisfies conditions (h-0) and (j-0) is

$$\theta_0(\xi,\tau) = \frac{\xi}{\xi_{i0}(\tau)} - 1. \qquad\qquad \text{(n)}$$

The interface position $\xi_{i0}(\tau)$ is obtained from the solution to (l-0). Substituting (n) into (l-0)

$$\frac{1}{\xi_{i0}} = \frac{d\xi_{i0}}{d\tau}.$$

Solving this equation and using initial condition (m-0) gives

$$\xi_{i0}(\tau) = \sqrt{2\tau}. \qquad\qquad \text{(o)}$$

Thus the zero-order temperature solution (n) becomes

$$\theta_0(\xi,\tau) = \frac{\xi}{\sqrt{2\tau}} - 1. \qquad\qquad \text{(p)}$$

First-order solution: Substituting (p) into (g-1)

$$\frac{\partial^2 \theta_1}{\partial \xi^2} = -\frac{\xi}{2\sqrt{2}}\tau^{-3/2}.$$

Integrating twice and using conditions (h-1) and (j-1)

$$\theta_1(\xi,\tau) = -\frac{\xi^3}{12\sqrt{2}}\tau^{-3/2} - \xi_{i1}\frac{\xi}{2}\tau^{-1} + \frac{\xi}{6\sqrt{2}}\tau^{-1/2}. \qquad (q)$$

The interface position $\xi_{i1}(\tau)$ is obtained from the solution to (l-1). Substituting (o) and (q) into (l-1)

$$-\frac{\xi_{i1}}{2}\tau^{-1} - \frac{1}{3\sqrt{2}}\tau^{-1/2} = \frac{d\xi_{i1}}{d\tau}.$$

Solving this equation and using initial condition (m-1) gives

$$\xi_{i1} = -\frac{\tau^{1/2}}{3\sqrt{2}}. \qquad (r)$$

Thus the first-order solution (q) becomes

$$\theta_1(\xi,\tau) = -\frac{\xi^3}{12\sqrt{2}}\tau^{-3/2} + \frac{\xi}{3\sqrt{2}}\tau^{-1/2}. \qquad (s)$$

Second-order solution. Substituting (s) into (g-2)

$$\frac{\partial^2 \theta_2}{\partial \xi^2} = \frac{\xi^3}{8\sqrt{2}}\tau^{-5/2} - \frac{\xi}{6\sqrt{2}}\tau^{-3/2}.$$

Integrating twice and using conditions (h-2) and (j-2)

$$\theta_2(\xi,\tau) = \frac{\xi^5}{160\sqrt{2}}\tau^{-5/2} - \frac{\xi^3}{36\sqrt{2}}\tau^{-3/2} + \frac{\xi}{12\sqrt{2}}\tau^{-1/2} - \xi_{i2}\frac{\xi}{2}\tau^{-1}. \qquad (t)$$

The interface position $\xi_{i2}(\tau)$ is obtained from the solution to (l-2). Substituting (o), (r) and (t) into (l-2)

$$\frac{2}{9\sqrt{2}}\tau^{-1/2} - \frac{\xi_{i2}}{2}\tau^{-1} = \frac{d\xi_{i2}}{d\tau}.$$

Solving this equation and using initial condition (m-2) gives

$$\xi_{i2} = \frac{\sqrt{2}}{9}\tau^{1/2}. \tag{u}$$

Thus the second-order solution (t) becomes

$$\theta_2(\xi,\tau) = \frac{\xi^5}{160\sqrt{2}}\tau^{-5/2} - \frac{\xi^3}{36\sqrt{2}}\tau^{-3/2} - \frac{\xi}{36\sqrt{2}}\tau^{-1/2}. \tag{v}$$

The perturbation solution for θ becomes

$$\theta(\xi,\tau) = \frac{\xi}{\sqrt{2\tau}} - 1 - \varepsilon\left[\frac{\xi^3}{12\sqrt{2}}\tau^{-3/2} - \frac{\xi}{3\sqrt{2}}\tau^{-1/2}\right] +$$

$$\varepsilon^2\left[\frac{\xi^5}{160\sqrt{2}}\tau^{-5/2} - \frac{\xi^3}{36\sqrt{2}}\tau^{-3/2} - \frac{\xi}{36\sqrt{2}}\tau^{-1/2}\right] + \dots (9.23)$$

With ξ_{i0}, ξ_{i1} and ξ_{i2} determined, the solution to the interface motion is obtained by substituting (o), (r) and (u) into (e)

$$\xi_i(\tau) = \sqrt{2\tau}\left[1 - (1/6)\,\varepsilon + (1/9)\,\varepsilon^2 + \dots\right]. \tag{w}$$

Substituting (b) into (w) gives the dimensional interface location $x_i(t)$

$$x_i(t) = \sqrt{2\alpha t}\left[\varepsilon^{1/2} - (1/6)\varepsilon^{3/2} + (1/9)\varepsilon^{5/2} + \dots\right]. \tag{9.24}$$

(4) Checking. *Limiting check*: (i) According to the definition of ε in (b), the special case of $\varepsilon = 0$ corresponds to a material of infinite latent heat $\mathcal{L}$. The interface for this limiting case remains stationary. Setting $\varepsilon = 0$ in eq. (9.24) gives $x_i(t) = 0$. (ii) If the specific heat is infinite or if the conductivity vanishes, the interface remains stationary. These limiting conditions give $\alpha = 0$. Setting $\alpha = 0$ in eq. (9.24) gives $x_i(t) = 0$.

Boundary and initial conditions check: Evaluating eq. (9.23) at $\xi = 0$ gives $\theta(0,\tau) = -1$. Thus this condition is satisfied.

(5) Comments. (i) The exact solution to Stefan's problem is given in Section 6.7. The exact interface position, x_{ie}, is

$$x_{ie}(t) = \lambda\sqrt{4\alpha t}, \tag{9.25}$$

where λ is the root of eq. (6.24)

$$\sqrt{\pi} \, \lambda \, e^{\lambda^2} \operatorname{erf} \lambda = \varepsilon .$$ (6.24)

Table 9.3

Table 9.3 compares the two solutions. The perturbation solution is very accurate even at values of ε as high as 0.8 where the error is 4.6%. (ii) The perturbation solution is expressed in a simple explicit form while the exact solution involves a transcendental equation whose solution requires iteration.

ε	λ	x_i / x_{ie}
0.1	0.2200	1.0005
0.2	0.3064	1.0022
0.3	0.3699	1.0054
0.4	0.4212	1.0097
0.5	0.4648	1.0160
0.6	0.5028	1.0241
0.7	0.5365	1.0341
0.8	0.5669	1.0463

9.5 Useful Expansions

To equate terms of identical powers of ε, the perturbation parameter must appear explicitly as a coefficient in a governing equation or boundary condition. This rules out problems in which the perturbation parameter appears as an exponent, such as e^{ε}, or in the argument of a function, such as $\cos \varepsilon$. However, in such cases the use of an appropriate expansion, valid for small ε, may resolve the difficulty. In Example 9.4, a Taylor series expansion was used. The following are other examples in which an expansion is used to provide the desired form for ε.

Exponential expansion: $e^{\varepsilon} = 1 + \dfrac{\varepsilon^2}{2!} + \dfrac{\varepsilon^3}{3!} + \dfrac{\varepsilon^4}{4!} + \ldots$ (x)

Trigonometric expansion: $\cos \varepsilon = 1 - \dfrac{\varepsilon^2}{2!} + \dfrac{\varepsilon^4}{4!} + \ldots$ (y)

Binomial expansion: $(1 - \varepsilon)^{-1/2} = 1 + \dfrac{\varepsilon}{2} + \dfrac{3}{4} \dfrac{\varepsilon^2}{2!} + \dfrac{15}{8} \dfrac{\varepsilon^3}{3!} + \ldots$ (z)

In each of these expansions, ε appears as a coefficient and thus can be equated to other terms containing ε raised to the same power.

REFERENCES

[1] Aziz, A. and T.Y. Na, *Perturbation Methods in Heat Transfer*, Hemisphere Publishing Corporation, New York, 1984.

[2] Van Dyke, M. *Perturbation Methods in Fluid Mechanics*, Academic Press, New York, 1964.

PROBLEMS

9.1   An electric wire of radius r_o is initially at uniform temperature T_i. At time $t \geq 0$ current is passed through the wire resulting in volumetric energy generation q'''.   The wire loses heat to the surroundings by convection and radiation. The ambient fluid and the surroundings temperatures are at zero kelvin.   The heat transfer coefficient is h and surface emissivity is e. Assume that the Biot number is small compared to unity and that radiation is small compared to convection. Consider the case where energy generation is such that the wire temperature does not exceed T_i. Formulate the governing equations in dimensionless form and identify an appropriate perturbation parameter.   Construct a   first-order perturbation solution to the transient temperature.

9.2   A foil is initially at a uniform temperature T_i.   At time $t \geq 0$ the lower surface is heated with uniform flux q''_o and the upper surface is allowed to lose heat to the surroundings by convection and radiation. The ambient fluid and the surroundings are at zero kelvin. The heat transfer coefficient is h and surface emissivity is e. Assume that the Biot number is small compared to unity and that radiation is small compared to convection. Consider the case where the heat flux q''_o is such that foil temperature does not exceed T_i. Formulate the governing equations in dimensionless form and identify an appropriate perturbation parameter. Construct a first-order perturbation solution to the transient temperature.

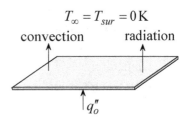

9.3   One side of a plate of thickness L and conductivity k is maintained at T_o while the other side loses heat by convection and radiation. The ambient fluid temperature and the surroundings temperature are at absolute zero. The emissivity of the surface is e. Use a simplified radiation model to obtain a second-order perturbation solution to the temperature distribution in the plate.

9.4   A rectangular sheet of length L, width W and thickness δ is initially
at uniform temperature T_i. At time $t = 0$ the plate begins to lose heat
to the surroundings by convection. The ambient temperature is T_∞.
The heat transfer coefficient h varies with temperature according to

$$h(T) = h_\infty [1 + \beta(T - T_\infty)],$$

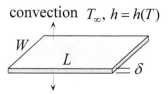

where h_∞ and β are constant. Assume
that the Biot number is small compared
to unity. Formulate the governing
equations in dimensionless form and
identify an appropriate perturbation parameter based on
$\beta(T - T_\infty) \ll 1$. Construct a second-order perturbation solution to
the transient temperature distribution. Compare your result with the
exact solution for $\varepsilon = 0.1$ and 0.5.

9.5   Consider one-dimensional steady state conduction in a wall of
thickness L. One side is at T_i while the opposite side exchanges heat
by convection. The heat transfer coefficient is h and the ambient
temperature is $T_\infty = 0\,^\circ C$. The thermal conductivity varies with
temperature according to

$$k(T) = k_i [1 + \beta(T - T_i)],$$

where k_i is the conductivity at the reference temperature T_i
and β is constant. Formulate the dimensionless governing equations
and identify an appropriate perturbation parameter based
on $\beta T_i \ll 1$. Construct a second-order perturbation solution to the
temperature distribution in the wall.

9.6   A metal sheet of surface area A_s, volume V and density ρ is initially
at a uniform temperature T_i. At time $t \geq 0$ the surface is allowed to
exchange heat by convection with the ambient. The heat transfer
coefficient is h and the ambient temperature is T_∞. The specific heat
varies with temperature according to

$$c(T) = c_\infty [1 + \beta(T - T_\infty)],$$

where c_∞ is the specific heat at T_∞ and β is constant. Assume that
the Biot number is small compared to unity. Formulate the governing

equations in dimensionless form and identify an appropriate perturbation parameter based on $\beta(T_i - T_\infty) \ll 1$. Construct a second-order perturbation solution to the transient temperature.

9.7 An electric wire of radius r_o is initially at a uniform temperature T_i. At time $t \geq 0$ current is passed through the wire resulting in volumetric energy generation q'''. Simultaneously, the wire is allowed to lose heat to the ambient by convection. The heat transfer coefficient is h and the ambient temperature is T_∞. The specific heat varies with temperature according to

$$c(T) = c_\infty \left[1 + \beta(T - T_\infty) \right],$$

where c_∞ is the specific heat at T_∞ and β is constant. Assume that the Biot number is small compared to unity. Formulate the governing equations in dimensionless form and identify an appropriate perturbation parameter based on $\beta(T_i - T_\infty) \ll 1$. Construct a first-order perturbation solution to the transient temperature.

9.8 Under certain conditions the heat transfer coefficient varies with surface temperature. Consider an electric wire of radius r_o which is initially at a uniform temperature T_i. At time $t \geq 0$ current is passed through the wire resulting in volumetric energy generation q'''. Simultaneously, the wire is allowed to lose heat to the ambient by convection. The heat transfer coefficient is h and the ambient temperature is T_∞. The heat transfer coefficient varies with temperature according to

$$h(T) = h_\infty \left[1 - \beta(T - T_\infty) \right],$$

where h_∞ and β are constant. Assume that the Biot number is small compared to unity. Formulate the governing equations in dimensionless form and identify an appropriate perturbation parameter based on $\beta(T_i - T_\infty) \ll 1$. Construct a second-order perturbation solution to the transient temperature.

9.9   Consider steady state heat transfer in a semi-infinite fin of radius r_o which generates energy at volumetric rate q'''. The fin exchanges heat by convection and radiation with the surroundings. The ambient

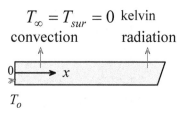

$T_\infty = T_{sur} = 0$ kelvin
convection   radiation

and surroundings temperatures are at zero kelvin. The heat transfer coefficient is h and surface emissivity is e. The base is maintained at constant temperature T_o. Assume that the Biot number is small compared to unity and that radiation is small compared to convection heat transfer. Formulate the governing equations in dimensionless form and identify an appropriate perturbation parameter. Construct a first-order perturbation solution to the heat transfer rate from the fin.

9.10 A large plate of thickness L, conductivity k and diffusivity α is initially at uniform temperature T_i. At $t \geq 0$ one surface is insulated while the other is allowed to exchange heat by convection and radiation. The heat transfer coefficient is h, emissivity e, ambient temperature T_∞ and surroundings temperature is T_{sur}. Consider one-dimensional transient conduction and assume that radiation is small compared to convection. Formulate the heat equation and boundary and initial conditions using the following dimensionless variables:

$$\theta = T/T_i, \quad \xi = x/L, \quad \tau = \alpha t/L^2$$

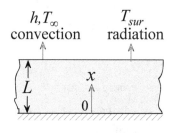

h, T_∞ T_{sur}
convection   radiation

Identify an appropriate perturbation parameter and set up the governing equations and boundary and initial conditions for the zero, first and second-order perturbation problems.

9.11 A thin plastic sheet of thickness t, width w and surface emissivity e is heated in a furnace to temperature T_o. The sheet moves on a conveyor belt with velocity U. It is cooled by convection and radiation along its upper surface while the lower surface is insulated.

The heat transfer coefficient is h and the surroundings and ambient temperatures are at zero kelvin. Assume that radiation is small compared to convection heat transfer and that the Biot number is small compared to

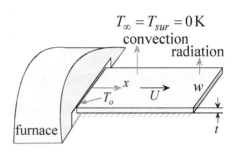

unity. Formulate the governing equations in dimensionless form and identify an appropriate perturbation parameter. Construct a first-order perturbation solution to the steady state surface heat transfer rate.

9.12 Consider a plate with a slightly tapered side. The length is L and the variable height $H(x)$ is given by

$$H(x) = H_o\left[1 + \varepsilon(x/L)\right],$$

where H_o is constant. The perturbation parameter ε depends on the taper angle β according to

$$\varepsilon = \frac{L}{H_o}\tan\beta.$$

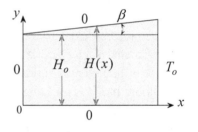

Three sides of the plate are at zero temperature while the fourth side is at uniform temperature T_o. Assume two-dimensional steady state conduction and construct a first-order perturbation solution to the temperature distribution.

9.13 A semi infinite solid region is initially at the fusion temperature T_f. The surface at $x = 0$ is suddenly maintained at constant temperature $T_o > T_f$. Assume that the Stefan number is small compared to unity, construct a first-order perturbation solution to the liquid-solid interface location.

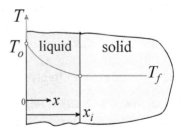

Heat Transfer in Living Tissue

10.1 Introduction

The determination of temperature distribution in blood perfused tissue is important in many medical therapies and physiological studies. Examples are found in cryosurgery, frost bite, hyperthermia, skin burns and body thermal regulation and response to environmental conditions and during thermal stress. The key to thermal modeling of blood perfused tissue is the formulation of an appropriate heat transfer equation. Such an equation must take into consideration three factors: (1) blood perfusion, (2) the vascular architecture, and (3) variation in thermal properties and blood flow rate. The problem is characterized by anisotropic blood flow in a complex network of branching arteries and veins with changing size and orientation. In addition, blood is exchanged between artery-vein pairs through capillary bleed-off along vessel walls, draining blood from arteries and adding it to veins. Energy is transported between neighboring vessels as well as between vessels and tissue. Thus, heat transfer takes place in a blood perfused inhomogeneous matrix undergoing metabolic heat production. The search for heat equations modeling this complex process began over half a century ago and remains an active topic among current investigators. Over the years, several bioheat transfer equations have been formulated. A brief description of some of these equations will be presented and their limitations, shortcomings and applicability outlined. Before these equations are presented, related aspects of the vascular circulation network and blood flow and temperature patterns are summarized.

10.2 Vascular Architecture and Blood Flow

Blood from the heart is distributed to body tissues and organs through a system of vessels (arteries) that undergo many generations of branching accompanied by diminishing size and flow rate. Because of their small size, vessels are measured in micrometers, $\mu = 10^{-6}$ m. This unit is also known as micron. Fig. 10.1 is a schematic diagram showing a typical vascular structure.

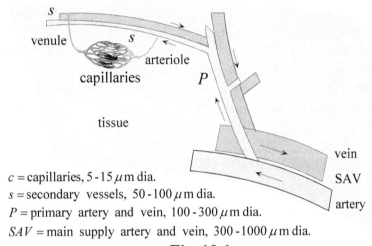

c = capillaries, 5 - $15\,\mu$m dia.
s = secondary vessels, 50 - $100\,\mu$m dia.
P = primary artery and vein, 100 - $300\,\mu$m dia.
SAV = main supply artery and vein, 300 - $1000\,\mu$m dia.

Fig. 10.1

Blood leaves the heart through the *aorta*, which is the largest artery (diameter $\approx 5{,}000$ μm). Vessels supplying blood to muscles are known as *main supply arteries* and *veins* (SAV, $300 - 1000$ μm diameter). They branch into *primary arteries,* (P, $100 - 300$ μm diameter) which feed the *secondary arteries* (s, $50 - 100$ μm diameter). These vessels deliver blood to the *arterioles* ($20 - 40$ μm diameter) which supply blood to the smallest vessels known as *capillaries* ($5 - 15$ μm diameter). Blood is returned to the heart through a system of vessels known as *veins*. For the most part they run parallel to the arteries forming pairs of counter current flow channels. The veins undergo confluence as they proceed from the capillaries to *venules, secondary veins* and to *primary veins*. Blood is returned to the heart through the *vena cava*, which is the largest vessel in the circulatory system. It should be noted that veins are larger than arteries by as much as 100%.

10.3 Blood Temperature Variation

Blood leaves the heart at the arterial temperature T_{a0}. It remains essentially at this temperature until it reaches the main arteries where equilibration with surrounding tissue begins to take place. Equilibration becomes complete prior to reaching the arterioles and capillaries where blood and tissue are at the same temperature T. Tissue temperature can be higher or lower than T_{a0}, depending on tissue location in the body. Blood returning from capillary beds near the skin is cooler than that from deep tissue layers. Blood mixing due to venous confluence from different tissue sources brings blood temperature back to T_{a0} as it returns to the heart via the vena cava. Fig. 10.2 is a schematic of blood temperature variation along the artery-vein paths.

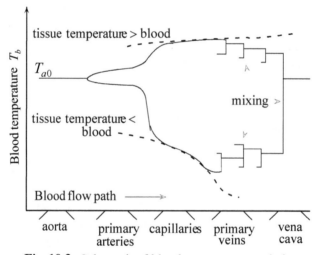

Fig. 10.2   Schematic of blood tempearature variation
in vessels. (From [5] with permission.)

To maintain constant body temperature heat production due to metabolism must be continuously removed from the body. Blood circulation is a key mechanism for regulating body temperature. During conditions of thermal stress, blood flow to tissues under the skin increases. This results in higher rates of cooler venous blood which is used to bring the temperature of blood returning to the heart to its normal level.

10.4 Mathematical Modeling of Vessels-Tissue Heat Transfer

The complex nature of heat transfer in living tissue precludes exact mathematical modeling. Assumptions and simplifications must be made to make the problem tractable while capturing the essential features of the process. The following is an abridged review of various heat equations for the determination of temperature distribution in living tissues. We will begin with the Pennes bioheat equation which was published in 1948. What is attractive and remarkable about this equation is its simplicity and applicability under certain conditions. Nevertheless, to address its shortcomings, several investigators have formulated alternate equations to model heat transfer in living tissues. In each case the aim was to refine the Pennes equation by accounting for factors that are known to play a role in the process. Improvements have come at the expense of mathematical complexity and/or dependency on vascular geometry data, blood perfusion and thermal properties. Detailed derivation and discussion of all equations is beyond the scope of this chapter. Instead five selected models will be presented and their main features identified.

10.4.1 Pennes Bioheat Equation [1]

(a) Formulation

Pennes bioheat equation is based on simplifying assumptions concerning the following four central factors:

(1) *Equilibration Site.* The principal heat exchange between blood and tissue takes place in the capillary beds, the arterioles supplying blood to the capillaries and the venules draining it. Thus all pre-arteriole and post-venule heat transfer between blood and tissue is neglected.

(2) *Blood Perfusion.* The flow of blood in the small capillaries is assumed to be isotropic. This neglects the effect of blood flow directionality.

(3) *Vascular Architecture.* Larger blood vessels in the vicinity of capillary beds play no role in the energy exchange between tissue and capillary blood. Thus the Pennes model does not consider the local vascular geometry.

(4) *Blood Temperature.* Blood is assumed to reach the arterioles supplying the capillary beds at the body core temperature T_{a0}. It instantaneously exchanges energy and equilibrates with the local tissue temperature T. Based on these assumptions, Pennes [1]

modeled blood effect as an isotropic heat source or sink which is proportional to blood flow rate and the difference between body core temperature T_{a0} and local tissue temperature T. In this model, blood originating at temperature T_{a0} does not experience energy loss or gain as it flows through long branching arteries leading to the arterioles and capillaries. Using this idealized process the contribution of blood to the energy balance can be quantified.

Consider the blood perfused tissue element shown in Fig. 10.3. The element is large enough to be saturated with arterioles, venules and capillaries but small compared to the characteristic dimension of the region under consideration. This tissue-vessels matrix is treated as a continuum whose collective temperature is T. Following the formulation of the heat conduction equation of Section 1.4, energy conservation for the element is given by eq. (1.6)

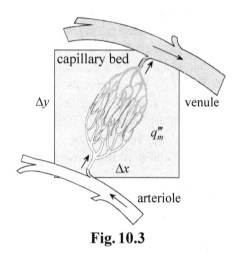

Fig. 10.3

$$\dot{E}_{in} + \dot{E}_g - \dot{E}_{out} = \dot{E} . \qquad (1.6)$$

In Section 1.4 the rate of energy added to the element is by conduction and convection (mass motion). Here the convection component is eliminated and replaced by energy added due to blood perfusion. The simplest way to account for this effect is to treat it as energy generation $\dot{E}_g$. Let

q_b''' = net rate of energy added by the blood per unit volume of tissue
q_m''' = rate of metabolic energy production per unit volume of tissue

Thus equation (b) of Section 1.4 becomes

$$\dot{E}_g = q''' dx\,dy\,dz = (q_b''' + q_m''')dx\,dy\,dz . \qquad (a)$$

To formulate an expression for q_b''' consider the elements shown in Fig. 10.3. According to Pennes, blood enters the element at the body core temperature T_{a0}, it instantaneously equilibrates and exists at the temperature of the element T. Thus

$$q_b''' = \rho_b c_b \dot{w}_b (T_{a0} - T),\qquad(10.1)$$

where

c_b = specific heat of blood
$\dot{w}_b$ = blood volumetric flow rate per unit tissue volume
ρ_b = density of blood

Substituting eq. (10.1) into (a)

$$q''' = q_m''' + \rho_b c_b \dot{w}_b (T_{a0} - T).\qquad(10.2)$$

Returning to the heat conduction equation (1.7), we eliminate the convection terms (set $U = V = W = 0$) and use eq. (10.2) to obtain

$$\nabla \cdot k\nabla T + \rho_b c_b \dot{w}_b (T_{a0} - T) + q_m''' = \rho c \frac{\partial T}{\partial t},\qquad(10.3)$$

where

c = specific heat of tissue
k = thermal conductivity of tissue
ρ = density of tissue

The first term in eq. (10.3) represents conduction in the three directions. It takes the following forms depending on the coordinate system:

Cartesian coordinates:

$$\nabla \cdot k\nabla T = \frac{\partial}{\partial x}\left(k\frac{\partial T}{\partial x}\right) + \frac{\partial}{\partial y}\left(k\frac{\partial T}{\partial y}\right) + \frac{\partial}{\partial z}\left(k\frac{\partial T}{\partial z}\right).\qquad(10.3a)$$

Cylindrical coordinates:

$$\nabla \cdot k\nabla T = \frac{1}{r}\frac{\partial}{\partial r}\left(kr\frac{\partial T}{\partial r}\right) + \frac{1}{r^2}\frac{\partial}{\partial \theta}\left(k\frac{\partial T}{\partial \theta}\right) + \frac{\partial}{\partial z}\left(k\frac{\partial T}{\partial z}\right).\qquad(10.3b)$$

Spherical coordinates:

$$\nabla \cdot k\nabla T = \frac{1}{r^2}\frac{\partial}{\partial r}\left(kr^2\frac{\partial T}{\partial r}\right) +$$

$$\frac{1}{r^2 \sin\phi}\frac{\partial}{\partial \phi}\left(k\sin\frac{\partial T}{\partial \phi}\right) + \frac{1}{r^2\sin^2\phi}\frac{\partial}{\partial \theta}\left(k\frac{\partial T}{\partial \theta}\right).\qquad(10.3c)$$

Equation (10.3) is known as the *Pennes bioheat equation*. Note that the mathematical role of the perfusion term in Pennes's equation is identical to the effect of surface convection in fins, as shown in equations (2.5), (2.19), (2.23) and (2.24). The same effect is observed in porous fins with coolant flow (see problems 5.12, 5.17, and 5.18).

(b) Shortcomings

The Pennes equation has been the subject of extensive study and evaluation [2-11]. The following gives a summary of critical observations made by various investigators. Attention is focused on the four assumptions made in the formulation of the Pennes equation. Discrepancy between theoretical predictions and experimental results is traced to these assumptions.

(1) *Equilibration Site*. Thermal equilibration does not occur in the capillaries, as Pennes assumed. Instead it takes place in pre-arteriole and post-venule vessels having diameters ranging from $70 - 500$ μm [3-5]. This conclusion is based on theoretical determination of the *thermal equilibration length*, L_e, which is the distance blood travels along a vessel for its temperature to equilibrate with the local tissue temperature. For arterioles and capillaries this distance is much shorter than their length L. Vessels for which $L_e / L > 1$ are commonly referred to as *thermally significant*.

(2) *Blood Perfusion*. Directionality of blood perfusion is an important factor in the interchange of energy between vessels and tissue. The Pennes equation does not account for this effect. Capillary blood perfusion is not isotropic but proceeds from arterioles to venules.

(3) *Vascular Architecture*. Pennes equation does not consider the local vascular geometry. Thus significant features of the circulatory system are not accounted for. This includes energy exchange with large vessels, countercurrent heat transfer between artery-vein pairs and vessel branching and diminution.

(4) *Blood Temperature*. The arterial temperature varies continuously from the deep body temperature of the aorta to the secondary arteries supplying the arterioles, and similarly for the venous return. Thus, contrary to Pennes' assumption, pre-arteriole blood temperature is not equal to body core temperature T_{a0} and vein return temperature is not

equal to the local tissue temperature T. Both approximations overestimate the effect of blood perfusion on local tissue temperature.

(c) Applicability

Despite the serious shortcomings of the Pennes equation it has enjoyed surprising success in many applications such as hyperthermia therapy, blood perfusion measurements, cryosurgery and thermal simulation of whole body. In some cases analytical results are in reasonable agreement with experimental data. Studies have shown that the Pennes equation is applicable in tissue regions where vessel diameters are greater than $500\,\mu m$ and for which equilibration length to total length L_e / L is greater than 0.3 [6].

Example 10.1: Temperature Distribution in the Forearm

Model the forearm as a cylinder of radius R with volumetric blood perfusion rate per unit tissue volume $\dot{w}_b$ and metabolic heat production q'''_m. The arm surface exchanges heat with the surroundings by convection. The heat transfer coefficient is h and the ambient temperature is T_∞. Use Pennes bioheat equation to determine the steady state one-dimensional temperature distribution in the arm.

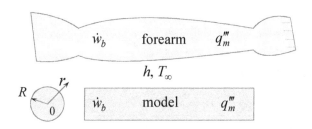

Fig. 10.4

(1) Observations. (i) The forearm can be modeled as a cylinder with uniform energy generation. (ii) Heat is transported to the surface by conduction and removed from the surface by convection. (iii) In general, temperature variation in the forearm is three-dimensional.

(2) Origin and Coordinates. Fig. 10.4 shows the origin and the radial coordinate r.

(3) Formulation.

(i) **Assumptions.** (1) Steady state, (2) the forearm can be modeled as a constant radius cylinder, (3) bone and tissue have the same properties and are uniform throughout, (4) uniform metabolic heat production, (5) uniform blood perfusion, (5) no variation in the angular direction, (6) negligible axial conduction, (7) skin layer is neglected and (8) Pennes bioheat equation is applicable.

(ii) **Governing Equations.** Pennes bioheat equation (10.3) for one-dimensional steady state radial heat transfer simplifies to

$$\frac{1}{r}\frac{d}{dr}\left(r\frac{dT}{dr}\right) + \frac{\rho_b c_b \dot{w}_b}{k}(T_{a0} - T) + \frac{q_m'''}{k} = 0. \qquad (a)$$

(iii) **Boundary Conditions.** Temperature symmetry and convection at the surface give the following two boundary conditions:

$$\frac{dT(0)}{dr} = 0, \text{ or } T(0) = \text{finite}, \qquad (b)$$

$$-k\frac{dT(R)}{dr} = h\left[T(R) - T_\infty\right]. \qquad (c)$$

(4) **Solution.** Eq.(a) is rewritten in dimensionless form using the following dimensionless variables

$$\theta = \frac{T - T_{a0}}{T_\infty - T_{a0}}, \quad \xi = \frac{r}{R}. \qquad (d)$$

Substituting (d) into (a)

$$\frac{1}{\xi}\frac{d}{d\xi}\left(\xi\frac{d\theta}{d\xi}\right) - \frac{\rho_b c_b \dot{w}_b R^2}{k}\theta - \frac{q_m''' R^2}{k(T_{a0} - T_\infty)} = 0. \qquad (e)$$

The coefficient of the second term in (e) is a dimensionless parameter representing the effect of blood flow. Let

$$\beta = \frac{\rho_b c_b \dot{w}_b R^2}{k}. \qquad (f)$$

The last term in (e) is a parameter representing the effect of metabolic heat. Let

$$\gamma = \frac{q_m''' R^2}{k(T_{a0} - T_\infty)}. \tag{g}$$

Substituting (f) and (g) into (e)

$$\frac{1}{\xi}\frac{d}{d\xi}\left(\xi\frac{d\theta}{d\xi}\right) - \beta\theta - \gamma = 0. \tag{h}$$

Boundary conditions (b) and (c) are similarly expressed in dimensionless form

$$\frac{d\theta(0)}{d\xi} = 0, \text{ or } \theta(0) = \text{finite}, \tag{i}$$

$$-\frac{d\theta(1)}{d\xi} = Bi[\theta(1) - 1], \tag{j}$$

where Bi is the Biot number defined as

$$Bi = \frac{hR}{k}.$$

The homogeneous part of equation (h) is a Bessel differential equation. The solution to (h) is

$$\theta(\xi) = C_1 I_0(\sqrt{\beta}\,\xi) + C_2 K_0(\sqrt{\beta}\,\xi) - \frac{\gamma}{\beta}, \tag{k}$$

where C_1 and C_2 are constants of integration. Boundary conditions (i) and (j) give

$$C_1 = \frac{Bi[1 + (\gamma/\beta)]}{\sqrt{\beta}\,I_1(\sqrt{\beta}) + Bi\,I_0(\sqrt{\beta})}, \quad C_2 = 0. \tag{m}$$

Substituting (m) into (k) gives

$$\theta(r) = \frac{T(r) - T_{a0}}{T_\infty - T_{a0}} = \frac{Bi[1 + (\gamma/\beta)]}{\sqrt{\beta}\,I_1(\sqrt{\beta}) + Bi\,I_0(\sqrt{\beta})} I_0(\sqrt{\beta}\,r/R) - \frac{\gamma}{\beta}. \tag{n}$$

(5) Checking. *Dimensional check*: The parameters Bi, β and γ are dimensionless. Thus the arguments of the Bessel functions and each term in solution (n) are dimensionless.

Limiting check: If no heat is removed by convection ($h = 0$), the entire arm reaches a uniform temperature T_o and all energy generation due to metabolic heat is transferred to the blood. Conservation of energy for the blood gives

$$q_m''' = \rho_b c_b \dot{w}_b (T_o - T_{a0}).$$

Solving for T_o

$$T_o = T_{a0} + \frac{q_m'''}{\rho_b c_b \dot{w}_b}. \qquad (o)$$

Setting $h = Bi = 0$ in (n)

$$T(r) = T_{a0} + \frac{q_m'''}{\rho_b c_b \dot{w}_b}. \qquad (p)$$

This agrees with (o).

(6) Comments. (1) The solution is characterized by three parameters: surface convection Bi, metabolic heat γ, and blood perfusion parameter β. (2) Setting $r = 0$ and $r = R$ in (n) gives the center and surface temperatures, respectively. (3) The solution corresponding to zero metabolic heat production is obtained by setting $q_m''' = \gamma = 0$. However, the solution for zero blood perfusion rate can not be deduced from (n) since setting $\beta = 0$ in (n) gives terms of infinite magnitude. This is due to the fact that β appears in differential equation (h) as a coefficient of the variable θ. To obtain a solution for zero perfusion one must first set $\beta = 0$ in (h) and then solve the resulting equation. This procedure gives

$$\frac{T - T_\infty}{(R^2 q_m''' / k)} = \frac{1}{2Bi} + \frac{1}{4}\left[1 - (r / R)^2\right]. \qquad (q)$$

The two solutions, (n) and (q), make it possible to examine the effect of blood perfusion relative to metabolic heat production on the temperature distribution.

10.4.2 Chen-Holmes Equation [5]

An important development in the evolution of bioheat transfer modeling is the demonstration by Chen and Holmes that blood equilibration with the tissue occurs prior to reaching the arterioles. Based on this finding they modified Pennes perfusion term, taking into consideration blood flow directionality and vascular geometry, and formulated the following equation

$$\nabla \cdot k\nabla T + \overset{*}{w}_b \rho_b c_b (T_a^* - T) - \rho_b c_b \overline{u} \cdot \nabla T + \nabla \cdot k_p \nabla T + q_m = \rho c \frac{\partial T}{\partial t}.$$

(10.4)

Although the second term in this equation appears similar to Pennes' perfusion term, it is different in two respects: (1) $\overset{*}{w}_b$ is the perfusion rate at the *local* generation of vessel branching and (2) T_a^* is not equal to the body core temperature. It is essentially the temperature of blood upstream of the arterioles. The third term in eq. (10.4) accounts for energy convected due to equilibrated blood. Directionality of blood flow is described by the vector $\overline{u}$, which is the volumetric flow rate per unit area. This term is similar to the convection term encountered in moving fins and in flow through porous media (see equations (2.19) and (5.6)). The fourth term in eq. (10.4) describes conduction mechanisms associated with small temperature fluctuations in equilibrated blood. The symbol k_p denotes "perfusion conductivity". It is a function of blood flow velocity, vessel inclination angle relative to local temperature gradient, vessel radius and number density.

In formulating eq. (10.4) mass transfer between vessels and tissue is neglected and thermal properties k, c and ρ are assumed to be the same as those of the solid tissue. Other assumptions limit the applicability of this equation to vessels that are smaller than $300 \mu m$ in diameter and equilibration length ratio $L_e / L < 0.6$.

Although the Chen-Holmes model represents a significant improvement over Pennes' equation, its application requires detailed knowledge of the vascular network and blood perfusion. This makes it difficult to use. Nevertheless, the model met with some success in predicting temperature distribution in the pig kidney.

10.4.3 Three-Temperature Model for Peripheral Tissue [7]

Since blood perfused tissue consists of three elements: arteries, veins and tissue, it follows that to rigorously analyze heat transfer in such a medium it is necessary to assign three temperature variables: arterial temperature T_a, venous T_v and tissue T. This approach was followed in analyzing the peripheral tissue of a limb schematically shown in Fig. 10.5 [7]. Three vascular layers were identified: deep layer, intermediate and cutaneous layers. The deep layer is characterized by countercurrent artery-vein pairs that are thermally significant. The intermediate layer is modeled as a porous media exchanging heat with pairs of thermally significant vessels. The thin cutaneous layer just below the skin is independently supplied by countercurrent artery-vein vessels called *cutaneous plexus*. Blood supply to the cutaneous layer is controlled through vasodilation and vasoconstriction of the cutaneous plexus. This is an important mechanism for regulating surface heat flux. This layer is divided into two regions; an upper region with negligible blood effect and a lower region having uniform blood perfusion heat source similar to the Pennes term.

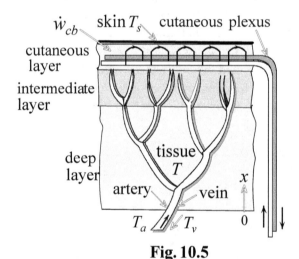

Fig. 10.5

A total of seven governing equations were formulated: three coupled equations for the deep layer, two for the intermediate and two for the cutaneous. Although this model accounts for the vascular geometry and blood flow directionality, solutions to the set of seven equations require numerical integration as well as detailed vascular data. Nevertheless, it was applied to peripheral tissue to examine the effect of various parameters [12]. It was also used to evaluate the performance of other models [6].

Formulation of the three-temperature equations for the deep layer will not be presented here since a simplified form will be outlined in the following section. However, attention is focused on the cutaneous layer. The one-dimensional steady state heat equation for the lower region is given by

$$\frac{d^2 T_1}{dx^2} + \frac{\rho_b c_b \dot{w}_{cb}}{k}(T_{c0} - T_1) = 0,$$ (10.5)

where

T_1 = temperature distribution in the lower region of the cutaneous layer
T_{c0} = temperature of blood supplying the cutaneous pelxus
$\dot{w}_{cb}$ = cutaneous layer volumetric blood perfusion rate per unit tissue volume
x = coordinate normal to skin surface

The upper region of the cutaneous layer just under the skin surface is governed by pure conduction. Thus the one-dimensional steady state heat equation for this region is

$$\frac{d^2 T_2}{dx^2} = 0,$$ (10.6)

where

T_2 = temperature distribution in the upper region of the cutaneous layer

10.4.4 Weinbaum-Jiji Simplified Bioheat Equation for Peripheral Tissue [8]

Recognizing the complexity of the three-temperature model, Weinbaum and Jiji [8] introduced simplifications reducing the three coupled deep layer equations for T_a, T_v and T to a single equation for the tissue temperature. Although the simplified form retains the effect of vascular geometry and accounts for energy exchange between artery, vein and tissue, the added approximations narrow its applicability. A brief description of this bioheat equation follows.

Fig. 10.6 shows a control volume containing a finite number of artery-vein pairs. Blood flow in each pair is in opposite direction (countercurrent). In addition, artery blood temperature T_a is different from vein temperature T_v. Thus, these vessels are *thermally significant* (not in thermal equilibrium). Not shown in Fig.10.6 are numerous *thermally insignificant* capillaries, arterioles, and venules that saturate the tissue. In formulating conservation of energy for the tissue within the control volume, one must take into consideration the following: (1) Conduction through the tissue. (2) Energy exchange by conduction between vessel pairs and tissue. Note that heat conduction from the artery to the

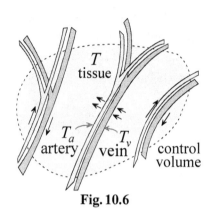

Fig. 10.6

tissue is not equal to conduction from the tissue to the vein. This imbalance is described as *incomplete countercurrent exchange*. (3) Energy exchange between vessels and tissue due to capillary blood bleed-off from artery to vein.

(a) Assumptions. Key assumptions in the simplified bioheat equation are: (1) blood bleed-off rate leaving the artery is equal to that returning to vein and is uniformly distributed along each pair, (2) bleed-off blood leaves the artery at T_a and enters the vein at the venous blood temperature T_v, (3) artery and vein have the same radius, (4) negligible axial conduction through vessels, (5) equilibration length ratio $L_e / L \ll 1$ and (6) tissue temperature T is approximated by the average of the local artery and vein temperatures. That is

$$T \approx (T_a + T_v)/2 . \qquad (10.7)$$

(b) Formulation. Based on the above assumptions, application of conservation of mass for the artery and vein and conservation of energy for the artery, vein and tissue in the control volume, give the simplified bioheat equation for tissue temperature. For the special one-dimensional case where blood vessels and temperature gradient are in the same direction x, the equation reduces to [8]

$$\rho c \frac{\partial T}{\partial t} = \frac{\partial}{\partial x}\left(k_{eff} \frac{\partial T}{\partial x}\right) + q_m''', \qquad (10.8)$$

where k_{eff} is the *effective conductivity*, defined as

$$k_{eff} = k\left[1 + \frac{n}{k^2\sigma}(\pi\rho_b c_b a^2 u)^2\right], \tag{10.9}$$

were

a = vessel radius

n = number of vessel pairs crossing control volume surface per unit area

u = average blood velocity in countercurrent artery or vein

σ = shape factor, defined in eq. (10.10)

The shape factor σ is associated with the resistance to heat transfer between two parallel vessels embedded in an infinite medium. For the case of vessels at uniform surface temperatures with center to center spacing l, the shape factor is given by [13]

$$\sigma = \frac{\pi}{\cosh(l/2a)}. \tag{10.10}$$

Equation (10.9) shows that k_{eff} reflects the effects of vascular geometry and blood perfusion on tissue temperature. It is useful to separate these two effects so that their roles can be analyzed individually. The variables $a, \sigma,$ n and u depend on the vascular geometry. Using conservation of mass, the local blood velocity u can be expressed in terms of the inlet velocity u_o to the tissue layer and the vascular geometry. Thus, eq. (10.9) is rewritten as

$$k_{eff} = k\left[1 + \frac{(2\rho_b c_b a_o u_o)^2}{k_b^2}V(\xi)\right], \tag{10.11}$$

where

a_o = vessel radius at the inlet to the tissue layer at $x = 0$

$V(\xi)$ = dimensionless vascular geometry function

$\xi = x/L$ = dimensionless distance

L = tissue layer thickness

u_o = blood velocity at the inlet to the tissue layer at $x = 0$

Given the vascular data the function $V(\xi)$ can be constructed. It is important to note that this function is independent of blood perfusion. In addition, the coefficient $(2\rho_b c_b a_o u_o/k_b)$ which characterizes blood flow rate is independent of the vascular geometry. It represents the inlet Peclet number, which is the product of Reynolds and Prandtl numbers, defined as

$$Pe_o = \frac{2\rho_b c_b a_o u_o}{k_b}. \tag{10.12}$$

Substituting eq. (10.12) into eq. (10.11) gives

$$k_{eff} = k\left[1 + Pe_o^2 V(\xi)\right]. \tag{10.13}$$

The following observations are made regarding the definition of effective conductivity k_{eff}

(1) For the more general three-dimensional case, the orientation of vessel pairs relative to the direction of the local tissue temperature gradient gives rise to a tensor conductivity that has similar properties to an anisotropic material [8].

(2) The second term on the right hand side of eqs. (10.11) and (10.13) represents the enhancement in tissue conductivity. This enhancement is due to countercurrent convection in the thermally significant microvessel pairs and capillary blood bleed-off.

The two regions of the cutaneous layer shown in Fig. 10.5 are governed by equations (10.5) and (10.6). However, for consistency with the formulation of k_{eff} in the tissue layer, eq. (10.5) will be expressed in terms of the Peclet number Pe_0. Blood perfusion rate $\dot{w}_b$ in the tissue layer is given by

$$\dot{w}_b = \frac{\pi n_o a_o^2 u_o}{L}, \tag{10.14}$$

where n_o is the number of arteries entering the tissue layer per unit area. Substituting eq. (10.12) into eq. (10.14) gives

$$\dot{w}_b = \frac{\pi n_o a_o k_b}{2L\rho_b c_b} Pe_o. \tag{10.15}$$

Define R as the ratio of *total* rate of blood supplied to the cutaneous layer to the *total* rate of blood supplied to the tissue layer. Thus

$$R = \frac{L_1 \dot{w}_{cb}}{L \dot{w}_b}, \tag{10.16}$$

where L_1 is the thickness of the cutaneous layer. Substituting eqs. (10.15) and (10.16) into eq. (10.5) gives

$$\frac{d^2 T_1}{dx^2} + \frac{\pi n_o a_o k_b}{2kL_1} RPe_0 (T_{c0} - T_1) = 0. \qquad (10.17)$$

(c) Limitation and Applicability. To test the validity of assumptions made in formulating eq. (10.8), numerical results using the solution to this equation were compared with those obtained from the solution to the three-temperature model developed in [7] and summarized in Section 10.4.3. The comparison was made for a simplified representation of one-dimensional heat transfer in the limb [6]. Results indicate that eq. (10.8) gives accurate predictions of tissue temperature for vessels smaller than 200 μm in diameter and equilibration length ratio $L_e / L < 0.3$. Experimental measurements on the rat spinotrapezius muscle showed that eq. (10.8) is valid for $L_e / L < 0.2$ [14]. It should be noted that the upper limit on vessel size decreases sharply under conditions of exercise due to an increase in blood flow rate.

Formulation of eq. (10.8) is based on vascular architecture characteristics of peripheral tissues less than 2 cm thick. It does not apply to deeper layers and to skeletal muscles where the vascular geometry takes on a different configuration.

Example 10.2: Temperature Distribution in Peripheral Tissue.

Consider the peripheral tissue shown schematically in Fig.10.5. Tissue thickness is L and skin surface is maintained at uniform temperature T_s. Blood at T_{a0} is supplied to the thermally significant arteries at $x = 0$. During resting state and neutral environment the effect of blood flow through the cutaneous layer is negligible. Consequently heat transfer through this layer is essentially by conduction. A representative vascular geometry function $V(\xi)$, shown in Fig.10.7, for a typical peripheral tissue can be approximated by [15]

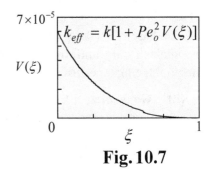

Fig. 10.7

$$V(\xi) = A + B\xi + C\xi^2,$$

where A, B and C are constants:

$$A = 6.32 \times 10^{-5}, \ B = -15.9 \times 10^{-5} \ and \ C = 10 \times 10^{-5}.$$

Use the Weinbaum-Jiji simplified bioheat equation to obtain a solution to the temperature distribution in the tissue. Express the results in dimensionless form in terms of the following dimensionless quantities:

$$\xi = x/L, \quad \theta = \frac{T - T_s}{T_{a0} - T_s}, \quad Pe_0 = \frac{2\rho_b c_b a_0 u_0}{k_b}, \quad \gamma = \frac{q_m''' L^2}{k(T_{a0} - T_s)}.$$

Construct a plot showing the effect of blood flow rate (Peclet number Pe_0) and metabolic heat production γ on tissue temperature $\theta(\xi)$. Note that an increase in γ brings about an increase in Pe_0. Compare $\theta(\xi)$ for $Pe_0 = 60$ and $\gamma = 0.02$ (resting state) with $Pe_0 = 180$ and $\gamma = 0.6$ (moderate exercise).

(1) Observations. (i) The variation of conductivity with distance along the three layers shown in Fig. 10.5 is known. (ii) The tissue can be modeled as a single layer with variable effective conductivity and constant energy generation due to metabolic heat. (iii) Tissue temperature increases as blood perfusion and/or metabolic heat are increased.

(2) Origin and Coordinates. Fig. 10.8 shows the origin and coordinate x.

(3) Formulation.

(i) **Assumptions.** (1) All assumptions leading to eqs. (10.8) and (10.9) are applicable, (2) steady state, (3) one-dimensional, (4) tissue temperature at the base $x = 0$ is equal to T_{a0}, (5) skin is maintained at uniform temperature and (6) negligible blood perfusion in the cutaneous layer.

(ii) **Governing Equations.** The bioheat equation for this model is obtained from eq. (10.8)

$$\frac{d}{dx}\left(k_{eff} \frac{dT}{dx}\right) + q_m''' = 0, \qquad \text{(a)}$$

where k_{eff} is defined in eq. (10.13) as

$$k_{eff} = k\left[1 + Pe_0^2 \ V(\xi)\right], \qquad \text{(b)}$$

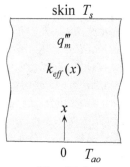

skin T_s

q_m'''

$k_{eff}(x)$

x

$0 \quad T_{a0}$

Fig. 10.8

where k is tissue conductivity corresponding to zero blood perfusion enhancement and $V(\xi)$ is specified as

$$V(\xi) = A + B\xi + C\xi^2 . \tag{c}$$

(iii) Boundary Conditions. The two required boundary conditions are

$$T(0) = T_{a0} , \tag{d}$$

$$T(L) = T_s . \tag{e}$$

(4) Solution. To express the problem in non-dimensional form the following dimensionless quantities are defined:

$$\xi = \frac{x}{L}, \quad \theta = \frac{T - T_s}{T_{a0} - T_s}, \quad \gamma = \frac{q_m''' L^2}{k(T_{a0} - T_s)} . \tag{f}$$

Substituting (b), (c) and (f) into (a) gives

$$\frac{d}{d\xi}\left[\{1 + Pe_0^2(A + B\xi + C\xi^2)\}\frac{d\theta}{d\xi}\right] + \gamma = 0 . \tag{g}$$

Boundary conditions (d) and (e) transform to

$$\theta(0) = 1 , \tag{h}$$

$$\theta(1) = 0 . \tag{i}$$

Integrating (g) once

$$\left[1 + Pe_0^2(A + B\xi + C\xi^2)\right]\frac{d\theta}{d\xi} = C_1 - \gamma\xi .$$

Separating variables and integrating again

$$\theta = C_1 \int\frac{d\xi}{1 + Pe_0^2(A + B\xi + C\xi^2)} - \gamma\int\frac{\xi d\xi}{1 + Pe_0^2(A + B\xi + C\xi^2)} + C_2 , \tag{j}$$

where C_1 and C_2 are constants of integration. The two integrals in (j) are of the form

$$\int\frac{d\xi}{a + b\xi + c\xi^2} \quad \text{and} \quad \int\frac{\xi d\xi}{a + b\xi + c\xi^2} , \tag{k}$$

where the coefficients a, b and c are defined as

$$a = 1 + APe_0^2, \quad b = BPe_0^2, \quad c = CPe_0^2 . \tag{m}$$

Evaluating the two integrals analytically and substituting into (j) give the solution to θ

$$\theta = \frac{2C_1}{\sqrt{d}}\tan^{-1}\frac{b+2c\xi}{\sqrt{d}} - \frac{\gamma}{c}\left[\frac{1}{2}\ln(a+b\xi+c\xi^2) - \frac{b}{\sqrt{d}}\tan^{-1}\frac{b+2c\xi}{\sqrt{d}}\right] + C_2,$$

(n)

where

$$d = 4ac - b^2.$$

(o)

Boundary conditions (h) and (i) give the constants C_1 and C_2. The solution to θ becomes

$$\theta = 2\frac{C_1}{\sqrt{d}}\left[\tan^{-1}\frac{b+2c\xi}{\sqrt{d}} - \tan^{-1}\frac{b+2c}{\sqrt{d}}\right] -$$
$$\frac{\gamma}{c}\left\{\frac{1}{2}\ln\frac{a+b\xi+c\xi^2}{a+b+c} - \frac{b}{\sqrt{d}}\left[\tan^{-1}\frac{b+2c\xi}{\sqrt{d}} - \tan^{-1}\frac{b+2c}{\sqrt{d}}\right]\right\}, \text{(p)}$$

where the constant C_1 is given by

$$C_1 = \frac{1 - \frac{\gamma}{c}\left\{\frac{1}{2}\ln\frac{a+b+c}{a} + \frac{b}{\sqrt{d}}\left[\tan^{-1}\frac{b}{\sqrt{d}} - \tan^{-1}\frac{b+2c}{\sqrt{d}}\right]\right\}}{\frac{2}{\sqrt{d}}\left[\tan^{-1}\frac{b}{\sqrt{d}} - \tan^{-1}\frac{b+2c}{\sqrt{d}}\right]}.$$

(q)

Table 10.1		
	Pe_o	
	60	180
a	1.2275	3.0477
b	-0.5724	-5.1516
c	0.36	3.24
d	1.44	12.96

Table 10.2		
	k_{eff}/k	
ξ	$Pe_o = 60$	$Pe_o = 180$
0	1.44	3.05
0.2	1.13	2.15
0.4	1.06	1.51
0.6	1.01	1.12
0.8	1.00	1.02
1.0	1.02	1.14

The constants a, b, c and d depend on Pe_0. They are listed in Table 10.1. The constant C_1 depends on both Pe_o and γ. For $Pe_0 = 60$ and $\gamma = 0.02$

eq.(q) gives $C_1 = -1.047$ and for $Pe_0 = 180$ and $\gamma = 0.6$ it gives $C_1 = -1.0176$. Table 10.2 lists the enhancement in conductivity for the two values of Pe_0. Fig. 10.9 shows the corresponding temperature distribution.

(5) Checking. *Dimensional check*: Solution θ, metabolic heat parameter γ, Peclet number Pe_0 and the arguments of $\tan^{-1}$ and log are dimensionless.

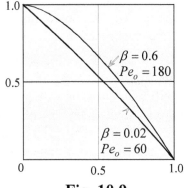

Boundary conditions check: Solution (p) satisfies boundary conditions (h) and (i).

Qualitative check: As anticipated, Fig. 10.9 shows that tissue temperature increases as blood perfusion and metabolic heat are increased.

Fig. 10.9

(6) Comments. (i) Table 10.2 shows the enhancement in k_{eff} due to blood perfusion. Increasing perfusion rate (Pe_0) increases k_{eff}. However, the enhancement diminishes rapidly towards the end of the tissue layer ($\xi = 0.8$). (ii) Fig. 10.9 shows that temperature distribution for $Pe_0 = 60$ and $\gamma = 0.02$ is nearly linear. The slight departure from linearity is due to metabolic heat production. For $Pe_0 = 180$ and $\gamma = 0.6$ tissue temperature is higher. The increase in temperature is primarily due to the increase in blood flow rate Pe_0 rather than metabolic heat γ. For example, at the mid-section $\xi = 0.5$, $\theta(0.5) = 0.688$ for $Pe_0 = 180$ and $\gamma = 0.6$. For zero metabolic heat ($\gamma = 0$), the mid-section temperature drops slightly to 0.64. (iii) Although solution (p) shows that tissue temperature is governed by the parameters Pe_0 and γ, the two are physiologically related. (iv) The increase in metabolic heat production during exercise results in an increase in cutaneous blood flow. Thus neglecting blood perfusion in the cutaneous layer during vigorous exercise is not a reasonable assumption.

10.4. 5 The *s*-Vessel Tissue Cylinder Model [16]

In the Pennes bioheat equation the arterial supply temperature is set equal to the body core temperature and venous return temperature is approximated by the local tissue temperature. Both approximations have been questioned. On the other hand the Chen-Holmes equation and

Weinbaum-Jiji equation are more complex, requiring detailed vascular data and have limited range of applicability. Furthermore, the Weinbaum-Jiji equation is limited to peripheral tissue less than 2 cm thick. These shortcomings motivated further search for a more tractable equation. The *s-vessel tissue cylinder model* addresses some of these issues [16]. An important contribution of this model is the development of a rational theory for the determination of the venous return temperature.

(a) The Basic Vascular Unit.   Comprehensive anatomical studies on the vascular geometry of different types of skeletal muscles have identified significant common arrangements [17]. Fig. 10.10 is a schematic representation of the vascular structure in the cat tenuissimus muscle. The sizes of the thermally significant vessels are indicated. The main supply artery and vein *SAV* branch into primary pairs *P*. The *P* vessels branch into secondary pairs *s* which run roughly parallel to the surface. Terminal arterioles and venules *t* branch off the *s* vessels to feed and drain the capillary beds *c* in the tissue. Blood flow in the *SAV*, *P*, and *s* vessels is countercurrent.

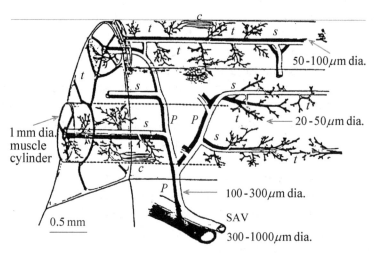

Fig. 10.10 Schematic of a representative vascular arrangement
(From [17] with permission.]

The vascular arrangement shown in Fig. 10.10 has an important feature which is central to the formulation of the new model for tissue heat transfer. Each countercurrent *s* pair is surrounded by a cylindrical tissue which is approximately 1000 μm in diameter. The length of the cylinder depends on the spacing of the P vessels along the SAV pair, which is typically 10-15 mm. The tissue is a matrix of numerous fibers, arterioles, venules and capillary beds. Attention is focused on this repetitive tissue cylinder as the basic heat exchange unit found in most skeletal muscles. Formulation of a bioheat equation for this basic unit can be viewed as the governing equation for the temperature distribution in the aggregate of all cylinders comprising the muscle.

(b) Assumptions. To formulate a bioheat equation for the tissue cylinder, the following assumptions are made: (1) bleed-off blood flow rate in vessels leaving the artery is equal to that returning to the vein and is uniformly distributed along each pair of s vessels, (2) negligible axial conduction through vessels and tissue cylinder, (3) radii of the s vessels do not vary along the tissue cylinder, (4) changes in temperature between the inlet to the P vessels and the inlet to the tissue cylinder is negligible, (5) the temperature field in the tissue cylinder is based on pure radial conduction with a heat-source pair representing the s vessels, and (6) the outer surface of the cylinder is at uniform temperature T_{local}.

(c) Formulation. The capillaries, arterioles and venules (t vessels) are essentially in local thermal equilibrium with the surrounding tissue. However, the pair of s vessels within the cylinder is thermally significant having blood temperatures that are different from tissue temperature. Three temperature variables are needed to describe the temperature distribution in the tissue cylinder: arterial temperature T_a, venous temperature T_v and tissue temperature T. Fig. 10.11 shows a tissue cylinder and its cross

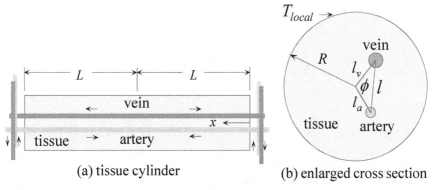

(a) tissue cylinder   (b) enlarged cross section

Fig. 10.11

section containing a pair of s vessels. Three governing equations for the temperature field in the s vessels and tissue cylinder were formulated [12]. The two-dimensional velocity field of the axially changing Poiseuille flow in the s vessels was independently determined from the solution to the Navier-Stokes equations. Continuity of temperature and flux at the surfaces of the vessels provides two boundary conditions in the radial direction r. Tissue temperature at the outer radius R of the cylinder is assumed uniform equal to T_{local}. Turning to the boundary conditions in the axial direction x, note that each cylinder extends a distance $x = 2L$ between two neighboring P vessels (see Fig. 10.11). Thus there is symmetry about the mid-plane $x = L$. At the inlet to the cylinder, $x = 0$, the bulk temperature of the artery T_{ab0} is specified. At $x = L$ the flow in the s vessels vanishes and the artery, vein and tissue are in thermal equilibrium at the local tissue temperature T_{local}.

(d) Solution. The three differential equations for the artery, vein and tissue temperature were solved analytically [16]. The most important aspect of the solution is the determination of T_{vb0}, the outlet bulk temperature of the vein at $x = 0$. The importance of this finding will become clear later. For simplicity, we will present only the results for the special case of equal size blood vessels symmetrically positioned relative to the center of the cylinder, i.e., $l_a = l_v$ (see Fig. 10.11). For this case the dimensionless artery-vein temperature difference, ΔT^* at $x = 0$ is given by

$$\Delta T^* = \frac{T_{ab0} - T_{vb0}}{T_{ab0} - T_{local}} = 1 + \frac{A_{11}}{A_{12}} + \sqrt{\frac{A_{11}^2}{A_{12}^2} - 1}, \qquad (10.18)$$

where

$$A_{11} = -\frac{1}{4}\left\{ \ln\left[R\left(1 - \frac{l_a^2}{R^2}\right) \right] + \frac{11}{24} \right\}, \qquad (10.19)$$

$$A_{12} = \frac{1}{4}\ln\left[\frac{R}{l}\sqrt{1 - \frac{2l_a^2 \cos\phi}{R^2} + \frac{l_a^4}{R^4}} \right]. \qquad (10.20)$$

where ϕ, l_a and l (vessel center to center spacing) are defined in Fig. 10.11.

(d) Modification of Pennes Perfusion Term. With the venous return temperature T_{vb0} determined, application of conservation of energy to the blood at $x = 0$ gives the total energy q_b delivered by the blood to the tissue cylinder

$$q_b = \rho_b c_b \pi a_a^2 u_a (T_{ab0} - T_{vb0}),$$ (10.21)

where

a_a = artery radius
u_a = average artery blood velocity at $x = 0$

Using eq. (10.18) to express the temperature difference $(T_{ab0} - T_{vb0})$ in terms of ΔT^*, eq. (10.21) becomes

$$q_b = \rho_b c_b \pi a_a^2 u_a \Delta T^* (T_{ab0} - T_{local}).$$

Dividing through by the volume of the cylinder gives

$$\frac{q_b}{\pi R^2 L} = \rho_b c_b \frac{\pi a_a^2 u_a}{\pi R^2 L} \Delta T^* (T_{ab0} - T_{local}).$$ (10.22)

Energy generation due to blood flow per unit tissue volume, q_b''', and volumetric blood flow per unit tissue volume $\dot{w}_b$ are

$$q_b''' = \frac{q_b}{\pi R^2 L},$$ (10.23)

and

$$\dot{w}_b = \frac{\pi a_a^2 u_a}{\pi R^2 L}.$$ (10.24)

Substituting eq.(10.23) and eq.(10.24) into eq.(10.22) gives

$$q_b''' = \rho_b c_b \dot{w}_b \Delta T^* (T_{ab0} - T_{local}).$$ (10.25)

Since $R \gg l$, T_{local} is approximately equal to the local average tissue temperature T. Thus, eq. (10.25) becomes

$$q_b''' = \rho_b c_b \dot{w}_b \Delta T^* (T_{ab0} - T).$$ (10.26)

This result replaces eq. (10.1) which was introduced in the formulation of the Pennes equation. It differs from eq. (10.1) in two respects: First, the artery supply temperature is not set equal to the body core temperature. Second, it includes the ΔT^* factor. Using eq.(10.26) to replace the blood perfusion term in the Pennes equation (10.3) gives the bioheat equation for the s-vessel tissue cylinder model as

$$\nabla \cdot k\nabla T + \rho_b c_b \dot{w}_b \Delta T^* (T_{ab0} - T) + q_m''' = \rho c \frac{\partial T}{\partial t}. \qquad (10.27)$$

The following observations are made regarding this result:

(1) The dimensionless factor ΔT^* is identified as a *correction coefficient*. Its definition in eq.(10.18) shows that it depends only on the vascular geometry of the tissue cylinder. More importantly, it is independent of blood flow rate. The analysis provides a closed-form solution for the determination of this factor. Its value for most muscle tissues ranges from 0.6 to 0.8. Thus the description of microvascular structure needed to apply this equation is much simpler than that required by Chen-Holmes and Weinbaum-Jiji equations.

(2) The model analytically determines the venous return temperature and accounts for contribution of countercurrent heat exchange in the thermally significant vessels.

(3) The artery temperature T_{ab0} appearing in eq. (10.27) is unknown. It is approximated by the body core temperature in the Pennes bioheat equation. Its determination involves countercurrent heat exchange in SAV vessels which have diameters ranging from 300 μm to 1000 μm. The determination of T_{ab0} is presented in Reference [18].

(4) While equations (10.5) and (10.6) apply to the cutaneous layer of peripheral tissue, eq. (10.23) applies to the region below the cutaneous layer.

Example 10.3: Surface Heat Loss from Peripheral Tissue.

Fig. 10.12 is a schematic of a peripheral tissue of thickness L with a cutaneous layer of thickness L_1. Blood at a volumetric rate $\dot{w}_b$ per unit tissue volume and temperature T_{ab0} is supplied to the tissue by the primary arteries. Cutaneous plexus supplies blood to the cutaneous layer at a volumetric rate $\dot{w}_{bc}$ and temperature T_{cb0}. Skin surface is maintained at uniform temperature T_s. Assume that blood perfusion $\dot{w}_{cb}$ is uniformly distributed throughout the cutaneous layer, including the skin. Assume

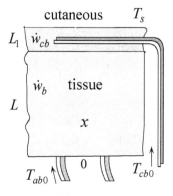

Fig. 10.12

further that metabolic heat production per unit tissue volume is q_m''' and neglect metabolic heat in the cutaneous layer. Using the s-vessel tissue cylinder model, determine surface heat flux for a specified correction coefficient ΔT^*.

(1) Observations. (i) Surface heat flux can be determined once the temperature distribution in the tissue is known. (ii) This is a two layer composite problem: tissue and cutaneous.

(2) Origin and Coordinates. Fig. 10.12 shows the origin and coordinate x.

(3) Formulation

(i) Assumptions. (1) All assumptions leading to bioheat equations (10.5) and (10.27) are applicable, (2) steady state, (3) one-dimensional, (4) constant properties, (5) uniform metabolic heat in the tissue layer, (6) negligible metabolic heat in the cutaneous layer, (7) uniform blood perfusion throughout the cutaneous layer, (8) tissue temperature at the base $x = 0$ is equal to T_{ab0} and (9) specified surface temperature.

(ii) Governing Equations. Application of Fourier's law at the surface gives surface heat flux

$$q_s'' = -k\frac{\partial T_1(L+L_1)}{\partial x}, \tag{a}$$

where

k = tissue conductivity
q_s'' = surface heat flux
T_1 = temperature distribution in the cutaneous layer

Two equations are needed to determine the temperature distribution: one for the tissue layer and one for the cutaneous layer. For the tissue layer eq. (10.27) gives

$$\frac{d^2T}{dx^2} + \frac{\rho_b c_b \dot{w}_b \Delta T^*}{k}(T_{ab0} - T) + \frac{q_m'''}{k} = 0, \qquad 0 \le x \le L, \tag{b}$$

where T is the temperature distribution in the tissue layer. Treating the cutaneous layer as a single region having uniform blood perfusion throughout, equation (10.5) gives

$$\frac{d^2T_1}{dx} + \frac{\rho_b c_b \dot{w}_{cb}}{k}(T_{cb0} - T_1) = 0, \qquad L \le x \le L + L_1. \tag{c}$$

(iii) Boundary Conditions. Four boundary conditions are required for eqs. (b) and (c)

$$T(0) = T_{ab0}, \tag{d}$$

$$T(L) = T_1(L), \tag{e}$$

$$\frac{dT(L)}{dx} = \frac{dT_1(L)}{dx}, \tag{f}$$

$$T_1(L + L_1) = T_s. \tag{g}$$

(4) Solution. Equations (b) and (c) are rewritten in dimensionless form using the following dimensionless variables

$$\theta = \frac{T - T_{ab0}}{T_s - T_{ab0}}, \quad \phi = \frac{T_1 - T_{ab0}}{T_s - T_{ab0}}, \quad \xi = \frac{x}{L}. \tag{h}$$

Substituting (h) into (b) and (c) gives

$$\frac{d^2\theta}{d\xi^2} - \frac{\rho_b c_b \dot{w}_b L^2}{k} \Delta T^* \theta - \frac{q_b''' L^2}{k(T_{ab0} - T_s)} = 0, \quad 0 \le \xi \le 1, \tag{i}$$

and

$$\frac{d^2\phi}{d\xi^2} - \frac{\rho_b c_b \dot{w}_{cb} L^2}{k} \left[\phi - \frac{T_{ab0} - T_{cb0}}{T_{ab0} - T_s} \right] = 0, \quad 1 \le \xi \le 1 + \xi_0. \tag{j}$$

Introduce the following definitions of the dimensionless parameters in (i) and (j)

$$\beta = \frac{\rho_b c_b \dot{w}_b L^2}{k}, \quad \beta_c = \frac{\rho_b c_b \dot{w}_{cb} L^2}{k},$$

$$\gamma = \frac{q_m''' L^2}{k(T_{ab0} - T_s)}, \quad \lambda = \frac{T_{ab0} - T_{cbo}}{T_{ab0} - T_s}. \tag{k}$$

Substituting (k) into (i) and (j)

$$\frac{d^2\theta}{d\xi^2} - \beta \Delta T^* \theta - \gamma = 0, \quad 0 \le \xi \le 1, \tag{m}$$

$$\frac{d^2\phi}{d\xi^2} - \beta_c \phi + \beta_c \lambda = 0, \quad 1 \le \xi \le 1 + \xi_0, \tag{n}$$

where

$$\xi_0 = \frac{L_1}{L}. \tag{o}$$

Boundary conditions (d)-(g) transform to

$$\theta(0) = 0, \tag{p}$$

$$\theta(1) = \phi(1), \tag{q}$$

$$\frac{d\theta(1)}{dx} = \frac{d\phi(1)}{dx}, \tag{r}$$

$$\phi(1 + \xi_0) = 1. \tag{s}$$

The solutions to (m) and (n) are

$$\theta = A \sinh \sqrt{\beta \Delta T^*} \, \xi + B \cosh \sqrt{\beta \Delta T^*} \, \xi - \frac{\gamma}{\beta \Delta T^*}, \tag{t}$$

$$\phi = C \sinh \sqrt{\beta_c} \, \xi + D \cosh \sqrt{\beta_c} \, \xi + \lambda. \tag{u}$$

where A, B, C and D are constants of integration. Application of boundary conditions (p)-(s) gives the four constants

$$A = \frac{\lambda + \dfrac{(1-\lambda)\cosh\sqrt{\beta_c}}{\cosh\sqrt{\beta_c}\,(1+\xi_0)} + \dfrac{\lambda \left|1 - \cosh\sqrt{\gamma\Delta T^*}\right|}{\beta \Delta T^*} + C_1 C_2 C_3}{\sinh\sqrt{\beta \Delta T^*} - C_2 C_3 \sqrt{\beta \Delta T^*} \cosh\sqrt{\beta \Delta T^*}}, \tag{v}$$

$$B = \frac{\gamma}{\beta \Delta T^*}, \tag{w}$$

$$C = \left[A\sqrt{\beta \Delta T^*} \cosh\sqrt{\omega \beta \Delta T^*} + C_1 \right] C_2, \tag{x}$$

$$D = \frac{1-\gamma}{\cosh\sqrt{\beta_c}\,(1+\xi_0)} - C_2 \left[A\sqrt{\beta \Delta T^*} \cosh\sqrt{\beta \Delta T^*} + C_1 \right] \tanh\sqrt{\beta}(1+\xi_0). \tag{y}$$

The constants C_1, C_2 and C_3 are given by

$$C_1 = \frac{\gamma}{\sqrt{\beta \Delta T^*}} \sinh \sqrt{\beta \Delta T^*} - \frac{(1-\lambda)\sqrt{\beta_c}\, \sinh \sqrt{\beta_c}}{\cosh \sqrt{\beta_c}\,(1+\xi_0)},$$

$$C_2 = \left\{ \sqrt{\beta_c} \left[\cosh \sqrt{\beta_c} - \sinh \sqrt{\beta_c}\, \tanh \sqrt{\beta_c}\,(1+\xi_0) \right] \right\}^{-1},$$

$$C_3 = \sinh \sqrt{\beta_c} - \cosh \sqrt{\beta_c}\, \tanh \sqrt{\beta_c}\,(1+\xi_0).$$

Surface heat flux is determined by substituting (u) into (a)

$$\frac{q_s'' L}{k\,(T_{ab0} - T_s)} = \sqrt{\beta_c}\, \left[C \cosh \sqrt{\beta_c}\,(1+\xi_0) + D \sinh \sqrt{\beta_c}\,(1+\xi_0) \right]. \quad (z)$$

(5) Checking. *Dimensional check*: The parameters $\beta, \beta_c, \gamma, \lambda$, and ξ_0 and the arguments of the sinh and cosh are dimensionless.

Limiting check: For the special case of $T_s = T_{ab0} = T_{ac0}$ and $q_b''' = 0$, solutions (t), (u) and (z) reduce to the expected results of $T(x) = T_1(x) = T_{ab0}$ and $q_s'' = 0$.

(5) Comments. (i) The solution is characterized by five parameters: $\beta, \beta_c, \gamma, \lambda$, and ξ_0. (ii) Equation (z) can be used to examine the effect of cutaneous blood perfusion on surface heat flux. Changing blood flow rate through the cutaneous layer is an important mechanism for regulating body temperature. (iii) The solution does not apply to the special case of zero blood perfusion rate since β and β_c appear in the differential equations as coefficients of the variables θ and ϕ, respectively. To obtain the solution to this case, β and β_c must be set equal to zero in equations (m) and (n) prior to solving them.

REFERENCES

[1] Pennes, H.H., "Analysis of Tissue and Arterial Blood Temperatures in the Resting Forearm," *J. Applied Physiology*, vol. 1, pp. 93-122, 1948.

[2] Wulff, W. "The Energy Conservation Equation for Living Tissues," *IEEE Transactions of Biomedical Engineering*, BME-21, pp. 494-495, 1974.

[3] Weinbaum, S., and Jiji, L.M., "A Two Phase Theory for the Influence of Circulation on the Heat Transfer in Surface Tissue," *In, 1979 Advances in Bioengineering* (M.K. Wells, ed.), pp. 179-182, ASME, New York, 1979.

[4] Klinger, H.G., "Heat Transfer in Perfused Biological Tissue-1: General Theory," *Bulletin of Mathematical Biology,* vol. 36, pp. 403-415, 1980.

[5] Chen, M.M., and Holmes, K.R., "Microvascular Contributions in Tissue Heat Transfer," *Annals of the New York Academy of Sciences,* vol., 335, pp. 137-150, 1980.

[6] Charny, C.K., Weinbaum, S., and Levin, R.L., "An Evaluation of the Weinbaum-Jiji Bioheat Equation for Normal and Hyperthermic Conditions," *ASME J. of Biomechanical Engineering,* vol. 112, pp. 80-87, 1990.

[7] Jiji, L.M., Weinbaum, S., and Lemons, D.E., "Theory and Experiment for the Effect of Vascular Microstructure on Surface tissue Heat Transfer-Part II: Model Formulation and Solution," *ASME J. of Biomechanical Engineering,* vol., 106, pp. 331-341, 1984.

[8] Weinbaum, S., and Jiji, L.M., "A New Simplified Bioheat Equation for the Effect of Blood Flow on Local Average Tissue Temperature," *ASME J. of Biomechanical Engineering,* vol. 107, pp. 131-139, 1985.

[9] Weinbaum, S., Jiji, L.M., and Lemons, D.E., "Theory and Experiment for the Effect of Vascular Microstructure on Surface Tissue Heat Transfer-Part I: Anatomical Foundation and Model Conceptualization," *ASME J. of Biomechanical Engineering,* vol., 106, pp. 321-330, 1984.

[10] Weinbaum, S., and Lemons, D.E., "Heat Transfer in Living Tissue: The Search for a Blood-Tissue Energy Equation and the Local Thermal Microvascular Control Mechanism," *BMES Bulletin,* vol. 16, no. 3, pp. 38-43, 1992

[11] Arkin, H., Xu, L.X., and Holmes, K.R., "Recent Development in Modeling Heat Transfer in Blood Perfused Tissue," *IEEE Transactions on Biomedical Engineering,* vol. 41, no. 2, pp.97-107, 1994.

[12] Dagan, Z., Weinbaum, S., and Jiji, L.M., "Parametric Studies on the Three-Layer Microcirculatory Model for Surface Tissue Energy

Exchange," *ASME J. Biomechanical Engineering*, vol. 108, pp. 89-96, 1986.

[13] Chato, J.C. "Heat Transfer to Blood Vessels," *ASME J. Biomechanical Engineering,* vol. 102, pp. 110-118, 1980.

[14] Song, J., Xu, L.X., Lemons, D.E., and Weinbaum, S., "Enhancement in the Effective Thermal Conductivity in Rat Spinotrapezius Due to Vasoregulation," *ASME J. of Biomechanical Engineering*, vol. 119, pp. 461-468, 1997.

[15] Song, W.J., Weinbaum, S., and Jiji, L.M., "A Theoretical Model for Peripheral Tissue Heat Transfer Using the Bioheat Equation of Weinbaum and Jiji," *ASME J. of Biomechanical Engineering*, vol. 109, pp. 72-78, 1987.

[16] Weinbaum, S., Xu, L.X., Zhu, L., and Ekpene, A., " A New Fundamental Bioheat Equation for Muscle Tissue: Part I: Blood Perfusion Term," *ASME J. Biomechanical Engineering*, vol. 119, pp.278-288, 1997.

[17] Myrhage, R., and Eriksson, E., "Arrangement of the Vascular Bed in Different Types of Skeletal Muscles, " *Progress in Applied Microcirculation*, vol. 5, pp.1-14, 1984.

[18] Zhu, L., Xu, L.X., He, Q., and Weinbaum, S., "A New Fundamental Bioheat Equation for Muscle Tissue: Part II: Temperature of SAV Vessels," *ASME J. Biomechanical Engineering*, vol. 124, pp.121-132, 2002.

[19] Farlow, J.O., Thompson, C.V., and Rosner, D.E., "Plates of the Dinosaur Stegosaurus: Forced Convection Heat Loss Fins?", Science, vol. 192, pp.1123-1125, 1976.

PROBLEMS

10.1   Pennes[1] obtained experimental data on temperature distribution in the forearm of several subjects. The average center and skin surface temperatures were found to be $36.1\,^{\circ}C$ and $33.6\,^{\circ}C$, respectively. Use the Pennes model to predict these two temperatures based on the following data:

c_b = specific heat of blood = 3.8 J/g–$^\circ$C
h = heat transfer coefficient = 4.18 W/m^2–$^\circ$C
k_b = thermal conductivity of blood = 0.5 W/m–$^\circ$C
k = thermal conductivity of muscle = 0.5 W/m–$^\circ$C
q_m''' = metabolic heat production = 0.000418 W/cm^3
R = average forearm radius = 4 cm
T_{a0} = $36.3\,^\circ$C
T_∞ = $26.6\,^\circ$C
$\dot{w}_b$ = volumetric blood perfusion rate per unit tissue volume
  = 0.0003 cm^3/s/cm^3
ρ_b = blood density = 1050 kg/m^3

10.2   Blood perfusion rate plays an important role in regulating body temperature and skin heat flux. Use Pennes's data on the forearm of Problem 10.1 to construct a plot of skin surface temperature and heat flux for blood perfusion rates ranging from $\dot{w}_b = 0$ to $w_b = 0.0006$ (cm^3/s)/cm^3.

10.3   Certain clinical procedures involve cooling of human legs prior to surgery. Cooling is accomplished by maintaining surface temperature below body temperature. Model the leg as a cylinder of radius R with volumetric blood perfusion rate per unit tissue volume $\dot{w}_b$ and metabolic heat production rate q_m'''. Assume uniform skin surface temperature T_s. Use the Pennes bioheat equation to determine the steady state one-dimensional temperature distribution in the leg.

10.4   A manufacturer of suits for divers is interested in evaluating the effect of thermal conductivity of suit material on skin temperature. Use Pennes model for the forearm to predict skin surface temperature of a diver wearing a tight suit of thickness δ and thermal conductivity k_o. The volumetric blood perfusion rate per unit tissue volume is $\dot{w}_b$ and metabolic heat production rate is q_m'''. The ambient temperature is T_∞ and the heat transfer coefficient is h. Neglect curvature of the suit layer.

10.5   Consider the single layer model of the peripheral tissue of Example 10.2. Tissue thickness is L and blood supply temperature to the deep layer at $x = 0$ is T_{a0}. The skin surface exchanges heat by convection. The ambient temperature is T_∞ and the heat transfer coefficient is h. Assume that the vascular geometry function $V(\xi)$ can be approximated by

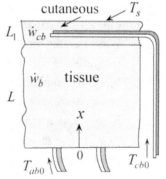

$$V(\xi) = A + B\xi + C\xi^2,$$

where

$$A = 6.32 \times 10^{-5}, \quad B = -15.9 \times 10^{-5}, \quad \text{and} \quad C = 10 \times 10^{-5}.$$

Use the Weinbaum-Jiji simplified bioheat equation to obtain a solution to the temperature distribution in the tissue. Express the result in non-dimensional form using the following dimensionless quantities:

$$\theta = \frac{T - T_\infty}{T_{a0} - T_\infty}, \quad \xi = \frac{x}{L}, \quad \gamma = \frac{q_m''' L^2}{k(T_{a0} - T_\infty)}, \quad Bi = \frac{hL}{k},$$

$$Pe_0 = \frac{2\rho_b c_b a_0 u_0}{k_b}.$$

Construct a plot showing the effect of Biot number (surface convection) on tissue temperature distribution $\theta(\xi)$ for $Pe_0 = 180$, $\gamma = 0.6$ and $Bi = 0.1$ and 1.0.

10.6   In Example 10.3 skin surface is maintained at specified temperature. To examine the effect of surface convection on skin surface heat flux, repeat Example 10.3 assuming that the skin exchanges heat with the surroundings by convection. The ambient temperature is T_∞ and the heat transfer coefficient is h.

10.7   The vascular geometry of the peripheral tissue of Example 10.2 is approximated by a polynomial function. To evaluate the sensitivity of temperature distribution to the assumed vascular geometry function, consider a linear representation of the form

$$V(\xi) = A + B\xi,$$

where $A = 6.32 \times 10^{-5}$ and $B = -6.32 \times 10^{-5}$. Determine k_{eff}/k and $\theta(\xi)$ for $\gamma = 0.6$ and $Pe_0 = 180$. Compare your result with Example 10.2.

10.8   A digit consists mostly of bone surrounded by a thin cutaneous layer. A simplified model for analyzing the temperature distribution and heat transfer in digits is a cylindrical bone covered by a uniform cutaneous layer. Neglecting axial and angular variations, the problem reduces to one-dimensional temperature distribution. Consider the case of a digit with negligible metabolic heat production. The skin surface exchanges heat with the ambient by convection. The heat transfer coefficient is h and the ambient temperature is T_∞. Using the Pennes equation determine the steady state temperature distribution and heat transfer rate. Note that in the absence of metabolic heat production the bone in this model is at uniform temperature.

10.9   Fin approximation can be applied in modeling organs such as the elephant ear, rat tail, chicken legs, duck beak and human digits. Temperature distribution in these organs is three-dimensional. However, the problem can be significantly simplified using fin approximation. As an example, consider the rat tail. Anatomical studies have shown that it consists of three layers: bone, tendon and cutaneous layer. There are three major axial artery-vein pairs: one ventral and two lateral. These pairs are located in the tendon near the cutaneous layer as shown. The ventral vein is small compared to the lateral veins, and the lateral arteries are small compared to the ventral artery. Blood perfusion from the arteries to the veins takes place mostly in the cutaneous layer through a network of small vessels. Assume that blood is supplied to the cutaneous layer at uniform temperature T_{a0} all along the tail. Blood equilibrates at the local cutaneous temperature T before returning to the veins. Assume further that (1) cutaneous layer, tendon and bone have the same conductivity, (2) negligible angular variation, (3) uniform blood perfusion along the tail, (4) negligible metabolic heat, (5) steady state, (6) uniform outer radius and (7) negligible temperature variation in the radial direction (fin approximation is valid, $Bi \ll 1$). Surface heat exchange is by convection. The heat transfer coefficient is h and the ambient temperature is T_∞. Using Pennes model for the cutaneous layer, show that the heat equation for the rat tail is given by

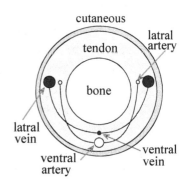

$$\frac{d^2\theta}{d\xi^2} - (m+\beta)\theta + m = 0,$$

where

$$\theta = \frac{T-T_{a0}}{T_\infty - T_{a0}}, \quad \xi = \frac{x}{L},$$

$$m = \frac{2hL^2}{kR}, \quad \beta = \frac{\dot{w}_b \rho_b c_b L^2}{k}.$$

Here L is tail length, R tail radius and x is axial distance along the tail.

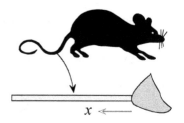

10.10 Consider the rat tail model described in Problem 10.9. Assume that the base of the tail is at the artery supply temperature T_{a0} and that the tip is insulated. Show that the axial temperature distribution and total heat transfer rate from the tail are given by

$$\theta(\xi) = \frac{m}{\beta+m}\left[(\tanh\sqrt{\beta+m})\sinh\sqrt{\beta+m}\,\xi - \cosh\sqrt{\beta+m}\,\xi + 1\right],$$

and

$$q = 2\pi hRL(T_{a0}-T_\infty)\left[1 - \frac{m}{\beta+m} + \frac{m\tanh\sqrt{\beta+m}}{(\beta+m)\sqrt{\beta+m}}\right],$$

Construct a plot of the axial temperature distribution and calculate the heat transfer rate for the following data:

$$c_b = 3.8 \frac{J}{g\text{-}^\circ C}, \quad h = 15.9 \frac{W\text{-}m^2}{^\circ C}, \quad k = 0.5 \frac{W\text{-}m}{^\circ C},$$

$$L = 22.5\,\text{cm}, \quad R = 0.365\,\text{cm}, \quad T_{a0} = 36\,^\circ C, \quad T_\infty = 22.5\,^\circ C,$$

$$\dot{w}_b = 0.01947 \frac{cm^3/s}{cm^3}, \quad \rho = 1.05 \frac{g}{cm^3}.$$

10.11 Studies have shown that blood perfusion along the rat tail is non-uniform.   This case can be analyzed by dividing the tail into sections and assigning different blood

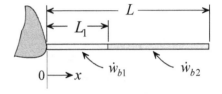

perfusion rate to each section. Consider the rat tail described in Problem 10.9. Model the tail as having two sections. The first section extends a length L_1 from the base and the second has a length of $(L - L_1)$. Blood perfusion rate in the first section is $\dot{w}_{b1}$ and in the second section $\dot{w}_{b2}$. Determine the axial temperature distribution in the tail in terms of the following dimensionless quantities

$$\theta = \frac{T - T_{a0}}{T_\infty - T_{a0}}, \quad \xi = \frac{x}{L}, \quad \xi_1 = \frac{L_1}{L}, \quad m = \frac{2hL^2}{kR},$$

$$\beta_1 = \frac{\dot{w}_{b1}\rho_b c_b L^2}{k}, \quad \beta_2 = \frac{\dot{w}_{b2}\rho_b c_b L^2}{k}.$$

10.12 The cutaneous layer of a peripheral tissue is supplied by blood at temperature T_{c0} and a *total* flow rate $\dot{W}_{cb}$. The tissue is supplied by blood at temperature T_{a0} and *total* flow rate $\dot{W}_b$. Tissue thickness is L and cutaneous layer thickness is L_1. One mechanism for regulating surface heat loss is by controlling blood flow through the cutaneous layer. Use the Weinbaum-Jiji simplified bioheat equation to examine the effect of blood flow ratio, $R = \dfrac{\dot{W}_{cb}}{W_b} = \dfrac{L_1 \dot{w}_{cb}}{L \dot{w}_b}$, on surface heat flux. The skin is maintained at uniform temperature T_s. Assume a linear tissue vascular geometry function of the form

$$V(\xi) = A + B\xi.$$

10.13 Studies have concluded that the plates on the back of the dinosaur Stegosaurus served a thermoregulatory function as heat dissipating fins [19]. There are indications that the network of channels within the plates may be blood vessels. Model the plate as a rectangular fin of width W, length L and thickness t. Use the Pennes model to formulate the heat equation for this blood

perfused plate. The plate exchanges heat with the ambient air by convection. The heat transfer coefficient is h and the ambient temperature is T_∞. Assume that blood reaches each part of the plate at temperature T_{a0} and that it

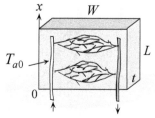

equilibrates at the local temperature T. Assume further that (1) blood perfusion is uniform, (2) negligible metabolic heat production, (3) negligible temperature variation along plate thickness t (fin approximation is valid, $Bi \ll 1$), (4) steady state and (5) constant properties. Show that the heat equation for this model is given by

$$\frac{d^2\theta}{d\xi^2} - (m + \beta)\theta + m = 0,$$

where

$$\theta = \frac{T - T_{a0}}{T_\infty - T_{a0}}, \quad \xi = \frac{x}{L}, \quad m = \frac{2h(W + t)L^2}{kWt}, \quad \beta = \frac{\dot{w}_b \rho_b c_b L^2}{k}.$$

10.14 Modeling the plates on the back of the dinosaur Stegosaurus as rectangular fins, use the bioheat fin equation formulated in Problem 10.13 to show that the enhancement in heat transfer, η, due to blood flow is given by

$$\eta = \frac{q}{q_o} = \frac{\sqrt{m}}{\beta + m} \cdot \frac{\beta + \dfrac{m}{\sqrt{\beta + m}}\tanh\sqrt{\beta + m}}{\tanh\sqrt{m}},$$

where

$$m = \frac{2h(W+t)L^2}{kWt}, \quad \beta = \frac{\dot{w}_b \rho_b c_b L^2}{k}.$$

q = heat transfer rate from plate with blood perfusion

q_o = heat transfer rate from plate with no blood perfusion

Compute the enhancement η and total heat loss from 10 plates for the following data:

c_b = specific heat of blood = $3800\,\text{J/kg–}^\circ\text{C}$
h = heat transfer coefficient = $14.9\,\text{W/m}^2\text{–}^\circ\text{C}$
k = thermal conductivity = $0.6\,\text{W/m–}^\circ\text{C}$
L = plate length = $0.45\,\text{m}$
t = plate thickness = $0.2\,\text{m}$
T_{a0} = blood supply temperature = $37\,^\circ\text{C}$
T_∞ = ambient temperature = $27\,^\circ\text{C}$
$\dot{w}_b$ = blood perfusion rate per unit tissue volume
    = $0.00045\,(\text{cm}^3/\text{s})/\text{cm}^3$
W = plate width = 0.7 m
ρ_b = blood density = $1050\,\text{kg/m}^3$

10.15 Elephant ears serve as a thermoregulatory device by controlling blood supply rate to the ears and by flapping them. Increasing blood perfusion increases heat loss due to an increase in surface temperature. Flapping results in an increase in the heat transfer coefficient as air flow over the ears changes from natural to forced convection. Consider the following data on a 2000 kg elephant which generates 1640 W:

c_b = specific heat of blood = $3800\,\text{J/kg–}^\circ\text{C}$

h_n = natural convection heat transfer coefficient = $2 \, W/m^2 - ^\circ C$

h_f = forced convection heat transfer coefficient (flapping)
    $= 17.6 \, W/m^2 - ^\circ C$

k = thermal conductivity = $0.6 \, W/m - ^\circ C$

L = equivalent length of square ear = $0.93 \, m$

t = average ear thickness = $0.6 \, cm$

T_{a0} = blood supply temperature = $36 \, ^\circ C$

T_∞ = ambient temperature = $24 \, ^\circ C$

$\dot{w}_b$ = blood perfusion rate per unit tissue volume
    $= 0.0015 \, (cm^3/s)/cm^3$

ρ_b = blood density = $1050 \, kg/m^3$

Neglecting metabolic heat production in the ear, model the ear as a square fin using the bioheat equation formulated in Problem 10.13 to determine the total heat transfer rate from two sides of two ears with and without flapping. In addition compute the enhancement in heat transfer for the two cases. Define enhancement η as

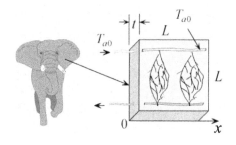

$$\eta = \frac{q}{q_o},$$

where

q = heat transfer rate from ear

q_o = heat transfer rate from ear with no blood perfusion and no flapping

10.16 Cryosurgical probes are used in medical procedures to selectively freeze and destroy diseased tissue. The cryoprobe surface is maintained at a temperature below tissue freezing temperature causing a frozen front to form at the surface and propagate outward. Knowledge of the maximum frozen layer or lesion size is helpful to the surgeon in selecting the proper settings for the cryoproble. Maximum lesion size corresponds to the steady state temperature distribution. Consider a planar probe which is inserted in a large region of tissue. Probe thickness is L and its surface temperature is T_o. Using the s-vessel tissue cylinder model, determine the maximum lesion size for the following data:

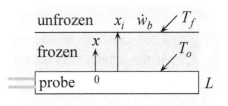

c_b = specific heat of blood = 3800 J/kg–$^\circ$C
L = probe thickness = 4 mm
k = thermal conductivity of unfrozen tissue = 0.6 W/m–$^\circ$C
k_s = thermal conductivity of frozen tissue = 1.8 W/m–$^\circ$C
q'''_m = metabolic heat production = 0.021 W/cm^3
T_{ab0} = artery blood temperature = 36.5 $^\circ$C
T_f = tissue freezing temperature = 0 $^\circ$C
T_o = probe surface temperature = –42 $^\circ$C
$\dot{w}_b$ = blood perfusion rate per unit tissue volume
      = 0.0032 (cm^3/s)/cm^3
ρ_b = blood density = 1050 kg/m^3
$\Delta T^* = 0.75$

10.17 A cylindrical cryosurgical probe consists of a tube whose surface temperature is maintained below tissue freezing temperature T_f. The frozen region or lesion around the cryoprobe reaches its maximum size at steady state. Predicting the maximum lesion size is important in avoiding damaging healthy tissue and in targeting diseased areas. Consider a cylindrical probe which is inserted in a large tissue region. Probe radius is r_o and its surface temperature is T_o. Metabolic heat production is q'''_m and blood perfusion rate is $\dot{w}_b$. Let T be the temperature distribution in the unfrozen tissue and T_s the

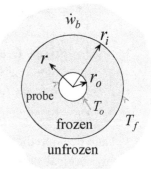

temperature distribution in the frozen tissue. Using the s-vessel tissue cylinder model, determine the steady state temperature distribution in the two regions and the maximum lesion size. Note that the conductivity of frozen tissue k_s is significantly different from the conductivity k of unfrozen tissue. Express the result in dimensionless form using the following dimensionless quantities:

$$\theta = \frac{T - T_{ab0}}{T_f - T_{ab0}}, \quad \theta_s = \frac{T_s - T_o}{T_f - T_o}, \quad \xi = \frac{r}{r_o}, \quad \xi_i = \frac{r_i}{r_o},$$

$$\beta = \frac{\dot{w}_b \rho_b c_b \Delta T^* r_o^2}{k}, \quad \gamma = \frac{q_m^{'''} r_o^2}{k(T_{ab0} - T_f)}.$$

10.18 A brain surgical procedure requires the use of a spherical cryosurgical probe to create a frozen region (lesion) 6 mm in radius. A 3 mm radius spherical cryoprobe is selected for insertion into the diseased area. Using the Pennes model, determine the required probe surface temperature such that the maximum lesion size does not exceed 6 mm. Note that maximum size corresponds to the steady state temperature distribution in the frozen and unfrozen regions around the probe. The following data is given

c_b = specific heat of blood = $3800 \text{ J/kg-}^\circ\text{C}$
k = thermal conductivity of unfrozen tissue = $0.6 \text{ W/m-}^\circ\text{C}$
k_s = thermal conductivity of frozen tissue = $1.8 \text{ W/m-}^\circ\text{C}$
$q_m^{'''}$ = metabolic heat production = 0.011 W/cm^3
T_{a0} = artery blood temperature = $36.5 \ ^\circ\text{C}$
T_f = tissue freezing temperature = $0 \ ^\circ\text{C}$
$\dot{w}_b$ = blood perfusion rate per unit tissue volume
$\quad = 0.0083 \ (\text{cm}^3/\text{s})/\text{cm}^3$
ρ_b = blood density = 1050 kg/m^3

10.19 Analytical prediction of the growth of the frozen region around cryosurgical probes provides important guidelines for establishing probe application time. Consider the planar probe described in Problem 10.16. Use the s-vessel tissue cylinder model and assume a quasi-steady approximation to show that the dimensionless interface location ξ_i is given by:

$$\tau = \frac{1}{\lambda^2}\left[\lambda\xi_i - \ln(1 + \lambda\xi_i)\right],$$

where

$$\tau = \frac{k_s(T_f - T_o)}{\rho L^2 \mathcal{L}}t, \quad \lambda = \sqrt{\beta}\,\frac{k(T_f - T_{ab0})}{k_s(T_f - T_o)}\left[1 + \frac{\gamma}{\beta}\right], \quad \xi_i = \frac{x_i}{L},$$

$$\beta = \frac{\dot{w}_b\rho_b c_b \Delta T^* L^2}{k}, \quad \gamma = \frac{q_m''' L_o^2}{k(T_{ab0} - T_f)}.$$

where $\mathcal{L} = 333{,}690$ J/kg is the latent heat of fusion and $\rho_s = 1040\,\text{kg/m}^3$ is the density of frozen tissue. How long should the probe of Problem 10.16 be applied so that the frozen layer is 3.5 mm thick?

10.20 Consider the cylindrical cryosurgical described in Problem 10.17. Using the s-vessel tissue cylinder model and assuming quasi-steady interface motion, determine lesion size as a function of time.

10.21 The spherical cryosurgical probe of Problem 10.18 is used to create a lesion corresponding to 95% of its maximum size. Using the Pennes model and assuming quasi-steady interface motion, determine the probe application time. Probe temperature is $T_o = -29.6°\text{C}$, latent heat of fusion is $\mathcal{L} = 333{,}690$ J/kg and the density of frozen tissue is $\rho_s = 1040\,\text{kg/m}^3$.

10.22 Prolonged exposure to cold environment of elephants can result in frost bite on their ears. Model the elephant ear as a sheet of total surface area (two sides) A and uniform thickness δ. Assume uniform blood perfusion $\dot{w}_b$ and uniform metabolic heat q_m'''. The ear loses heat by convection. The ambient temperature is T_∞ and the heat transfer coefficient is h. Using lumped capacity approximation and the Pennes model, show that the dimensionless transient heat equation is given by

$$\frac{d\theta}{d\tau} = (1+\gamma) - (1+\beta)\theta,$$

where

$$\theta = \frac{T - T_{a0}}{T_\infty - T_{a0}}, \quad \tau = \frac{2h}{\delta \rho c}t, \quad \beta = \frac{\dot{w}_b \rho_b c_b \delta}{2h}, \quad \gamma = \frac{q_m''' \delta}{2h(T_\infty - T_{a0})}.$$

Here ρ is tissue density and c tissue specific heat. The subscript b refers to blood. Determine the maximum time a zoo elephant can remain outdoors on a cold winter day without resulting in frost bite when the ambient temperature is lower than freezing temperature T_f. Assume that initially the ears are at uniform temperature T_i.

MICROSCALE CONDUCTION

Chris Dames
Department of Mechanical Engineering
University of California, Riverside

11.1   Introduction

Heat conduction at the microscale can be dramatically different than at the macroscale. The differences become clear by comparing the thermal conductivity of a "bulk" material (that is, a large sample that is by definition free of microscale effects) to the effective thermal conductivity of a microscopic sample of the same material. The values of thermal conductivity that are readily available in textbooks and standard reference books (so-called "handbook values") apply to bulk samples, but must be used with great caution for microscale samples. As one example, according to a standard reference [1], the thermal conductivity k of pure silicon at room temperature is $148 \, W/m\text{-}^\circ C$. This value is appropriate for silicon samples with characteristic lengths ranging from meters to millimeters to microns. However, a silicon nanowire of diameter $56 \, nm$ ($1 \, nm = 10^{-9} \, m$) has thermal conductivity of only $26 \, W/m\text{-}^\circ C$ [2], a reduction by more than a factor of five compared to the bulk value. The reductions are even more dramatic at low temperature: at $20 \, K$ the values are $4940 \, W/m\text{-}^\circ C$ for bulk Si [1], and $0.72 \, W/m\text{-}^\circ C$ for the Si nanowire [2], a difference of more than a factor of 6000! By the end of this chapter, readers should understand the physical reasons for this tremendous reduction, and be able to evaluate it numerically with approximate calculations.

Although the great majority of microsystems follow this same pattern of having thermal conductivity less than their bulk counterparts, there are also certain materials that exhibit nanoscale thermal conductivity greater than their bulk forms. Carbon is the most prominent example. At 300 K the thermal conductivity of bulk diamond is $2310\,W/m\text{-}^\circ C$, and the value for graphite (in-plane) is $2000\,W/m\text{-}^\circ C$ [1], whereas single-walled carbon nanotubes have been measured with thermal conductivity of $3500\,W/m\text{-}^\circ C$ and up [3,4].

These dramatic variations in the thermal conductivity between macro- and micro-systems are of great importance for many modern applications. The feature sizes in microelectronics (transistors and memory) and some optoelectronics (certain solid-state lasers) are now in the range of 100 nm or less, and the corresponding reduction in the thermal conductivity of the constituent materials presents a serious obstacle to dissipating heat away from the device regions. Of course, there are also applications where reduced thermal conductivity is desirable, for example, in thermoelectric energy converters [5] and thermal barrier coatings.

11.1.1   Categories of Microscale Phenomena

Heat transfer at the microscale is a rich and challenging research field that encompasses many different phenomena [6-9]. The classical Fourier law of diffusive heat conduction will break down for processes that are too fast, or systems that are too small. In this chapter we will limit ourselves to the second category, by focusing exclusively on steady-state problems. The essential questions are:

- What are the physical mechanisms by which the classical macroscopic Fourier's law will fail for small systems?
- At what length scales does this happen?
- How can we modify the macroscopic Fourier's law to still be useful at the microscale? What is the effective thermal conductivity k for a microstructure such as a nanowire or thin film?

Our framework for answering these questions is shown in Fig. 11.1. A useful concept for discussing the fundamentals of microscale heat transfer is the "wavepacket." As shown in Fig. 11.1(a), a wavepacket is a wave-like disturbance that is localized within a small volume of space. The wavepacket represents an energy carrier propagating through a medium. A wavepacket is like a particle in that it has a small size and propagates in a certain direction, and a wavepacket is like a wave in that it has an average wavelength λ. For heat conduction in an insulator like diamond, the wavepacket can be thought of as a small pulse of a sound wave, whereas for heat conduction in a metal like copper the wavepacket can be thought of as the wavefunction of an electron. (These energy carriers, and others, will be discussed in detail later in the chapter.) Figure 11.1(b) depicts the wavepacket traveling through a medium for an extended period of time. The characteristic length of the medium is denoted by L, which might be the thickness of a thin film, or the diameter of a nanowire. Occasionally the wavepacket collides with something in the medium (for example, an impurity atom, a grain boundary, or even another wavepacket) which causes the packet to scatter and change direction. These collisions impede the packet's progress through the medium, thereby slowing down the energy transport from bottom to top in Fig. 11.1. The average distance between such scattering events is known as the "mean free path," denoted by Λ. As long as $\lambda \ll \Lambda \ll L$, we can generally ignore the wave nature of the packet, and treat it simply as a particle [10].

In Fig. 11.1(c) the characteristic length has been decreased to be smaller

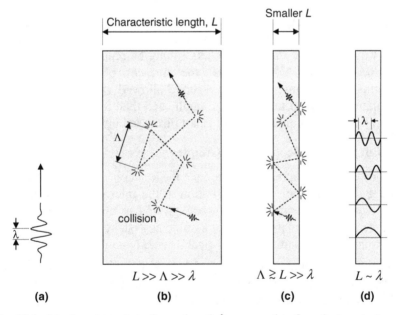

Fig. 11.1 (a) A wavepacket of wavelength λ propagating from bottom to top. (b) Bulk behavior: wavepacket in a large system with characteristic length L. The average distance between collisions is the mean free path, Λ. (c) Classical size effect (the focus of this chapter): If the characteristic length is reduced to be comparable to or smaller than Λ, the frequency of collisions is greatly increased by scattering on the boundaries of the system. The wavepacket can be treated as a particle. (d) Quantum size effect: If the characteristic length is further reduced to be comparable to the wavelength, the wavepacket can no longer be treated as a particle, and a more sophisticated analysis is required.

than Λ. In this case the wavepacket has numerous collisions with the boundaries of the system (for example, with the surfaces of a thin film), and these collisions further impede the energy flow from bottom to top. This is the essence of the *"classical size effect,"* which is the focus of this chapter. Because the wavelength is still small compared to L and Λ, we again ignore the wave nature of the packet and approximate it as a particle. (The reader is advised that there are certain systems where this particle approximation breaks down even though $\lambda \ll \Lambda, L$ [6]. The most common failure is transport perpendicular to multilayer structures with interfaces that are very sharp compared to λ, most notably the constructive and destructive interference of light rays in thin film optics. The propagation of sound waves through multilayer thin-film "superlattices" is less commonly a concern, because the roughness of the layer interfaces is usually not negligible compared to λ. When the particle approximation does break down, such systems require a more careful treatment of multiple coherent scattering events [6,7], which is beyond the scope of this chapter.)

Finally, Fig. 11.1(d) depicts the *"quantum size effect."* If the system length is so small as to be comparable to the wavelength of the energy carrier, then the wave nature of the wavepackets cannot be ignored. For example, due to the boundary conditions at the surfaces of a thin film, the wavelengths may be quantized into discrete allowed values. In this case the calculations become much more complicated than for the classical size effect [6,7]. The quantum size effect is not discussed in this chapter.

11.1.2 Purpose and Scope of this Chapter

The purpose of this chapter is to introduce the key concepts of the classical size effect in steady-state heat conduction at the microscale, in a manner accessible to readers coming from a background in classical engineering heat transfer. In particular, we do not assume any background in solid state physics, quantum mechanics, or statistical thermodynamics.

It is hoped that this chapter will help the dedicated reader develop the skills to think critically about the major issues in the field, to be confident reading a specialized journal paper, and to make realistic numerical estimates for the heat transfer in a range of microscale systems.

Readers interested in pursuing this subject at a more advanced level can choose from several excellent resources, including comprehensive textbooks by Chen [6] and Zhang [7] and a volume edited by Volz [8]. Also of particular value are an extensive chapter by Majumdar [9], and a review article by Cahill *et al.* [11]. Truly dedicated students may also want to strengthen their background in the supporting subjects of solid state physics [12-14], quantum mechanics [15,16], and statistical thermodynamics [17,18].

The chapter is organized as follows. In Section 11.2 we introduce the framework of kinetic theory and derive a key equation for the thermal conductivity of a substance in terms of other, more fundamental properties. These fundamental properties are outlined in Section 11.3 for four important types of energy carriers, with coverage of ideal gases, metals, insulators, and radiation. Finally, in Section 11.4 we introduce theoretical tools to model the thermal conductivity reduction in nanostructures including nanowires and thin films.

11.2   Understanding the Essential Physics of Thermal Conductivity Using the Kinetic Theory of Gases

Microscale heat transfer in solids can be understood using a variety of theoretical approaches, including the Boltzmann transport equation, molecular dynamics, the Landauer formalism, the Green-Kubo formalism, radiation analogies, and other techniques [6-9,11]. In this chapter we focus on the theoretical framework of *kinetic theory*, because this approach offers maximum physical insight for minimal complexity, and can be applied fruitfully to a wide range of realistic problems in microscale conduction. Furthermore, we shall see that in many cases even the results of more complicated analyses (e.g. Boltzmann equation, Landauer formalism) can still be understood and explored using kinetic theory.

11.2.1 Derivation of Fourier's Law and an Expression for the Thermal Conductivity

Figure 11.2(a) depicts a gas of particles, such as the individual molecules in an ideal gas. Later we will generalize this derivation to other types of particles, such as free electrons in a solid piece of metal. Some particles travel faster than others, but their representative speed is denoted by v. Consider the particle labeled "1" in Fig. 11.2(a,b). At the instant in time depicted in Fig. 11.2(a), it is clear that particle 1 is on a collision course with particle 2. Thus, a short time later these two particles collide, as indicated in Fig. 11.2(b). The collision changes the direction of both particles, and also causes some energy to be exchanged between them. Thus, if particle 1 was initially "hot" compared to particle 2, after the collision in Fig. 11.2(b), particle 1 will have given up some of its energy to particle 2. This energy exchange is an essential feature of the kinetic theory framework.

Such collisions happen repeatedly over time, causing each particle to undergo a "random walk" between collisions (Fig. 11.2(c)), exchanging energy in every collision. The distance that a particle travels between any two consecutive collisions is called the free path, and the average value of this free path over many collisions is called the "mean free path," which we denote by Λ. It is sometimes helpful to consider the corresponding "mean free time" between collisions, denoted by τ, and defined through the simple relation

$$\Lambda = v\tau . \tag{11.1}$$

Now let this gas of particles be placed in a large box with a negative temperature gradient in the x-direction. That is, the particles are hotter towards the left (small x) and colder towards the right (large x). As shown in Fig. 11.2(d), we will evaluate the *net* energy flow crossing an imaginary control surface defined by the plane $x = x_0$, by evaluating the difference between energy flows from left-to-right and from right-to-left.

Consider two control volumes each of thickness Λ in the x-direction, and with some large and arbitrary area A in the $y - z$ direction. The "left" control volume extends from x_0 to $x_0 - \Lambda$, and the "right" control volume

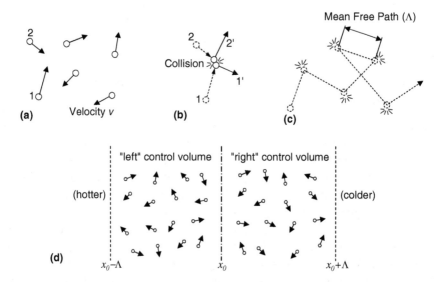

Fig. 11.2 (a) Schematic of a gas of particles. (b) A collision between two particles. (c) Each particle undergoes many random collisions over time. The average distance between these collisions is known as the "mean free path." (d) Control volumes used to analyze heat transfer.

extends from x_0 to $x_0 + \Lambda$. If we wait a time $\tau = \Lambda / v$, approximately half of the particles in the left control volume will exit that control volume through the plane $x = x_0$, and the other half will exit through the plane $x = x_0 - \Lambda$. Similarly, in that same time τ approximately half of the particles in the right control volume will exit through $x = x_0$, and the other half will exit through $x = x_0 + \Lambda$.

Due to the imposed temperature gradient, the particles crossing $x = x_0$ from the left control volume are hotter on average than those particles crossing $x = x_0$ from the right control volume. Thus there is a net energy exchange from left to right. We estimate the energy exchange as follows. The total energy contained in the left control volume is

$$\hat{U}_{left} = \Lambda A U_{left,avg},$$

where $\hat{U}$ is the internal energy in J (joules), and $U_{left,avg}$ is the internal energy density per unit volume, in J/m^3, averaged over the left control volume. (Note that it will prove more convenient to work with energy per unit volume [J/m^3] rather than energy per unit mass [J/kg], even though the latter quantity is more common in engineering thermodynamics. Properties per kg are easily converted to a per m^3 basis simply by multiplying by the mass density ρ. For example, $U_{left,avg} = \rho \hat{u}_{left,avg}$.) Similarly, the energy in the right control volume is

$$\hat{U}_{right} = \Lambda A U_{right,avg}.$$

As a result of the energy leaving each control volume, the net energy crossing $x = x_0$ is thus

$$\Delta \hat{U}_{LR} \approx \tfrac{1}{2} \hat{U}_{left} - \tfrac{1}{2} \hat{U}_{right} = \tfrac{1}{2} \Lambda A \left(U_{left,avg} - U_{right,avg} \right).$$

We next assume that U is a smoothly-varying function of x which can be approximated as a straight line over very small distances of the order of Λ. This means that

$$U_{left,avg} \approx U(x_0 - \tfrac{1}{2}\Lambda),$$

namely, U evaluated at the location $x_0 - \tfrac{1}{2}\Lambda$. Similarly,

$$U_{right,avg} \approx U(x_0 + \tfrac{1}{2}\Lambda).$$

Now we use a Taylor series expansion to write

$$U_{left,avg} - U_{right,avg} \approx -\left[\frac{dU}{dx}\right]_{x_0} \Lambda. \qquad (11.2)$$

Our assumption that the energy density of the particles conforms to the local temperature, even at the microscale, is also known as the assumption of local thermodynamic equilibrium.

The exact analysis of this problem shows that the correct expression is only $\tfrac{2}{3}$ as large as eq. (11.2),

$$U_{left,avg} - U_{right,avg} = -\tfrac{2}{3}\left[\frac{dU}{dx}\right]_{x_0} \Lambda.$$

This $\tfrac{2}{3}$ coefficient is found from a more careful treatment of the full three-dimensional problem including the varying locations of the particles' most recent collisions. For the rest of this chapter we proceed using the correct $\tfrac{2}{3}$ coefficient.

We wish to present our results in terms of temperature rather than internal energy, so we use the chain rule of calculus to write

$$\frac{dU}{dx} = \frac{dU}{dT}\frac{dT}{dx} = C\frac{dT}{dx},$$

where $C = \rho c_v$ is the specific heat capacity at constant volume, per unit volume, in $J/m^3-^\circ C$. Notice again the normalization per unit volume rather than per unit mass.

Finally, we collect our results, and divide by the area and the elapsed time to evaluate the heat flux in W/m^2:

$$q'' = \frac{\Delta \hat{U}_{LR}}{A\tau} = -\tfrac{1}{3}Cv\Lambda\frac{dT}{dx}. \qquad (11.3)$$

Notice the proportionality between q'' and dT/dx: here we have *derived* Fourier's law of heat conduction, an equation which is usually presented as an empirical experimental observation. Furthermore, we have derived an equation for the thermal conductivity of a gas of particles:

$$k = \tfrac{1}{3}Cv\Lambda .$$ (11.4)

These last two equations are the most important theoretical results in the study of heat conduction by kinetic theory. Equation (11.4) in particular forms the framework for the rest of this chapter. To understand heat conduction in any system, we need to methodically evaluate each term in eq. (11.4): what are the specific heat capacity, the particle velocity, and the mean free path? In microscale systems, we will see that the small size of the microstructure will generally reduce Λ without affecting C or v, thus leading to reductions in the thermal conductivity.

11.3 Energy Carriers

Having derived the kinetic theory expression eq. (11.4), we now discuss the relevant properties C, v, and Λ for a range of heat-conducting materials. We will begin with an ideal gas, the system for which kinetic theory is most easily understood. However, one of the most powerful aspects of our kinetic theory derivation is that the key results, eqs. (11.3) and (11.4), are also directly applicable to heat conduction in *solids*, as long as we can identify the corresponding "particles" that carry the heat. Thus we will extend our analysis to metals, such as gold and copper, as well as insulators and semiconductors, such as diamond and silicon. Finally, we will further extend this kinetic theory approach to heat transfer by radiation. In all cases (ideal gas, metal, insulator, radiation), we first identify the important energy carrier, and then ask the same three questions: For these carriers, what is C? What is v? And what is Λ?

11.3.1 Ideal Gases: Heat is Conducted by Gas Molecules

The kinetic theory of ideal gases has been thoroughly studied and is well-understood [19-22]. As a starting point, the reader is certainly familiar with the ideal gas law,

$$pV = m_{tot}RT ,$$ (11.5)

where p is absolute pressure, V is volume, m_{tot} is the total mass of the gas, T is absolute temperature in Kelvin, and R is the gas constant for the particular gas being studied, found from

$$R = R_U / M ,$$ (11.6)

where $R_U = 8.314$ J/mol-K is the universal gas constant, and M is the molecular weight of the gas molecule. It will also prove convenient to express R_U as

$$R_U = N_A k_B,$$ (11.7)

where $N_A = 6.022 \times 10^{23}$ mol^{-1} is Avogadro's number, and $k_B = 1.381 \times 10^{-23}$ J/K is Boltzmann's constant.

Note that in eq. (11.5) it is essential to express T in units of Kelvin (K) rather than centigrade (°C). This need to use absolute units is a recurring theme in this chapter, for all types of energy carriers. In this chapter the reader is strongly encouraged to use absolute units throughout, although conversion back to centigrade units can be made as the final step if desired when solving any particular problem. In particular note that W/m-°C and W/m-K are equivalent, as are J/m^3-°C and J/m^3-K.

In the rest of this section we summarize the main results that are useful for understanding the thermal conductivity of ideal gases. It is important to distinguish between monatomic gases (such as Ar and He) and diatomic/polyatomic gases (such as N_2, O_2, CO_2, and H_2O). The key distinction between these two categories is that a molecule of a monatomic gas can only store energy in its translational kinetic energy, whereas a molecule of diatomic/polyatomic gas has additional mechanisms for energy storage, namely rotational kinetic energy and interatomic vibrations within the molecule. These additional mechanisms make an exact theory much more complicated. Therefore, for simplicity we will focus our discussion on monatomic gases, while briefly commenting on the connections to diatomic/polyatomic gases.

Specific heat. Recall from the basic thermodynamics of an ideal gas that the specific heats at constant pressure (c_p) and constant volume (c_v), expressed on a per-unit-mass basis, are always related through

$$c_p = c_v + R \qquad [\text{J/kg-°C}].$$

To convert to a per-unit-volume basis, we multiply by the density $\rho = p/RT$ to find

$$C_p = C_v + p/T \qquad [\text{J/m}^3\text{-°C}].$$ (11.8)

For a monatomic ideal gas, the specific heat has the particularly simple form

$$c_v = \tfrac{3}{2} R \,,$$

$$C_v = \frac{3p}{2T} \,. \tag{11.9}$$

The specific heats of diatomic and polyatomic gases have a similar form but the numerical coefficient is in general a function of temperature that is larger than or equal to $\tfrac{3}{2}$, a reflection of the additional modes of energy storage.

Speed. Even at fixed pressure and temperature, the molecular speeds are distributed over a broad range of values. The distribution of speeds depends on the temperature and is known as the Maxwellian velocity distribution. For simplicity we would like to represent that entire distribution by a single representative "thermal velocity" v_{th}, for use in the kinetic theory framework of eq. (11.4). We will represent v_{th} by the root-mean-square (rms) velocity v_{rms}, given by

$$v_{th} = v_{rms} = \sqrt{3RT} \,. \tag{11.10}$$

Other relevant velocities that are sometime used to characterize a Maxwellian distribution are the most-probable speed ($v_{mp} = \sqrt{2RT}$), the mean speed ($v_{mean} = \sqrt{\tfrac{8}{\pi} RT}$), and the speed of sound ($v_{sound} = \sqrt{\gamma RT}$), where $\gamma = c_p / c_v$. Note that all of these velocities are proportional to $\sqrt{RT}$, with numerical prefactors varying by only tens of percent.

Mean free path. For ideal gas molecules the mean free path between *collisions* is

$$\Lambda_{coll} = \frac{1}{\pi \sqrt{2} \eta d^2} \,,$$

where $\eta = \rho N_A / M = p / k_B T$ is the average number of molecules per unit volume and d is the effective diameter of the gas molecule. This expression accounts for the relative motion between the different molecules [19]. However, when molecules collide, they do not exchange 100% of their energy. The effectiveness of the energy exchange depends on whether the molecules are monatomic or diatomic, and how the intermolecular forces are modeled. Based on several different detailed calculations for a monatomic ideal gas modeled as elastic spheres [19,22], the mean free path for *energy exchange* is given to within about 2% by

$$\Lambda_{en} = \frac{12}{5\sqrt{3\pi}\,\eta d^2} = \left(\frac{12}{5\sqrt{3\pi}}\right)\frac{k_B T}{p d^2}\,. \tag{11.11}$$

Note that Λ_{en} is about 3.5× larger than Λ_{coll}, and that it is Λ_{en} (not Λ_{coll}) that is the correct choice for evaluating thermal conductivity using eq. (11.4). For diatomic and polyatomic ideal gases, the energy-exchange mean free path has a similar form as eq. (11.11), but there is no such tidy theory for the numerical coefficient.

Example 11.1: Thermal Conductivity of an Ideal Gas.

The molecular diameter of He at $0\,°C$ and atmospheric pressure was determined to be $d = 0.2193\,nm$ using viscosity measurements [22]. Calculate the corresponding thermal conductivity, and compare with the experimental value listed in Appendix D.

(1) Observations. Helium is a monatomic gas, so we expect the theory outlined above to work well here.

(2) Formulation.

(i) Assumptions. (1) Helium can be modeled as a monatomic ideal gas. (2) Helium atoms can be treated as elastic spheres.

(ii) Governing Equation. We will evaluate the kinetic theory expression eq. (11.4) using the expressions for C, v, and Λ given above in eqs. (11.9), (11.10), and (11.11).

(3) Solutions.

Specific heat. Application of eq. (11.9) at 273.15 K and 101,300 Pa gives

$$C_v = \frac{3p}{2T} = \frac{3}{2}\frac{101{,}300\,\text{Pa}}{273.15\,\text{K}} = 556.3\frac{\text{J}}{\text{m}^3 - °\text{C}}\,,$$

where we have used the convenient identity that $1\,\text{Pa} = 1\,\text{J/m}^3$.

Speed. The atomic weight of He is $M = 4.003\,\text{g/mol}$, so from eq. (11.7) we have $R = R_U/M = 2077\,\text{J/kg - K}$. Thus, using eq. (11.10) we have

$$v_{th} = \sqrt{3RT} = \sqrt{3(2077\,\text{J/kg - K})(273.15\,\text{K})} = 1305\,\text{m/s}\,.$$

Mean free path. Applying eq. (11.11), we have

$$\Lambda_{en} = \left(\frac{12}{5\sqrt{3\pi}}\right)\frac{k_B T}{pd^2} = \left(\frac{12}{5\sqrt{3\pi}}\right)\frac{\left(1.381\times10^{-23}\,\text{J/K}\right)\left(273.15\,\text{K}\right)}{\left(101,300\,\text{Pa}\right)\left(0.2193\times10^{-9}\,\text{m}\right)^2}$$

$$\Lambda_{en} = 605\times10^{-9}\,\frac{\text{J}}{\text{Pa - m}^2} = 605\,\text{nm}\,.$$

Thermal conductivity. Combining these values for C, v, and Λ, we find

$$k = \tfrac{1}{3}Cv\Lambda = \tfrac{1}{3}\left(556.3\frac{\text{J}}{\text{m}^3\text{-}^{\circ}\text{C}}\right)\left(1305\frac{\text{m}}{\text{s}}\right)\left(605\text{nm}\right) = 0.1464\frac{\text{W}}{\text{m}-^{\circ}\text{C}}\,.$$

For comparison, from Appendix D the experimental value of k at $0°C$ is $0.142\,\text{W/m-}^{\circ}\text{C}$. The error in our calculation is therefore

$$\text{error} = \frac{0.1464 - 0.142}{0.142} = 3.1\%\,.$$

(4) Checking. *Dimensional check.* The units of $Cv\Lambda$ are $(\text{J-m}^{-3}\text{-}^{\circ}\text{C}^{-1})(\text{m}-\text{s}^{-1})(\text{m})$, correctly giving $\text{W/m-}^{\circ}\text{C}$, the required units of k.

(5) Comments. (1) The observed 3.1% error in k is only slightly larger than the 2% uncertainty stated in our equation (11.11) for Λ, and also reflects experimental uncertainty in the values we used for d and k from Appendix D. Overall this agreement between theory and experiment is very good. (2) This mean free path of 605 nm may appear quite small, but because helium's d is smaller than all other gases, this value of Λ is actually a relatively large value for an ideal gas. As revealed by eq. (11.11), molecules with larger diameters will necessarily have smaller Λ. (3) You may wonder what would happen if this helium gas were used to conduct heat between parallel plates with a gap much less than the calculated mean free path of 605 nm. This very important question refers to a "classical size effect," which is discussed later in the chapter. (4) See also the end-of-chapter Problems 11.1-11.3.

11.3.2 Metals: Heat is Conducted by Electrons

For the purposes of heat conduction, the defining feature of a metal is that it has a very large concentration of electrons that are free to roam throughout the material. These "free electrons" are responsible for the very high electrical conductivity of metals such as copper and gold. In addition, besides carrying charge, these free electrons also carry energy, and thus are the dominant transporter of heat in metals.

A theoretical discussion of the thermal properties of electrons in metals requires some knowledge of solid-state physics and thus is beyond the scope of this chapter. However, the key theoretical results are simple and compact, and are now presented below.

Speed. The electrons that contribute to heat conduction in a given metal all travel at virtually the same speed, known as the Fermi velocity, v_F, for that metal. These electrons are very fast indeed: as shown in Table 11.1, typical values of v_F are around $1-2 \times 10^6$ m/s. The Fermi velocity for many metals can be found tabulated in handbooks and textbooks from the field of solid state physics. It is also common to characterize the free electrons by their Fermi energy, E_F, which is related to v_F simply through $E_F = \frac{1}{2} m_e v_F^2$, where $m_e = 9.110 \times 10^{-31}$ kg is the mass of an electron. Thus, the Fermi energy is simply the kinetic energy associated with an electron traveling at the Fermi velocity. The Fermi energy is often reported in units of electron volts, denoted eV, with the conversion $1\,\text{eV} = 1.602 \times 10^{-19}\,\text{J}$. The Fermi velocity can also be related to the concentration of free electrons η_e, through

$$v_F = \frac{\hbar}{m_e} \left(3\pi^2 \eta_e \right)^{1/3} \tag{11.12}$$

where $\hbar = 1.055 \times 10^{-34}$ J-s is the reduced Planck's constant.

Table 11.1 Free electron properties of selected metals

Metal	Concentration of free electrons, η_e [m^{-3}]	Fermi Velocity, v_F [m/s]	Fermi Energy, E_F [eV]	Fermi Temperature, T_F [K]	Specific heat coefficient, γ [J/m^3 – K^2]
Li	4.70×10^{28}	1.29×10^6	4.75	5.52×10^4	58.1
Na	2.65×10^{28}	1.07×10^6	3.24	3.76×10^4	48.0
K	1.40×10^{28}	8.63×10^5	2.12	2.46×10^4	38.8
Cu	8.45×10^{28}	1.57×10^6	7.03	8.15×10^4	70.6
Ag	5.85×10^{28}	1.39×10^6	5.50	6.38×10^4	62.5
Au	5.90×10^{28}	1.39×10^6	5.53	6.42×10^4	62.6
Al	1.81×10^{29}	2.03×10^6	11.66	1.35×10^5	90.9
Pb	1.32×10^{29}	1.82×10^6	9.46	1.10×10^5	81.9

Source: Kittel [14]

Specific heat. The specific heat of free electrons C_e also follows a very simple equation, which we express in three equivalent ways:

$$C_e = \tfrac{1}{2}\pi^2 \eta_e k_B^2 T / E_F$$
$$C_e = \tfrac{1}{2}\pi^2 \eta_e k_B T / T_F \qquad (11.13)$$
$$C_e = \gamma T.$$

The second form of eq. (11.13) defines a characteristic "Fermi Temperature",

$$T_F = E_F / k_B \qquad (11.14)$$

and the third form expresses the fact that the specific heat is simply proportional to temperature, where the coefficient γ has the units $J/m^3 \text{-} K^2$ and is also summarized in Table 11.1. Note that the electron specific heat C_e is typically several orders of magnitude smaller than the standard handbook values of C for metals, even though the electrons dominate the thermal conductivity [23].

Mean free path. The physics of electron scattering in metals is more complicated than our current treatment allows. Around room temperature and above, for most metals of reasonable purity the dominant mechanism of electron scattering is collisions with sound waves [12]. In practice the electron mean free path in many metals at 300 K and above is approximately proportional to T^{-1}, the inverse of absolute temperature. As explored in Problem 11.5 at the end of the chapter, the electron mean free path in metals is typically of the order of tens of nm at room temperature.

11.3.3 Electrical Insulators and Semiconductors: Heat is Conducted by Phonons (Sound Waves)

The defining feature of electrical insulators (also known as dielectrics) is the extreme scarcity of free electrons as compared to metals. Of course, many electrons are still present in such crystals, but due to the nature of the atomic bonding, essentially all of these electrons are tightly bound to the vicinity of their parent atomic nucleus: virtually no electrons are free to roam around the crystal. Note that it is also possible to intentionally add a small percentage of free electrons to materials that are normally insulating, by incorporating impurity atoms known as "dopants" into the crystal. Although technologically very important for modifying the electrical properties of semiconductors, the additional free electrons due to doping are usually only a minor contribution to the thermal conductivity, and thus are ignored in our treatment below [24].

If free electrons are unimportant, how is thermal energy stored and transported in dielectric crystals? The answer: atomic vibrations. For heat transport, the simplest and most important class of atomic vibrations is sound waves. Sound waves by definition have wavelengths that are much larger than the typical spacing between individual atoms (this interatomic spacing is intimately related to, but not always equal to, the "lattice constant"). In this limit, the wavelength λ and oscillation frequency ω (in rad/s) follow a particularly simple relationship:

$$\omega = \frac{2\pi v_s}{\lambda} .$$
(11.15)

Here v_s is the speed of sound, which is tabulated in Table 11.2 for several dielectric materials. You may recognize eq. (11.15) as being analogous to the equation for the frequency of a light wave from elementary physics. Indeed, the analogy is quite relevant, and in particular we now introduce the concept of a "phonon," referring to the quantum of a sound wave, in the same way that a "photon" is the quantum of a light wave.

Table 11.2 Acoustic phonon properties of selected solids

Material	Number density of primitive unit cells, η_{PUC} [m^{-3}]	Atoms per primitive unit cell	Sound velocity (effective average), v_s [m/s]	Debye temperature, θ_D [K]	Thermal conductivity of solid at 300 K, k [W/m-°C]
C	8.80×10^{28}	2	13400	1775	2310
Si	2.50×10^{28}	2	5880	512	148
Ge	2.21×10^{28}	2	3550	297	53
Ar	2.66×10^{28}	1	1030	92	(gas)
Kr	2.17×10^{28}	1	867	72	(gas)
Xe	1.64×10^{28}	1	846	64	(gas)

Notes: Debye temperature is based on number density of primitive unit cells. Carbon is diamond phase. Ar, Kr, and Xe are only solids below 84 K, 116 K, and 161 K, respectively.   Sources: Purdue University Thermophysical Properties Research Center [1], Kittel [14].

Phonons are the key to thermal energy storage and transport in insulators. Although a detailed discussion of phonons is beyond the scope of this chapter, we now give the essential concepts. As summarized in Table 11.3, there are two broad classes of phonons: acoustic and optical.

Acoustic phonons can largely be thought of as sound waves following eq. (11.15), with the modification that there is an upper limit on the allowed frequencies. That there should be such a limit is evident by noting that the phonon wavelengths in eq. (11.15) become shorter at higher frequencies. At sufficiently high frequency, the wavelengths become so short as to be comparable to the lattice constant. It is unphysical to speak of wavelengths shorter than twice the interatomic spacing, thus placing an upper limit on the allowed frequencies. Typical values of the maximum frequency are 10^{13} - 10^{14} rad/s, depending on the material. In fact, eq. (11.15) itself breaks down near these limiting frequencies. If the details of phonon behavior at high frequencies are important, eq. (11.15) is replaced with a more general function $\omega = \omega(\lambda)$, which can be found simply by solving Newton's equations of motion for a collection of masses and springs (representing the atomic nuclei and interatomic forces, respectively), a technique known as "lattice dynamics." Even a full lattice dynamics solution for $\omega(\lambda)$ will always reduce to the linear form of eq. (11.15) in the limit of small ω (large λ).

Table 11.3 Comparison between acoustic and optical phonons

	Acoustic Phonons	Optical Phonons
Present in simplest crystals (only 1 atom per primitive unit cell)	Yes	No
Present in crystals with more than 1 atom per primitive unit cell	Yes	Yes
Relationship between frequency and wavelength	$\omega = 2\pi v_s / \lambda$ (approximately)	$\omega \approx$ const., typically 10^{13} - 10^{14} rad/s
Minimum frequency	0	
Maximum frequency	Typically 10^{13} - 10^{14} rad/s	
Speed	Sound velocity (approximately)	Zero (common approximation)
Important for energy storage (specific heat)	Yes	Yes
Important for thermal conductivity	Yes	Usually neglected

Optical phonons are present in all but the simplest materials. Specifically, optical phonons are present if and only if the material's crystal structure has more than one atom per "primitive unit cell." A primitive unit cell is the smallest possible repeating unit that can be used to build up an entire crystal. In simple crystals including those that are "simple cubic", "body-centered cubic", and "face-centered cubic", the number of primitive unit cells is simply equal to the number of atoms. All of the metals in Table 11.1 fall in this category, as do solid Ar, Kr, and Xe (Table 11.2). However, many important materials (such as Si, Ge, GaAs, and NaCl) have two atoms per primitive unit cell, and some complicated materials (for example, Bi_2Te_3) have even more than two atoms per primitive unit cell.

In contrast to acoustic phonons, the dynamics of optical phonon vibrations are generally dominated by the oscillations of atoms locally against their nearest neighbors [12-14]. For example, in the simplest case of two atoms per primitive unit cell, in the limit of long wavelengths every atom oscillates 180° out of phase with its neighbors. If adjacent atoms have opposite electric charges (as in an ionic crystal like sodium chloride, made up of Na^+ and Cl^- ions), these oscillations lead to a local electromagnetic field inside the crystal that oscillates over time, making these crystals capable of interacting strongly with light of certain frequencies. This is the reason for the name "optical" phonon.

The oscillation frequencies of optical phonons are slightly higher than the maximum frequencies of the acoustic phonons, and again typically around 10^{13} -10^{14} rad/s , depending on the material. Importantly, and in contrast to acoustic phonons, the oscillation frequencies of optical phonons for any particular crystal are relatively insensitive to changes in the phonon wavelength. This leads to the very important fact that optical phonons propagate through the crystal at velocities that are far slower than the speed of sound – so slow, in fact, that the velocity of optical phonons is commonly set to *zero*. Referring to eq. (11.4), this approximation of zero velocity is of great significance: it implies that optical phonons make a negligible contribution to the thermal conductivity.

For the purposes of this chapter, when discussing thermal conductivity we will ignore optical phonons to focus exclusively on acoustic phonons and use the following approximations, which are generally known as the *Debye approximation*.

Speed. All acoustic phonons will be approximated as traveling at the speed of sound, v_s . Real crystals have different sound speeds for transverse waves (waves with atomic displacements perpendicular to the direction of wave propagation) and longitudinal waves (atomic displacements are parallel to the direction of wave propagation). There are always twice as many transverse waves as longitudinal waves. We will designate the former speed as $v_{s,T}$ and the latter as $v_{s,L}$. For calculations of the thermal conductivity of nanostructures, it is convenient to combine $v_{s,T}$ and $v_{s,L}$ into a single effective sound velocity, in which case the best averaging method is

$$v_s = \left(\tfrac{2}{3} v_{s,T}^{-2} + \tfrac{1}{3} v_{s,L}^{-2} \right)^{-1/2} . \tag{11.16}$$

Averaging in this inverse-squares sense ensures that the result becomes exact at low temperature (see problems 11.10 and 11.11 at the end of the chapter).

Specific Heat. The specific heat in the Debye model is given by an integral expression (not presented here) which in general must be evaluated using numerical methods [12-14]. Although the numerical integration is not difficult, a convenient and practical *approximation* to the exact Debye calculation is

$$C = \frac{3\eta_{PUC} k_B}{1 + \frac{5}{4\pi^4} \left(\dfrac{\theta_D}{T} \right)^3} , \tag{11.17}$$

which is exact in the limits of low and high temperature. In the intermediate temperature range, eq. (11.17) can be used with less than 12% error compared to the exact Debye calculation. In eq. (11.17) η_{PUC} is the number of primitive unit cells per unit volume, and θ_D is the "Debye temperature," defined through

$$\theta_D = \frac{\hbar v_s}{k_B} \left(6\pi^2 \eta_{PUC} \right)^{1/3} . \tag{11.18}$$

Both η_{PUC} and θ_D are included in Table 11.2.

Readers are advised that a slightly different definition of θ_D is also in common use [14], which is identical to eq. (11.18) except that η_{PUC} is replaced by the number density of atoms, η_{atoms}. For simple crystals with one atom per primitive unit cell, the two definitions are equivalent. For more complicated crystals, the alternate definition results in a higher value of θ_D. For example, referring to Table 11.2, because silicon has two atoms per primitive unit cell it must have $\eta_{atoms} = 2\eta_{PUC}$. Thus, the alternate definition would lead to a θ_D that is larger by a factor of $2^{1/3}$, namely, 645 K [14]. Although this alternate Debye approach defining θ_D using η_{atoms} is convenient for calculating the total specific heat, it is inappropriate for modeling the thermal conductivity because it treats optical phonons as if they were traveling at v_s. Therefore, in this chapter we will use exclusively θ_D as defined in eq. (11.18) and listed in Table 11.2.

In the limits of low and high temperature, eq. (11.17) reduces correctly to the well-known limiting expressions

$$C = \tfrac{12\pi^4}{5}\eta_{PUC}k_B\left(\frac{T}{\theta_D}\right)^3 \quad (T < \tfrac{3}{20}\theta_D), \tag{11.19}$$

$$C = 3\eta_{PUC}k_B \quad (T > \tfrac{1}{2}\theta_D) . \tag{11.20}$$

Equation (11.19) is exact for $T/\theta_D \to 0$, and can be used up to $T \approx \tfrac{3}{20}\theta_D$ with errors less than 20% compared to the exact Debye calculation. Similarly, eq. (11.20) is exact for $T/\theta_D \to \infty$, and can be used down to $T \approx \tfrac{1}{2}\theta_D$ with errors less than 20%. The transition between low and high temperature regimes occurs around $T \approx \theta_D/3$. The low-temperature result eq. (11.19) is known as the "Debye T^3 law", and the high-temperature result eq. (11.20) is known as the "Law of Dulong and Petit."

Note that all of the results in eqs. (11.16) – (11.20) are only for the specific heat of *acoustic* phonons. Optical phonons, if present, make their own contribution, which is commonly described using an "Einstein model" (not discussed here). The values of specific heat reported in handbooks are for the total specific heat, including optical and acoustic phonons.

Mean Free Path. The physics of phonon scattering is beyond our present treatment. At room temperature and above, the thermal resistance is usually dominated by phonons scattering with other phonons in such a way that the overall energy flux is impeded, a process known as "Umklapp" scattering. Like electrons, in practice the phonon mean free path in many insulators is approximately proportional to T^{-1} at room temperature and above. Alloy atoms, if present, can also result in strong phonon scattering (for example, the Ge atoms in a crystal with composition $Si_{0.9}Ge_{0.1}$). If a crystal is "doped" with impurities to generate additional free electrons, the dopant atoms can also scatter phonons, an effect which is sometimes significant in reducing the phonon thermal conductivity, especially if the dopant concentration is larger than $\sim 10^{17}$ cm^{-3} ($\sim 10^{23}$ m^{-3}) [25].   At temperatures around 300 K and below it may also be important to consider the effects of phonons scattering off of impurities, isotopes, defects, and grain boundaries. Phonons may also scatter off of the boundaries of the sample itself, an effect which is revisited later in the chapter as the "classical size effect."

Example 11.2: Thermal Conductivity Trend with Temperature for Silicon.

In many situations the temperature-dependence of the thermal conductivity of a material can be approximated with a power law of the form $k(T) = aT^b$, *where a and b are constants. Assuming that the mean free path is proportional to* T^{-1}, *use data from Table 11.2 to propose a power-law approximation for the thermal conductivity of bulk silicon for temperatures from 300 K to 1000 K.*

(1) Observations.   Silicon is extremely important for applications in the microelectronics industry, and its thermal conductivity has been very well studied over a broad range of temperatures. However, in this example we will limit ourselves to the information given in Table 11.2.

(2) Formulation.

(i) Assumptions. (1) The thermal conductivity is dominated by acoustic phonons, for which the Debye model is an adequate approximation. (2) The specific heat can be approximated by the high-T limit, because the temperatures of interest are always greater than $\theta_D / 2$. (3) The mean free path will be assumed to vary inversely proportional to temperature.

(ii) Governing Equations.   We will approximate the specific heat of acoustic phonons using eq. (11.20).

(3) Solutions.

Speed. From Table 11.2, the sound velocity in silicon is 5880 m/s, which is assumed independent of temperature.

Specific heat. Using eq. (11.20) and the values from Table 11.2, we have

$$C = 3\eta_{PUC}k_B = 3\left(2.5 \times 10^{28}\,\text{m}^{-3}\right)\left(1.381 \times 10^{-23}\,\text{J/K}\right)$$
$$= 1.04 \times 10^6\,\text{J/m}^3\text{-}^\circ\text{C},$$

which is assumed constant from 300 K to 1000 K.

Thermal conductivity at 300 K. From Table 11.2 we have $k = 148$ W/m-°C at 300 K.

Mean free path. Combining these values for k, C, and v, we find at 300 K

$$\Lambda = \frac{3k}{Cv} = \frac{3\left(148\,\text{W/m-}^\circ\text{C}\right)}{\left(1.04 \times 10^6\,\text{J/m}^3\text{-}^\circ\text{C}\right)\left(5880\,\text{m/s}\right)} = 73\,\text{nm}.$$

Because the mean free path is assumed to vary in proportion to T^{-1}, we have

$$\Lambda = \frac{const.}{T}$$

$$\Lambda T = const.$$

$$\Lambda(T) \times T = \Lambda(300\,\text{K}) \times 300\,\text{K}$$

Thermal conductivity power law. Consider the ratio $k(T)/k(300\,\text{K})$. From kinetic theory we have

$$\frac{k(T)}{k(300\,\text{K})} = \frac{\frac{1}{3}Cv\Lambda(T)}{\frac{1}{3}Cv\Lambda(300\,\text{K})}$$

Because we assume $C \approx const.$,

$$\frac{k(T)}{k(300\,\text{K})} = \frac{\Lambda(T)}{\Lambda(300\,\text{K})},$$

and using our result for Λ,

$$\frac{k(T)}{k(300\,\text{K})} = \frac{300\,\text{K}}{T}$$

$$k(T) = k(300\,\text{K}) \times \left[\frac{300\,\text{K}}{T}\right]$$

$$= 148\,\text{W/m-K} \times 300\,\text{K} \times T^{-1}$$

Thus we have our final result

$$k(T) = \left(44{,}400\,\text{W/m}\right) \times T^{-1}$$

Comparing to the given form $k(T) = aT^b$, we identify $a = 44{,}400$ W/m and $b = -1$, where T must be expressed in Kelvin.

(4) Checking. *Dimensional check.* The right-hand side of the last equation has units of K^{-1} - W/m, which is equivalent to the expected units of W/m - $°C$. *Magnitude check.* Our calculation can be compared to the standard reference values listed in Appendix D as follows

T	k (our model)	k (reference)	Error = (model-ref)/(ref)
300 K	148	148	0%
600 K	74	61.9	20%
1000 K	44.4	31.2	42%

(5) Comments. (1) Once again the mean free path is best measured in nanometers. Values in the range of tens to hundreds of nm are typical for phonons in dielectric crystals at room temperature. As discussed later in Example 11.5, a more realistic estimate for the average mean free path of acoustic phonons in Si at 300 K is around 200-300 nm. (2) Using the density of silicon $\rho = 2330\,kg/m^3$ from Appendix D, we can convert the acoustic phonon specific heat to a mass basis, finding $446\,J/kg$ - $°C$. For comparison, the handbook value listed in Appendix D is $712\,J/kg$ - $°C$. This indicates that the optical phonons are making a significant contribution to the total specific heat. (Indeed, a more exact calculation gives the Debye acoustic specific heat as $380\,J/kg$ - $°C$ and the optical contribution as $298\,J/kg$ - $°C$, for a total of $678\,J/kg$ - $°C$, within 5% of the value from Appendix D.) (3) Clearly the T^{-1} power law is only approximate. Over this temperature range the actual thermal conductivity varies by a factor of 4.74, compared to our expected variation by a factor of 3.33. A better power law in this temperature range would be $k \propto T^{-1.3}$. (4) The power law is strongly dependent on temperature. For example, below 10 K the power law for bulk silicon is approximately $k \propto T^3$.

Example 11.3: Bulk Mean Free Paths as a Function of Temperature.

Estimate Λ for acoustic phonons in bulk silicon as a function of temperature. Consider T ranging from 1 K to 1000 K. Use the approximate Debye model for the specific heat and sound velocity. The thermal conductivity of bulk Si is given in Appendix D.

(1) Observations. This problem spans a large temperature range both far above and far below θ_D. Above room temperature we expect the mean free path to be dominated by phonon-phonon scattering with a power law of approximately $\Lambda \propto T^{-1}$, or, based on the results of Example 11.2, $\Lambda \propto T^{-1.3}$. We are not sure what to expect at low temperature, but clearly there is something interesting because the thermal conductivity climbs as T is reduced from 300 K down to 20 K, but then k falls rapidly as T is further reduced below 20 K.

(2) Formulation.

(i) Assumptions. (1) The thermal conductivity is dominated by acoustic phonons over this entire temperature range. (2) The Debye model is an adequate approximation for the acoustic phonons.

(ii) Governing Equations. For the specific heat we will use eq. (11.17). The sound velocity will be taken from Table 11.2.

(3) Solutions.

The mean free path is given by

$$\Lambda = 3k / Cv .$$

From eq. (11.17),

$$C = \frac{3\eta_{PUC} k_B}{1 + \frac{5}{4\pi^4}\left(\dfrac{\theta_D}{T}\right)^3} ,$$

where from Table 11.2 we have $\eta_{PUC} = 2.5 \times 10^{28}$ m^{-3} and $\theta_D = 512\,\text{K}$. The sound velocity is $v = 5880\,\text{m/s}$. Combining these expressions we can calculate the mean free path from the data for k in Appendix D. The results are depicted in Fig. 11.3 and Table 11.4 on the following page.

(4) Checking. *Limiting behavior check.* The specific heat transitions from T^3 at low temperature to constant at high temperature, as expected from eqs. (11.19) and (11.20). Furthermore, the transition occurs at approximately 120 K, which is near the expected transition of $\theta_D / 3 \approx 170\,\text{K}$. The trends of mean free path and thermal conductivity are both approximately T^{-1} at high temperature, as expected from Example 11.2. *Value check.* At 300 K we can compare our values to those from Example 11.2. Our C is 9.73×10^5 J/m^3-$^\circ$C, which is close to the value of

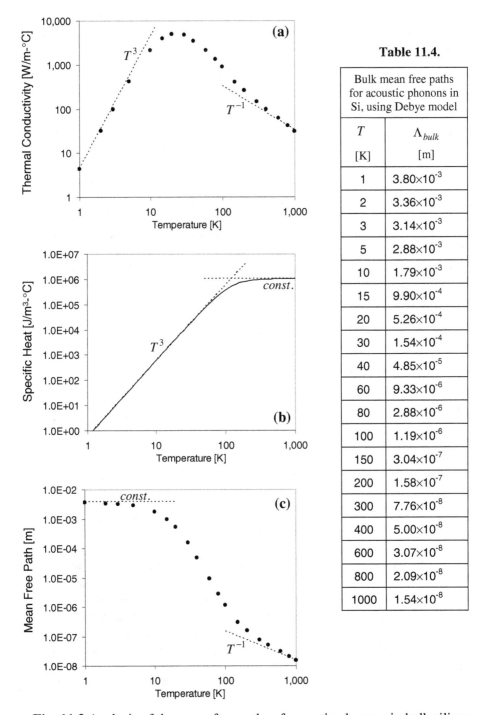

Table 11.4.

Bulk mean free paths for acoustic phonons in Si, using Debye model	
T [K]	Λ_{bulk} [m]
1	3.80×10^{-3}
2	3.36×10^{-3}
3	3.14×10^{-3}
5	2.88×10^{-3}
10	1.79×10^{-3}
15	9.90×10^{-4}
20	5.26×10^{-4}
30	1.54×10^{-4}
40	4.85×10^{-5}
60	9.33×10^{-6}
80	2.88×10^{-6}
100	1.19×10^{-6}
150	3.04×10^{-7}
200	1.58×10^{-7}
300	7.76×10^{-8}
400	5.00×10^{-8}
600	3.07×10^{-8}
800	2.09×10^{-8}
1000	1.54×10^{-8}

Fig. 11.3 Analysis of the mean free paths of acoustic phonons in bulk silicon, using the Debye model.

1.04×10^6 J/m^3-°C found in Example 11.2 using the high-temperature approximation. Our Λ is 78 nm, which is also close to the value of 73 nm found in Example 11.2 using the high temperature approximation.

(5) Comments. The low-temperature behavior is quite distinctive. The mean free path at low temperature appears to saturate at an almost constant value of around 4 mm. Because the low-T specific heat goes as T^3, this causes the low-T thermal conductivity to also go as T^3. This cubic trend is quite general for phonon thermal conductivity at very low temperature. As will become clear in Section 11.4 and Example 11.6, the reason is that even this nominally "bulk" sample transitions to behavior dominated by the classical size effect at sufficiently low temperatures. In fact, the low-temperature value of $\Lambda \approx 4$ mm corresponds to the characteristic length of the sample measured to generate the values of "bulk" k reported in Appendix D. It is important to recognize that other references for "bulk" silicon may report different values at low temperature, if their sample size is different, but the reported values at high temperature should be in very close agreement amongst different references.

11.3.4   Radiation: Heat is Carried by Photons (Light Waves)

Radiation heat transfer is usually studied as a phenomenon completely unrelated to conduction heat transfer. However, part of the beauty of the microscale perspective using kinetic theory is that radiation and conduction can be seen simply as two limiting cases of a single general phenomenon. This equivalence will become clearer in the sections below on boundary scattering and the classical size effect. For now we will simply review several major features of radiation heat transfer, focusing on the example of radiation between two parallel plates in a vacuum. We treat the radiation as a gas of photons.

Speed.   The speed of light in vacuum is a physical constant, $c = 2.998 \times 10^8$ m/s .

Specific Heat. Although not usually discussed in engineering thermodynamics or heat transfer, photons store energy just as do molecules, electrons, and phonons. Consider a vacuum chamber at absolute temperature T. The higher the temperature, the greater the number of photons inside the chamber, and the greater the energy per photon on average. Assuming a perfect vacuum inside the chamber, the specific heat of this photon gas is given by

$$C = 16\sigma T^3 / c , \qquad (11.21)$$

regardless of the emissivity of the walls of the chamber.   Here $\sigma = 5.670 \times 10^{-8}$ W/m^2-K^4 is the Stefan-Boltzmann constant. (Note that photons have no mass, so it would not be meaningful to attempt to convert the photon specific heat from J/m^3 - K to J/kg - K .)

Mean Free Path. The mean free path of photons varies tremendously but is usually much longer than the mean free path of molecules, electrons, and phonons. For example, photons emitted from the surface of the Sun travel over 90 million miles to reach the Earth without any scattering. In a typical vacuum chamber, on the other hand, the walls may be approximated as parallel plates with a gap of perhaps 10-100 mm. In this case the corresponding photon mean free path should also be of this approximate magnitude. Finally, heat transfer in crystals that are practically transparent in the infrared, such as glasses, may also have important contributions from the photons within the crystal itself, especially at high temperatures. In this case the photon mean free path usually depends strongly on the wavelength and material, with typical values of Λ in the range of microns to millimeters.

Example 11.4: Effective Thermal Conductivity for Radiation Heat Transfer between Two Parallel Plates (One Black, One Gray).

From elementary heat transfer, we know that the net radiation heat transfer from a gray plate "1" to a parallel black plate "2" is given by

$$Q = \varepsilon\sigma A\left(T_1^4 - T_2^4\right), \qquad\qquad (11.22)$$

where ε is the emissivity of plate 1, both plates have the same area A, and the gap L between the plates is much smaller than their length and width. Using the kinetic theory framework, re-express the heat transfer in terms of a "conduction thermal resistance" $R = L/k_{rad} A$, where k_{rad} will represent the effective thermal conductivity of the photon gas. Also derive an effective mean free path for the photons. You may assume that the temperature differences are much smaller than the average temperature. Evaluate your results numerically for a pair of plates with $A = 0.1\,m^2$, $L = 0.001\,m$, $\varepsilon = 0.2$, and $T_1 = 600\,K$, $T_2 = 500\,K$.

(1) Observations.

It is initially surprising that we might be able to express radiation as conduction, but considering that we have an expression for the speed and specific heat of photons, it should be possible. We will proceed carefully. Because in vacuum there are no photon scattering mechanisms other than at the plates themselves, we expect that the photon mean free path should be somehow proportional to the gap L.

(2) Formulation.

(i) Assumptions. (1) Small temperature differences:

$$\left|T_1 - T_2\right| << \tfrac{1}{2}\left(T_1 + T_2\right).$$

(2) Properties like the specific heat can be evaluated at the average temperature, $T = \tfrac{1}{2}\left(T_1 + T_2\right)$.

(ii) Governing Equations. We will use the radiation heat transfer equation (11.22), the kinetic theory equation (11.4), and the specific heat of photons from eq. (11.21).

(3) Solutions.

Conduction resistance. We need to rearrange eq. (11.22) into the familiar form

$$\left(T_1 - T_2\right) = RQ,$$

and the resulting coefficient R will be the desired conduction resistance. The challenge is obtaining an equation in $(T_1 - T_2)$ from our starting point of $\left(T_1^4 - T_2^4\right)$. First define the temperature difference $\Delta = T_1 - T_2$. Thus, $T_1 = T + \tfrac{1}{2}\Delta$, and $T_2 = T - \tfrac{1}{2}\Delta$. Substituting these into eq. (11.22), we have

$$Q = \varepsilon\sigma A\left[\left(T + \frac{1}{2}\Delta\right)^4 - \left(T - \frac{1}{2}\Delta\right)^4\right].$$

Factoring out T^4, we have

$$Q = \varepsilon\sigma A T^4\left[\left(1 + \tfrac{\Delta}{2T}\right)^4 - \left(1 - \tfrac{\Delta}{2T}\right)^4\right].$$

From the binomial theorem, or equivalently, a Taylor series expansion, we know that $(1 + x)^n \approx 1 + nx$, if $x << 1$. Thus, because $\Delta << T$, we have $\left(1 + \tfrac{\Delta}{2T}\right)^4 \approx 1 + \tfrac{2\Delta}{T}$. Then

$$Q = \varepsilon\sigma A T^4\left[\left(1 + \tfrac{2\Delta}{T}\right) - \left(1 - \tfrac{2\Delta}{T}\right)\right].$$

Simplifying,

$$Q = 4\Delta\varepsilon\sigma A T^3.$$

or

$$T_1 - T_2 = \frac{Q}{4\varepsilon\sigma AT^3}.$$

Thus, we have found the "conduction resistance"

$$R = \frac{T_1 - T_2}{Q} = \frac{1}{4\varepsilon\sigma AT^3}. \tag{11.23}$$

Effective thermal conductivity. By comparing eq. (11.23) to the standard form $R = L/kA$, we can solve for k_{rad}:

$$\frac{1}{4\varepsilon\sigma AT^3} = \frac{L}{k_{rad}A}$$

$$k_{rad} = 4\varepsilon\sigma T^3 L.$$

Effective mean free path. We now compare k_{rad} to the standard form $k = \frac{1}{3}Cv\Lambda_{eff}$, to solve for Λ_{eff}:

$$k_{rad} = 4\varepsilon\sigma T^3 L = \frac{1}{3}Cv\Lambda_{eff}$$

$$\Lambda_{eff} = \frac{12\varepsilon\sigma T^3 L}{Cv}$$

Substituting eq. (11.21) for C and recognizing that $v = c$, we have

$$\Lambda_{eff} = \frac{12\varepsilon\sigma T^3 L}{\left(16\sigma T^3/c\right)v} = \frac{3}{4}\varepsilon L. \tag{11.24}$$

Numerical calculation. From the exact radiation equation, we find that $Q = 76.1\,\text{W}$. From the conduction resistance equation, using $T = 550\,\text{K}$, we find $R = 1.33\,\text{K/W}$, resulting in $Q = 75.5\,\text{W}$, within 1% of the exact value.

(4) Checking. *Dimensional check.* Our conduction resistance has units of $[(\text{K}^4\text{-}\text{m}^2\text{-}\text{W}^{-1})(\text{m}^{-2})(\text{K}^{-3})]$, correctly yielding the expected units of $[\text{K/W}]$. The effective mean free path has dimensions of length. *Trend check: conduction resistance.* The equivalent conduction resistance is inversely proportional to area, which makes sense because reducing the area must reduce the heat transfer, thus increasing the thermal resistance. We also note that the conduction resistance is inversely proportional to the emissivity ε. This also makes sense, because from eq. (11.22) we know that reducing the emissivity should reduce the heat transfer in equal proportion. *Trend check: mean free path.* We see that the effective photon mean free path is simply proportional to the gap L: Doubling the gap will double the effective mean free path, which is physically satisfying.

(5) Comments.   (1) This example reveals how classical radiation heat transfer through a vacuum can sometimes be viewed as heat conduction, if k_{rad} is defined appropriately. (2) The effective thermal conductivity of this photon gas is proportional to the cube of temperature.   This can be understood by recalling that higher temperatures result in many more photons, of higher energy, which are available to transport the heat. You may also have heard the assertion in introductory heat transfer classes that "radiation is only important at high temperature," which is qualitatively consistent with our result here. (3) The effective mean free path requires further consideration. We see from eq. (11.24) that Λ_{eff} is proportional to the emissivity. How can a shinier plate result in a shorter Λ_{eff} even though the gap L remains constant? The answer to this apparent paradox is that Λ_{eff} is not only a function of the geometry of the system (in this case the gap between plates), but it also describes how effectively energy is *exchanged* between the two plates, per photon collision.   Thus the distinction here between L and Λ_{eff} is analogous to the distinction between Λ_{coll} and Λ_{en} made in Section 11.3.1 in the context of gas molecules. We know from eq. (11.22) that reducing the plate emissivity must reduce the radiation energy exchange. In the conduction framework, reducing the energy exchange must correspond to reducing the effective thermal conductivity. Because the specific heat and speed of light are independent of the plate emissivity, we are left with the effective mean free path as the only term which can capture the emissivity effect. (4) It is interesting to evaluate the numerical values of C and k separately. From eq. (11.21), $C = 5.03 \times 10^{-7} \text{ J/m}^3 \text{ -}^\circ\text{C}$.   Apparently the specific heat of photons is extremely small, which is why photons are usually ignored for energy storage. However, when evaluating heat transfer, we must recall that this small specific heat is countered by the very large speed, $c = 3.00 \times 10^8 \text{ m/s}$. For this particular problem, the effective mean free path is 0.150 mm, resulting   in   an   effective   thermal   conductivity   of $k = \frac{1}{3} Cv\Lambda_{eff} = 0.00754 \text{ W/m -}^\circ\text{C}$.   This is about 5 times lower than the thermal conductivity of air. However, the effective k_{rad} can vary by many orders of magnitude depending on the temperature and geometry of a particular problem. (5) The result of eq. (11.24) that $\Lambda_{eff} = \frac{3}{4}L$ for $\varepsilon = 1$ (that is, for perfectly absorbing, plane-parallel boundaries) will come up again later in the chapter in the context of heat conduction by electrons or phonons perpendicular to thin solid films. (6) If the vacuum between the plates is replaced by a dense gas or other medium that scatters the photons significantly, Λ_{eff} will be reduced, and so will the radiation heat transfer.

11.4   Thermal Conductivity Reduction by Boundary Scattering: The Classical Size Effect

With the exception of Example 11.4, our discussion thus far has focused on "bulk" materials: systems that are much larger than the mean free path of the energy carrier. In bulk materials the thermal conductivity is a property of the material, but is independent of the exact size and shape of the sample. In this case we can write

$$k_{bulk} = \frac{1}{3} C v \Lambda_{bulk} .$$

As we have seen above, typical Λ_{bulk} for molecules, electrons, and phonons are in the range of $10-1000\,\text{nm}$ at $300\,\text{K}$. For traditional applications with sample sizes ranging from millimeters to meters, the bulk mean free path is several orders of magnitude smaller than the sample, and the handbook values of thermal conductivity are appropriate. However, in many modern technologies the characteristic length of a structure may easily be as small as $10-100\,\text{nm}$. In this case the energy carriers will collide frequently with the boundaries of the structure. This is the essence of the "classical size effect" depicted in Fig. 1(c): the boundary collisions impede the transport of heat, which may lead to dramatic reductions in the thermal conductivity. The remainder of this chapter is dedicated to further exploring the classical size effect.

11.4.1 Accounting for Multiple Scattering Mechanisms: Matthiessen's rule

For a micro- or nano-structure where the classical size effect is important, we will still express the effective thermal conductivity k as

$$k = \frac{1}{3} C v \Lambda_{eff} ,$$

where Λ_{eff} is the effective mean free path. Note the very important fact that the speed and specific heat of the energy carriers in the nanostructure will be approximated as identical to the values in a bulk sample. That is,

$C_{eff} \approx C_{bulk}$ and $v_{eff} \approx v_{bulk}$, allowing us to focus our efforts purely on determining Λ_{eff} (which will be smaller than Λ_{bulk}). This tremendous simplification requires that quantum size effects be negligible, which means that situations such as Fig. 11.1(d) are forbidden.

An exact calculation of Λ_{eff} is a demanding task, usually involving the Boltzmann transport equation, that depends on the details of the geometry, boundary conditions, and other scattering mechanisms. Fortunately a powerful approximation is available that has good accuracy, provides excellent physical insight, is widely used, and can be adapted to represent the results of the more detailed solutions found from the Boltzmann equation. This approximation is known as Matthiessen's rule:

$$\Lambda_{eff}^{-1} = \sum_{\text{mechanism } i} \Lambda_i^{-1} . \qquad (11.25)$$

Matthiessen's rule simply states that the effective mean free path Λ_{eff} is made up of a sum of the mean free paths corresponding to each of the various scattering mechanisms, all summed in a reciprocal sense. The sum will include as many terms as there are distinct scattering mechanisms in the system. For example, phonons may scatter on impurities, on grain boundaries, on electrons, on other phonons, and on the sample boundaries, in which case there would be five terms in the sum of eq. (11.25):

$$\Lambda_{eff}^{-1} = \underbrace{\Lambda_{imp.}^{-1} + \Lambda_{g.b.}^{-1} + \Lambda_{ph-e}^{-1} + \Lambda_{ph-ph}^{-1}}_{\Lambda_{bulk}^{-1}} + \Lambda_{bdy}^{-1} .$$

However, for the purposes of this chapter we can partition all possible scattering mechanisms into two simple categories: all of those *mechanisms that are present in a large, bulk sample* (denoted Λ_{bulk}, representing for example the effect of the first 4 terms on the right-hand side of the above equation), and the *additional scattering mechanism(s) due to "boundary scattering"* (denoted Λ_{bdy}) *which is only present in small samples.*

Therefore, we will focus on Matthiessen's rule expressed as

$$\Lambda_{eff} = \left(\Lambda_{bulk}^{-1} + \Lambda_{bdy}^{-1} \right)^{-1} = \frac{\Lambda_{bulk}\Lambda_{bdy}}{\Lambda_{bulk} + \Lambda_{bdy}} . \qquad (11.26)$$

We have already seen in Section 11.3 how to evaluate Λ_{bulk} for various types of energy carriers. Thus, the major remaining task is to determine Λ_{bdy} for various nanostructures, such as nanowires and thin films. We will break the discussion into two common cases (Fig. 11.4): heat transport parallel to boundaries, such as heat flow along the plane of a thin film, and heat transport perpendicular to boundaries, such as heat flow perpendicular to a thin film.

11.4.2 Boundary Scattering for Heat Flow Parallel to Boundaries

As shown in Fig. 11.4(a), standard configurations for heat flow parallel to boundaries include nanowires and thin films.

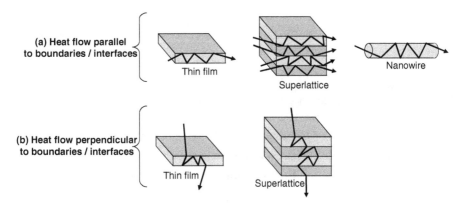

Fig. 11.4

Nanowires

For heat transport along a nanowire of diameter D, the mean free path due to boundary scattering is given by [13]

$$\Lambda_{bdy} = D\left(\frac{1+p}{1-p}\right),\tag{11.27}$$

where p is the *specularity*, defined as follows (Fig. 11.5):

   $p = 0$: scattering is 100% diffuse (rough surfaces)
   $p = 1$: scattering is 100% specular (smooth surfaces)
   $0 < p < 1$: scattering is p specular and $(1-p)$ diffuse

In this definition, "specular" scattering refers to mirror-like reflections, where the angle of incidence equals the angle of reflection. "Diffuse" scattering refers to reflections off of a very rough surface, such that incident particles are reflected with equal intensity in all directions (also

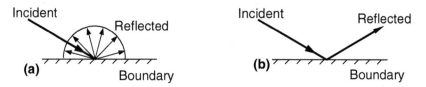

Fig. 11.5 Specularity. (a) Diffuse reflection ($p=0$): Reflection intensity is equal in all directions. (b) Specular reflection ($p=1$): Angle of reflection equals angle of incidence.

known as "Lambertian" or "cosine-law" reflection), regardless of the angle of incidence.

The specularity depends strongly on the surface roughness as compared to the wavelength of the energy carrier (molecule, electron, phonon, photon). For surface roughness much smaller than the wavelength, reflections will be specular ($p = 1$). For roughness larger than the wavelength, reflections will be diffuse ($p = 0$). For intermediate roughness, the specularity is commonly estimated using the following expression from Ziman [13]:

$$p = \exp\left(\frac{-16\pi^3\delta^2}{\lambda^2}\right),\qquad\qquad (11.28)$$

where δ is the root-mean-square (rms) roughness, and λ is the wavelength of the energy carrier [13,26]. At low temperatures, the wavelengths of most energy carriers become significantly longer than their 300 K values (see Problem 11.8).

Thin films (Diffuse: $p=0$)

Analysis of heat transport along a thin film of thickness L is best done using the Boltzmann transport equation. Fortunately, the results are easily expressed in the framework of kinetic theory by using equivalent mean free paths. For purely diffuse scattering ($p = 0$), the standard result is known as the Fuchs-Sondheimer solution [6,7,27-29]:

$$\frac{\Lambda_{eff}}{\Lambda_{bulk}} = 1 - \frac{3\Lambda_{bulk}}{8L}\left[1 - 4E_3\left(\frac{L}{\Lambda_{bulk}}\right) + 4E_5\left(\frac{L}{\Lambda_{bulk}}\right)\right], \qquad (11.29)$$

where E_3 and E_5 are "exponential integrals," special functions defined through

$$E_n(x) = \int_0^1 \mu^{n-2}\exp(-x/\mu)\,d\mu,$$

where μ is a dummy variable of integration. The exponential integrals E_3 and E_5 are depicted in Fig. 11.6, and these functions are also available tabulated and as functions in mathematical software packages such as Maple and Mathematica. Although eq. (11.29) can be combined with Matthiessen's Rule and solved for an effective boundary scattering mean free path Λ_{bdy}, it is usually more convenient to work with eq. (11.29) directly. For thick and thin films, the following asymptotic forms are also useful [28]

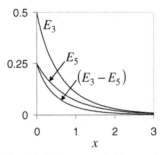

Fig. 11.6 Exponential integrals.

$$\frac{\Lambda_{eff}}{\Lambda_{bulk}} = \left(1 + \frac{3\Lambda_{bulk}}{8L}\right)^{-1} \qquad L \gg \Lambda_{bulk}, \qquad (11.30a)$$

$$\frac{\Lambda_{eff}}{\Lambda_{bulk}} = \frac{3L\ln(\Lambda_{bulk}/L)}{4\Lambda_{bulk}} \qquad L \ll \Lambda_{bulk}. \qquad (11.30b)$$

Compared to the exact solution, eq. (11.30a) has errors less than 10% for $L > 0.5\Lambda_{bulk}$, and errors less than 20% for L as small as $0.025\Lambda_{bulk}$. Similarly, eq. (11.30b) has errors less than 10% for $L < 0.019\Lambda_{bulk}$, and errors less than 20% for L as large as $0.125\Lambda_{bulk}$. Thus, for many practical calculations (tolerant of errors up to 20%), if evaluating the exponential integrals is inconvenient it would always be reasonable to use one of the limiting forms of eqs. (11.30).

Thin films (Some specularity: $p>0$)

When the film specularity becomes significant, the effective thermal conductivity increases. For $p>0$, the effective mean free path is given by the following integral [6,7]:

$$\frac{\Lambda_{eff}}{\Lambda_{bulk}} = 1 - \frac{3\Lambda_{bulk}(1-p)}{2L}\int_0^1 \left(\mu-\mu^3\right)\frac{1-\exp(-L/\mu\Lambda_{bulk})}{1-p\exp(-L/\mu\Lambda_{bulk})}d\mu . \quad (11.31)$$

This integral is readily evaluated numerically. The resulting Λ_{eff} are shown in Fig. 11.7 for a range of specularities. Note that the $p=0$ case reduces correctly to eq. (11.29), while in the $p=1$ case there is no reduction in k compared to bulk regardless of the value of L.

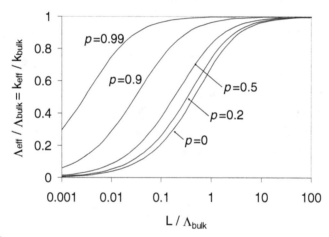

Fig. 11.7 The Fuchs-Sondheimer solution for heat transfer along a thin film, as a function of film thickness L and interface specularity p.

Example 11.5: Thermal conductivity of a silicon nanowire.

A silicon nanowire has diameter D=56 nm. What are the specularity, p, and the effective thermal conductivity, k, at $T = 300\,K$? You may assume that the surface roughness is approximately $\delta = 0.5\,nm$, and the average phonon wavelength at this temperature is $\lambda = 1\,nm$.

(1)   Observations.   (1) The roughness is slightly smaller than the wavelength, so we expect that the specularity may be somewhere in the transition regime $0<p<1$. (2) From Table 11.4 we know that the mean free path of acoustic phonons in bulk silicon using our Debye approximation is $\Lambda_{bulk}(300\,\text{K}) \approx 78\,\text{nm}$. Because the diameter of the nanowire is comparable to Λ_{bulk}, we expect that the thermal conductivity may be significantly reduced compared to bulk.

(2) Formulation.

(i) Assumptions. (1) The thermal conductivity is dominated by acoustic phonons, which are adequately represented by our Debye model. (2) The specific heat and phonon speed in the nanowire are identical to those in bulk.

(ii) Governing Equations. The specularity will be calculated using eq. (11.28), the boundary scattering mean free path from eq. (11.27), and the effective mean free path found from Matthiessen's Rule, eq. (11.26).

(3) Solutions.

Specularity. From eq. (11.28), for $\delta / \lambda = 0.5$, we have

$$p = \exp\left(\frac{-16\pi^3\delta^2}{\lambda^2}\right) = \exp\left(-16\pi^3(0.5)^2\right) = \exp(-124) = 1.37 \times 10^{-54}$$

which for all practical purposes is zero.

Boundary scattering mean free path. From eq. (11.27), with $p = 0$ we have

$$\Lambda_{bdy} = D\left(\frac{1+p}{1-p}\right) = D\left(\frac{1+0}{1-0}\right) = D = 56\,\text{nm} .$$

Effective mean free path. From Matthiessen's Rule, eq. (11.26), we have

$$\Lambda_{eff} = \frac{\Lambda_{bulk}\Lambda_{bdy}}{\Lambda_{bulk} + \Lambda_{bdy}} = \frac{(78\text{nm})(56\text{nm})}{78\text{nm} + 56\text{nm}} = 33\,\text{nm} .$$

Thermal conductivity. Although we could evaluate k from C, v, and Λ_{eff}, it is insightful to first consider the ratio of k / k_{bulk}:

$$\frac{k}{k_{bulk}} = \frac{\frac{1}{3}Cv\Lambda_{eff}}{\frac{1}{3}Cv\Lambda_{bulk}} .$$

Because we assume that C and v are independent of the structure size, we have simply

$$\frac{k}{k_{Bulk}} = \frac{\Lambda_{eff}}{\Lambda_{bulk}} = \frac{\Lambda_{bdy}}{\Lambda_{bulk} + \Lambda_{bdy}} = \frac{1}{1 + (\Lambda_{bulk} / \Lambda_{bdy})},$$

or, using $\Lambda_{bdy} = D$,

$$\frac{k}{k_{bulk}} = \frac{1}{1 + (\Lambda_{bulk} / D)}. \tag{11.32}$$

Thus, $k / k_{bulk} = 0.42$, or $k = 62$ W/m-°C.

(4) Checking. *Dimensional check.* Because we have expressed all equations as dimensionless ratios, there are no dimensions to check separately. *Limiting behaviors.* When the diameter is much larger than the bulk mean free path, eq. (11.32) reduces to $k = k_{bulk}$, correctly showing that the diameter no longer matters in such case. In the opposite limit, $D \ll \Lambda_{bulk}$, we have $k = k_{bulk} D / \Lambda_{bulk}$, which can be written $k = \frac{1}{3} CvD$. In this case further reductions in the nanowire diameter should reduce the thermal conductivity in direct proportion. This also makes sense: a smaller diameter results in increased scattering, reducing thermal conductivity.

(5) Comments. (1) A nanowire like this has actually been measured by Li and co-workers [2], and selected values of their $k(T)$ have been reproduced in Appendix D. Their result at 300 K is 25.7 W/m-°C, which is less than half of our estimate. The reason for the discrepancy is that our Debye approach to estimating the bulk mean free path of acoustic phonons (78 nm) is too conservative. (2) A better estimate for the average bulk mean free path in Si at 300 K is $\Lambda_{bulk} \approx 200\text{-}300$ nm [30-32], which takes into account the frequency dependence of the C and v contributions from the various acoustic phonons. Taking $\Lambda_{bulk} = 250$ nm as a representative average value and repeating the calculation, we would find $k = 27$ W/m-°C, in remarkably good agreement with the experimental value (in fact, it is fortuitous that the agreement should be this good). (3) It is common that the Debye estimate for the bulk mean free path is too low by a factor of 2 or more.   A better analytical approach is the Born-von Karman approximation [32], which in this case gives $\Lambda_{bulk} = 210$ nm and $k = 31$ W/m-°C.   However, the Born-von Karman approach requires a (simple) numerical integration to evaluate Cv, so we do not pursue it here.

Example 11.6: Temperature Dependence of the Thermal Conductivity of a Nanowire.

For Si nanowires of several diameters, plot the thermal conductivity as a function of temperature. Consider T from 1 K to 1000 K. Use the approximate Debye model for the specific heat. You will also need the bulk mean free path as a function of temperature: use the results from Example 11.3. Assume purely diffuse scattering. Consider three different diameters:

$$Macro\text{-}wire(D=1\ mm)$$

$$Micro\text{-}wire\ (D=1\ \mu m)$$

$$Nanowire\ (D=56\ nm)$$

Compare your calculations for the 56 nm wire with the experimental measurements of Ref. [2], some of which are included in Appendix D .

(1) Observations. Based on the results of Example 11.3, we know that boundary scattering dominates the thermal conductivity at low temperature. For small samples, boundary scattering becomes even stronger, so we expect the thermal conductivity to be reduced as D is reduced.

(2) Formulation.

(i) Assumptions. (1) The thermal conductivity is dominated by acoustic phonons, which are adequately represented by our Debye model. (2) The specific heat and phonon speed in the nanowire are identical to those in bulk. (3) Boundary scattering is perfectly diffuse $(p = 0)$.

(ii) Governing Equations. For the specific heat we will use eq. (11.17). The sound velocity will be taken from Table 11.2. The boundary scattering mean free path will be calculated from eq. (11.27), the bulk mean free paths will be taken from Table 11.4, and the effective mean free path calculated using Matthiessen's rule, eq. (11.26).

(3) Solutions.

From eq. (11.27), as in Example 11.5, with $p = 0$ we have $\Lambda_{bdy} = D$. The bulk mean free paths were tabulated as part of Example 11.3, allowing us to calculate Λ_{eff} from Matthiessen's Rule as

$$\Lambda_{eff} = \frac{\Lambda_{bulk}\Lambda_{bdy}}{\Lambda_{bulk} + \Lambda_{bdy}} = \frac{\Lambda_{bulk}D}{\Lambda_{bulk} + D} = \Lambda_{bulk}\left(1 + \frac{\Lambda_{bulk}}{D}\right)^{-1}.$$

Similarly, multiplying both sides by $\frac{1}{3}Cv$, we have

$$k_{eff} = k_{bulk}\left(1 + \frac{\Lambda_{bulk}}{D}\right)^{-1}.$$

The resulting mean free paths and thermal conductivities are shown in Fig. 11.8.

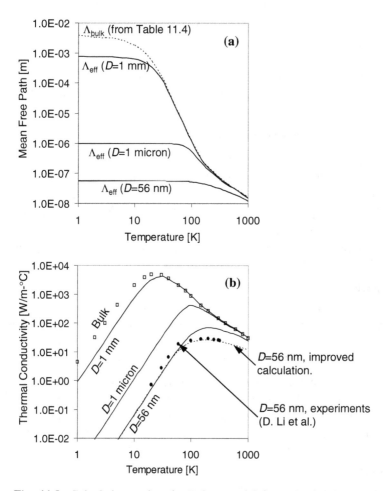

Fig. 11.8 Calculations using the Debye model for acoustic phonons in silicon wires of various diameters.   (a) Phonon mean free paths.   (b) Thermal conductivity. Also included in (b) are: reference values (squares) for bulk [1] from Appendix D; measured values (dots) for a 56 nm nanowire from [2] (given in Appendix D); and the result of an improved calculation (dashed line) for the 56 nm diameter wire [33].

(4) Checking. *Trend check.* The calculated thermal conductivities of the nanowires have the same general shape as a function of T as the recommended data for bulk Si: increasing as T^3 at low temperature, reaching a peak value, and then decreasing at high temperature. Similarly, the effective mean free paths become constant ($\Lambda_{eff} \approx D$) at low temperature, and decrease at high temperature. As expected, reducing the diameter leads to lower thermal conductivity. At very low temperature, we can see that $k \propto D$, as expected.

(5) Comments. (1) Note that the diameter effect is much more dramatic at low T than high T. This is because the bulk mean free paths are short at high T, so boundary scattering is relatively less important. (2) Reducing the diameter reduces the peak value of the $k(T)$ curve and also shifts the peak to a higher temperature. (3) It is interesting to consider the limiting behavior for even smaller diameters. At 1000 K the bulk mean free path in our Debye approximation for silicon is around 15 nm. If we consider a diameter 10× smaller than this, namely D=1.5 nm, then boundary scattering dominates over the entire temperature range from $1-1000\,\text{K}$. In this case the effective mean free path simply becomes a constant over the entire temperature range. Returning to eq. (11.4), if Λ and v are now both constants, then the shape of $k(T)$ would simply reproduce the shape of $C(T)$: cubic at low temperature, and constant at high temperature (Fig. 11.3b). (4) On the other hand, the limiting behavior for infinitely large diameters is that the thermal conductivity will diverge ("blow up") at low temperatures, because all of the bulk scattering mechanisms become continually weaker as $T \rightarrow 0\,\text{K}$. That is, if $D \rightarrow \infty$, by reducing T we find that Λ_{bulk} becomes larger even "faster" than C becomes smaller. (5) The comparison in Fig. 11.8(b) with the experimental results of Li *et al.* [2] (points) shows that our simple Debye model (solid lines) explains all the major trends of the experiments, including the reduction in k by more than a factor of 6000 at low T. However, the disagreement between model and experiment can be more than a factor of 2 at higher temperatures. As explained in the comments of Example 11.5, a more careful analysis using the Born-von Karman approach gives better agreement with experiment, especially if the frequency-dependence of the mean free paths is taken into account (dashed line in Fig. 11.8(b) [33].)

11.4.3 Boundary Scattering for Heat Flow Perpendicular to Boundaries

As shown in Fig. 11.4(b), the standard configurations for heat flow perpendicular to boundaries include a thin film or a superlattice (a stack of thin films with some repeating unit). Although the kinetic theory framework is in general less appropriate for these configurations (as compared to transport parallel to boundaries), kinetic theory still gives useful physical insight which, in simple cases such as a thin film with no heat generation, is also accurate.

Thin films with no heat generation

For heat transport perpendicular to a thin film of thickness L with no heat generation, the heat flow can be described using an effective thermal conductivity obtained from kinetic theory and Matthiessen's Rule, as long as the boundary scattering mean free path is expressed as

$$\Lambda_{bdy} = \frac{\frac{3}{4} L}{\alpha_1^{-1} + \alpha_2^{-1} - 1}, \qquad (11.33)$$

where α_1 and α_2 represent the absorptivities of the two bounding surfaces. Note that the concept of emissivity and absorptivity has been generalized from photon radiation to other types of energy carriers (recall also Example 11.4). The absorptivity must range from 1 ("black" surface) to 0 (perfectly reflecting surface), and represents the fraction of incident energy that is absorbed at the surface. In the context of heat conduction by gases, the absorptivities can be replaced by the "energy accommodation coefficients" [19-21]. We assume that emissivity and absorptivity are approximately equal after averaging over wavelength and direction (Kirchhoff's Law [34-36]).

Equation (11.33) is one expression of the so-called Rosseland diffusion approximation with Deissler jump boundary conditions, which is commonly derived in the field of radiation heat transfer in a participating medium [34-36], or using the Boltzmann transport equation [6,7]. We have simplified the result here to be expressed in the form of a mean free path for boundary scattering.

Example 11.7: Thermal Conductivity Perpendicular to a Silicon Thin Film.

Consider a silicon film at 300 K and with thickness L, sandwiched between two heat sinks. Assuming that the phonon absorptivities of the two contacts are both perfect, plot the effective thermal conductivity as a function of film thickness, for L ranging from 1 nm to 1 mm. Also plot the conduction thermal resistance for a sample of (1 cm x 1 cm) cross sectional area.

(1) Observations. Based on Table 11.4, we know that $\Lambda_{bulk} = 78$ nm for Si in our Debye approximation at 300 K, although the comments of Example 11.5 suggest that $\Lambda_{bulk} = 250$ nm would be a better choice. In either case, because the thinnest films of the problem $(L = 10$ nm) are significantly thinner than Λ_{bulk}, we expect that the effective thermal conductivity will be significantly reduced compared to the bulk value of k.

(2) Formulation.

(i) Assumptions. (1) The thermal conductivity is dominated by acoustic phonons, which can be approximated using the Debye model. (2) The specific heat and phonon speed in the thin film are identical to those in bulk. (3) Phonon absorptivities and emissivities at both contacts can be approximated as unity. No other contact resistance effects will be considered. (4) The bulk mean free path at 300 K will be approximated as $\Lambda_{bulk} = 250\,\text{nm}$.

(ii) Governing Equations. The specific heat is found from eq. (11.17). The boundary scattering mean free path is given by eq. (11.33), and the effective mean free path by Matthiessen's rule, eq. (11.26).

(3) Solutions. The functional dependencies of various quantities on the film thickness L are summarized in the plots of Fig. 11.9.

Specific heat. Same as bulk. Using eq. (11.17), with $\theta_D = 512\,\text{K}$ and $\eta_{PUC} = 2.50 \times 10^{28}\,\text{m}^{-3}$ from Table 11.2, at 300 K we have

$$C = \frac{3\eta_{PUC}k_B}{1+\frac{5}{4\pi^4}\left(\frac{\theta_D}{T}\right)^3} = \frac{3 \cdot 2.5\times10^{28}\,\text{m}^{-3} \cdot 1.381\times10^{-23}\,\text{J/K}}{1+\frac{5}{4\pi^4}\left(\frac{512\,\text{K}}{300\,\text{K}}\right)^3} = 9.73\times10^5\,\text{J/m}^3\text{-}^\circ\text{C}\,.$$

Speed. Same as bulk. From Table 11.2, $v = 5880\,\text{m/s}$.

Effective mean free path. For $\alpha_1 = \alpha_2 = 1$, eq. (11.33) simplifies to

$$\Lambda_{bdy} = \frac{\frac{3}{4}L}{\alpha_1^{-1}+\alpha_2^{-1}-1} = \frac{\frac{3}{4}L}{1+1-1} = \frac{3}{4}L\,.$$

Substituting this into eq. (11.26), we have

$$\Lambda_{eff} = \frac{\frac{3}{4}\Lambda_{bulk}L}{\Lambda_{bulk}+\frac{3}{4}L} = \frac{\Lambda_{bulk}}{1+\dfrac{4\Lambda_{bulk}}{3L}}\,.$$

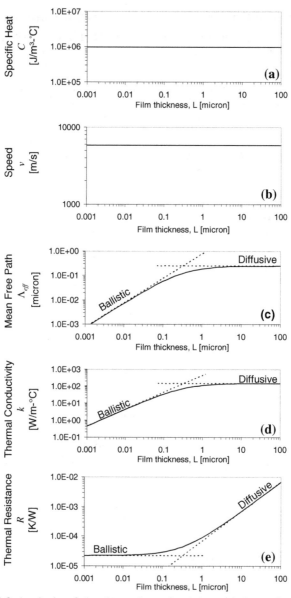

Fig. 11.9 Analysis of the thermal conductivity and thermal resistance perpendicular to a silicon thin film.

Thermal conductivity.   Combining C, v, and Λ_{eff} using eq. (11.4), we arrive at (Fig. 11.9)

$$k(L) = \tfrac{1}{3}Cv\Lambda_{eff} = \frac{\tfrac{1}{3}Cv\Lambda_{bulk}}{1 + \dfrac{4\Lambda_{bulk}}{3L}} = \frac{Cv\Lambda_{bulk}L}{3L + 4\Lambda_{bulk}}. \tag{11.34}$$

which can be rewritten

$$\frac{k(L)}{k_{bulk}} = \frac{1}{1 + \dfrac{4\Lambda_{bulk}}{3L}}. \tag{11.35}$$

Thermal resistance.   The conduction thermal resistance is defined as $R = L/kA$. Substituting from eq. (11.34), we have (Fig. 11.9)

$$R(L) = \frac{L}{kA} = \frac{L + \tfrac{4}{3}\Lambda_{bulk}}{k_{bulk}A}. \tag{11.36}$$

(4) Checking.   *Dimensional check.*   All terms in eq. (11.35) are dimensionless, as they must be. In eq. (11.36), both terms in the numerator have units of (m), and the denominator has units of $(W/m\text{-}^\circ C)(m^2)$, so that overall units are those of a conduction resistance: (K/W).   *Limiting behavior.*   In the limit $L \gg \Lambda_{bulk}$, eq. (11.35) correctly reduces to $k = k_{bulk} = 148\ W/m\text{-}^\circ C$ and eq. (11.36) correctly reduces to $R = L/k_{bulk}A$.

(5) Comments.   (1) The general sequence of plots in Fig. 11.9 is an excellent summary of the classical size effect. Although these plots were developed for heat conduction perpendicular to a thin film, the qualitative behavior and trends are quite general. For example, a calculation for the diameter-dependence of k in a copper nanowire would yield a similar collection of plots, except that the abscissa would be the nanowire diameter rather than the film thickness. (2) A subtle detail is that by eliminating the Cv product in eqs. (11.35) and (11.36), we have made the results for $k(L)$ and $R(L)$ more accurate. Because we know that the simplest Debye model estimate of $\Lambda_{bulk} \approx 78\ nm$ is too small, using the forms of eqs. (11.35) and (11.36) allows us to use the improved estimate $\Lambda_{bulk} \approx 250\ nm$, without worrying about the details of correcting for an effective C and v. (3) The thick-film limit, $L \gg \Lambda_{bulk}$, is also known as the *"diffusive"* regime, because the thermal resistance is dominated by the diffusion of energy carriers from the hot side to the cold side. In this case the thermal resistance is linearly proportional to the film thickness. (4) It is equally important to study the limiting behavior for a thin film, $L \ll \Lambda_{bulk}$. In this case eq. (11.35) shows that k is proportional to L. This is correct because further reducing the film thickness will increase the phonon scattering rate at the boundaries, thereby reducing the effective mean free path in equal proportion. Just as important is the fact that R becomes independent of L in the thin-film limit. This is known as the *"ballistic" regime*. In this limit there are effectively no scattering events within the film itself, and instead the thermal resistance is dominated by emission and absorption at the surfaces. This is identical to the traditional blackbody radiation resistance for photons traveling between two parallel plates in vacuum (recall Example 11.4). (5) In this example the transition between diffusive and ballistic regimes occurs when the film thickness is approximately equal to the bulk mean free path. More generally, it is clear from Matthiessen's rule, eq. (11.26), that the transition occurs when the bulk mean free path is comparable to the boundary scattering mean free path.

11.5 Closing Thoughts

Summary. We have used the kinetic theory framework of eq. (11.4) to introduce the essential concepts of conduction heat transfer by various energy carriers (gas molecules, electrons, phonons, and photons) in both bulk and nanostructures.   For any choice of energy carrier and nanostructure the key tasks are always the same: we must determine the specific heat C, the carrier velocity v, and the effective mean free path Λ_{eff}.

We have limited ourselves to the classical size effect: problems where the wavepacket of the energy carrier can be approximated as a particle, and the characteristic length L of the sample may be larger than, comparable to, or much smaller than the bulk mean free path Λ_{bulk} (Fig. 11.1(c)). In such problems we can use Matthiessen's Rule (eq. 11.26) to analyze the conduction heat transfer in terms of an effective thermal conductivity based on the effective mean free path Λ_{eff}. As summarized in Fig. 11.9(c-e) and Table 11.5 an important recurring theme is the transition from diffusive to ballistic behavior.

Suggestions for further study. This chapter has only addressed a tightly-focused subset of the rich pool of topics in microscale heat transfer [6-9,11].   Interested readers may like to pursue one or more of the following directions with natural links to the content of this chapter:

Improving kinetic theory by integrating over frequency. By expressing kinetic theory in the form of eq. (11.4), we have treated C, v, and Λ as lumped, frequency-independent quantities. This assumption captures the most important physics and all of the major trends. However, depending on the particular nanostructure and energy carrier, this assumption can sometimes lead to significant errors: it would not be surprising for this approach to result in an error of a factor of two or more when calculating the effective k. For example, one common scenario leading to significant errors is if the value of C for acoustic phonons is obtained from a handbook rather than by calculation. The problem arises because the handbook value of C includes the contribution of optical phonons as well as acoustic phonons, even though the former have a far slower velocity than the latter. This leads to a calculated Λ_{bulk} which is too small, which causes the effective k calculated for a nanostructure to be too big.

Table 11.5 Comparison between diffusive and ballistic limits

	Diffusive Behavior	Ballistic Behavior
Sample size	Large: $L \gg \Lambda_{bulk}$	Small: $L \ll \Lambda_{bulk}$
Thermal conductivity compared to handbook values k_{bulk}	$k = k_{bulk}$	$k < k_{bulk}$
Heat transfer *along a nanowire*: Dependence of thermal conductivity on diameter D	k independent of D	$k \propto D$
Heat transfer *along a thin film*: Dependence of thermal conductivity on film thickness L	k independent of L	k increases with L
Heat transfer *across a thin film*: Dependence of thermal conductivity on film thickness L	k independent of L	$k \propto L$
Heat transfer *across a thin film*: Dependence of thermal resistance on film thickness L	$R \propto L$	R independent of L

Fortunately, a simple solution exists: kinetic theory can be generalized to account for the frequency dependence of the properties of the energy carriers. In this way eq. 11.4 can be written as an integral over frequency:

$$k(T) = \tfrac{1}{3} \int C_{\omega}(\omega,T) v(\omega) \Lambda_{eff}(\omega,T) d\omega, \qquad (11.37)$$

where the arguments in parentheses indicate the functional dependencies of the various quantities, and now C_{ω} is the specific heat per unit volume per unit frequency, with units of $[(\text{J/m}^3 - °\text{C})/(\text{rad/s})]$. Equation (11.37) can be further generalized by treating the longitudinal and transverse waves separately. The quantities $C_\omega(\omega,T)$ and $v(\omega)$ are readily determined by studying the physics of each type of energy carrier, and Matthiessen's Rule can be applied on a frequency-dependent basis:

$$\Lambda_{eff}^{-1}(\omega,T) = \Lambda_{bulk}^{-1}(\omega,T) + \Lambda_{bdy}^{-1}(\omega) .$$

Typically $\Lambda_{bulk}(\omega,T)$ is parameterized by one or several fitting parameters, and Λ_{bdy} is calculated using the methods already discussed in this chapter. As shown by the dashed line in Fig. 11.8(b), calculations using the frequency-dependent eq. (11.37) are significantly more accurate than the calculations using the lumped eq. (11.4).

Alternative theoretical techniques. Several of the expressions given above for Λ_{bdy} have their origins in solutions of the Boltzmann transport equation, which is widely used in solid state physics [12,13]. Readers from a traditional engineering background may find it easier to approach these concepts from the perspective of radiation heat transfer in a participating medium, such as radiation through the atmosphere or through a cloud of soot [34-36]. One explicit link between these two similar approaches is the "equation of radiative transfer."

Superlattices. Solutions for the heat transfer in superlattices (Fig. 11.4) are generally derived using the Boltzmann transport equation [6,7], although again the results can sometimes be cast in the simple language of kinetic theory, especially if the superlattice interfaces are approximated as purely diffuse [32].

Ultrafast phenomena. Recall that classical size effects are important if the characteristic length of a sample is smaller than, or comparable to, the mean free path Λ_{bulk} of energy carriers in the bulk. In a very similar way, ultrafast effects are important if the characteristic *time* of a process is shorter than, or comparable to, the mean free time τ of the energy carriers (recall eq. 11.1). The time scale τ is also sometimes known as a relaxation time. The order-of-magnitude of τ depends on the energy carrier and temperature. For example, for phonons in Si at 300 K, using $\Lambda_{bulk} \approx 250\,\text{nm}$ and $v_s \approx 5880\,\text{m/s}$, we have $\tau \approx 4.3\times10^{-11}\,\text{s}$. These timescales are readily studied experimentally using pulsed lasers, with pulse widths typically measured in picoseconds $(1\,ps = 10^{-12}\,s)$ or femtoseconds $(1\,fs = 10^{-15}\,s)$. In most cases the energy carriers can still be treated as particles.

REFERENCES

[1]   Touloukian, Y. S., ed., Purdue University Thermophysical Properties Research Center, *Thermophysical Properties of Matter*. IFI/Plenum, New York, 1970.

[2]   Li, D., Wu, Y. Y., Kim, P., Shi, L., Yang, P. D., and Majumdar, A., "Thermal Conductivity of Individual Silicon Nanowires," *Applied Physics Letters*, vol. 83, pp. 2934-2936, 2003.

[3]   Yu, C. H., Shi, L., Yao, Z., Li, D. Y., and Majumdar, A., "Thermal Conductance and Thermopower of an Individual Single-Wall Carbon Nanotube," *Nano Letters*, vol. 5, pp. 1842-1846, 2005.

[4]   Pop, E., Mann, D., Wang, Q., Goodson, K., and Dai, H. J., "Thermal Conductance of an Individual Single-Wall Carbon Nanotube above Room Temperature," *Nano Letters,* vol. 6, 96-100, 2006.

[5]   Rowe, D. M., ed., *Thermoelectrics Handbook: Macro to Nano*, CRC Press, 2005.

[6]   Chen, G., *Nanoscale Energy Transport and Conversion: A Parallel Treatment of Electrons, Molecules, Phonons, and Photons*, Oxford University Press, 2005.

[7]   Zhang, Z. M., *Nano / Microscale Heat Transfer*, McGraw-Hill, New York, 2007.

[8]   Volz, S., ed., *Microscale and Nanoscale Heat Transfer,* Springer, 2007.

[9]   Tien, C. L., Majumdar, A., and Gerner, F., "Microscale Energy Transport in Solids," *Micro-scale Energy Transport*, Editors Taylor & Francis, 1998.

[10]   Strictly speaking, this requires that the sample size is much larger than the "phase coherence length," Λ_ϕ, to ensure that coherence effects are negligible. See also the comment at the end of the next paragraph.

[11]   Cahill, D. G., Ford, W. K., Goodson, K. E., Mahan, G. D., Majumdar, A., Maris, H. J., Merlin, R., and Phillpot, S. R., "Nanoscale Thermal Transport," *Journal of Applied Physics*, vol. 83, 793-818, 2003.

[12]   Ashcroft, N. W., and Mermin, N. D., *Solid State Physics*, Harcourt College Publishers, 1976.

[13]   Ziman, J. M., *Electrons and Phonons*, Clarendon Press, Oxford, 1960.

[14]   Kittel, C., *Introduction to Solid State Physics,* 8th edition, Wiley, 2004.

[15]   French, A. P., and Taylor, E. F., *An Introduction to Quantum Physics*, Norton, New York, 1978.

[16]   Hagelstein, P. L., Senturia, S. D., and Orlando, T. P., *Introductory Applied Quantum and Statistical Mechanics*, Wiley, 2004.

[17]   Kittel, C., and Kroemer, H., *Thermal Physics,* 2nd edition, W. H. Freeman, 1980.

[18]   Baierlein, R. *Thermal Physics*, Cambridge University Press, 1999.

[19]   Vincenti, W. G., and Kruger, C. H., *Introduction to Physical Gas Dynamics*, Wiley, 1965.

[20]   Springer, G. S., "Heat Transfer in Rarefied Gases," *Advances in Heat Transfer,* vol. 7, Edited by Irvine and James, Elsevier, 1971.

[21]   Cercignani, C., *Rarefied Gas Dynamics,* Cambridge University Press, 2000.

[22]   Chapman, S. and Cowling, T. G., *The Mathematical Theory of Non-Uniform gases*, Cambridge University Press, 1995.

[23]   The resolution to this apparent paradox is the fact that the specific heat of metals at typical temperatures is dominated by phonons, as discussed in the next section. This issue is explored further in Problem 11.6 at the end of the chapter.

[24]   However, the dopant atoms can still have a significant effect on the thermal conductivity because of the increased scattering of phonons. (One class of semiconductors where the electron contribution to the total thermal conductivity *is* important is thermoelectric materials. See Ref. 5.)

[25]   Asheghi, M., Kurabayashi, K., Kasnavi, R., and Goodson, K. E., "Thermal Conduction in Doped Single-Crystal Silicon Films," *Journal of Applied Physics*, vol. 91, 5079-5088, 2002.

[26]   The detailed form of this equation is in dispute. Some workers (Z. M. Zhang, Ref. [7]) argue that π should be squared rather than cubed.

[27]   Fuchs, K., "The Conductivity of Thin Metallic Films according to the Electron Theory of Metals," *Proceedings of the Cambridge Philosophical Society*, vol. 34, 100-108, 1938.

[28]   Sondheimer, E. H., "The Mean Free Path of Electrons in Metals," *Advances in Physics*, vol. 1, 1-42, 1952.

[29]   Majumdar, A., "Microscale Heat-Conduction in Dielectric Thin-Films," *Journal of Heat Transfer*, vol. 115, 7-16, 1993.

[30]   Ju, Y. S., and Goodson, K. E., "Phonon Scattering in Silicon Films with Thickness of Order 100 nm," *Applied Physics Letters*, vol. 74, 3005-3007, 1999.

[31]   Chen, G. "Thermal Conductivity and Ballistic-Phonon Transport in the Cross-Plane Direction of Superlattices," *Physical Review B*, vol. 57, 14958-14973, 1998.

[32]   Dames, C., and Chen, G., "Theoretical Phonon Thermal Conductivity of Si/Ge Superlattice Nanowires," *Journal of Applied Physics*, vol. 95, 682-693, 2004.

[33]   Dames, C. and Chen, G. "Thermal Conductivity of Nanostructured Materials," in *Thermoelectrics Handbook: Macro to Nano*, Edited by D. M. Rowe, CRC Press, 2005.

[34]   Modest, M., *Radiative Heat Transfer*, 2nd edition, Academic Press, 2003.

[35]   Siegel, R., and Howell, J., *Thermal Radiation Heat Transfer*, 4th edition, Taylor & Francis, 2001.

[36]   Brewster, M. Q., *Thermal Radiative Transfer and Properties*, Wiley-Interscience, 1992.

[37]   Mingo, N., and Broido, D. A., "Carbon Nanotube Ballistic Thermal Conductance and its Limits," *Physical Review Letters*, vol. 95, 096105, 2005.

[38]   Mingo, N., and Broido, D. A., "Length Dependence of Carbon Nanotube Thermal Conductivity and the 'Problem of Long Waves'," *Nano Letters*, vol. 5, 1221-1225, 2005.

PROBLEMS

11.1   Power-law dependencies for the thermal conductivity of an ideal gas in bulk. Use the ideal gas equations for C, v, and Λ to derive an equation expressing thermal conductivity in the form $k = \alpha T^\beta p^\gamma$, where α, β, and γ are constants (β and γ are dimensionless, while α is not). What happens to k if we increase temperature while holding pressure constant? What happens if we increase pressure while holding temperature constant?

11.2   Heat transfer through a gas at low pressure. Vacuum chambers are widely used to reduce the air pressure inside an experimental or processing chamber, in order to reduce undesirable heat losses by convection / air conduction. Consider the pressure-dependence of the heat transfer through helium gas between two parallel plates. The plates are separated by a gap $L = 10$ cm and have accommodation coefficients $\alpha_1 \approx \alpha_2 \approx 1$. The gas is at a temperature $T=300$ K. Calculate the thermal conductivity as a function of pressure for p ranging from one atmosphere down to 10^{-5} Pa (typical of a good high-vacuum pump). Repeat for a gap of 1 mm and 10 μm. In each case, what vacuum level is necessary to reduce the thermal conductivity to below 1% of its value at 1 atm?

11.3   Effect of accommodation coefficients. Repeat the previous problem assuming $\alpha_1 \approx \alpha_2 \approx 0.1$.

11.4   Estimate the mean free path of phonons in fused silica glass (polycrystalline silicon dioxide: SiO_2) at 300 K. How does this compare to the mean free path of phonons in silicon?

11.5   Estimate the mean free path for heat conduction by electrons in copper at 300 K. How does this compare to the mean free path of phonons in silicon?

11.6   Compare the specific heat of acoustic phonons in silicon at 300 K to the specific heat of electrons in copper at 300 K. Why is your calculated specific heat of copper so much smaller than the handbook value? If the specific heat of electrons in copper is so much smaller than that of phonons in silicon, how can the thermal conductivity of copper be higher than that of silicon?

11.7 Diameter-dependence of the thermal conductivity of a nanowire. For Si and Ge nanowires at 300 K, plot the thermal conductivity as a function of diameter. Consider D ranging from 1 mm down to 1 nm. Assume the specularity is approximately zero.

11.8 Temperature-dependence of effective diameter. Equation (11.27) can be thought of as defining an "effective diameter" D_{eff} in terms of the true diameter D and the specularity p, namely $D_{eff} = D\left(\dfrac{1+p}{1-p}\right)$. The specularity equation (11.28) links p to the roughness and average wavelength λ of an energy carrier. At lower temperatures, the average wavelengths become longer, so we anticipate that the specularity and effective diameter will both increase at low temperature. In analogy to Wien's displacement law of photon radiation, for phonons the average wavelength λ_{av} at a given temperature can be estimated from [33]

$$\lambda_{av} T \approx 50\,\text{nm - K} \times \left(\frac{v}{5000\,\text{m/s}}\right) \qquad (\lambda_{av} > 2b), \qquad (11.38)$$

$$\lambda_{av} \approx 2b \qquad\qquad\qquad \text{(otherwise)}, \qquad (11.39)$$

where v is the sound velocity and b is the spacing between nearest-neighbor atoms (typically $b \approx 0.2$ to $0.3\,\text{nm}$ for most solids). Consider a silicon nanowire of diameter $D = 100\,\text{nm}$. Assuming $b = 0.25\,\text{nm}$ and rms roughness of $\delta = 0.1\,\text{nm}$, plot the specularity p and effective diameter D_{eff} as functions of temperature, from 1000 K down to 1 K.   For simplicity in analysis it is usually desirable to approximate the boundary scattering as being perfectly diffuse $(p = 0)$. What is the lowest temperature for which this approximation is valid? That is, find the lowest temperature for which D_{eff} differs from D by less than 10%. Repeat for $\delta = 1\,\text{nm}$.

11.9 Single-walled carbon nanotubes (SWCNTs) are one-dimensional conductors with extremely high thermal conductivity, with reported values of $3500\,\text{W/m -}°\text{C}$ or higher at 300 K [3,4]. A

SWCNT of length L transporting heat between two contacts is in some ways similar to heat conduction perpendicular to a thin film of thickness L. Specifically, in both cases the presence of the contacts results in boundary scattering. (Note that, unlike a nanowire, an isolated SWCNT does not have boundary scattering at the sidewalls.) The simplest way to model the effective phonon mean free path is with Matthiessen's Rule. Assuming perfect contacts ($\alpha_1 \approx \alpha_2 \approx 1$), the mean free path for boundary scattering Λ_{bdy} can be approximated simply as L. At 300 K, assume the "bulk" phonon mean free path is $\Lambda_{bulk} \approx 1.5\,\mu m$, where here "bulk" refers to a nanotube with $L \to \infty$. (Estimates for Λ_{bulk} range from around 0.25 μm to 3 μm [3,4,37,38]). Make log-log plots of the thermal conductivity and thermal resistance of a SWCNT as functions of length, for L ranging from 1 mm down to 10 nm. The diameter of the tube is $d = 2.4\,nm$, and the most common definition for the cross-sectional area for heat conduction is $A = \pi db$, where the thickness of a single wall is $b = 0.34\,nm$. Assume that the thermal conductivity of a very long SWCNT at 300 K is $k \approx 5000\,W/m\text{-}°C$. You can compare your results with the results of a more comprehensive theoretical analysis by Mingo and Broido [38], selected values of which are given in Appendix D.

11.10 Velocity averaging for thermal conductivity. Equation (11.16) states that the correct way to calculate an average sound velocity from $v_{s,T}$ and $v_{s,L}$ is to average them in an inverse-squares sense.

Why don't we just do the simple average $v_s = \frac{2}{3}v_{s,T} + \frac{1}{3}v_{s,L}$?

Answer this question by focusing on the low temperature limit. Considering the three subsystems separately (there are always two polarizations of transverse acoustic waves and one polarization of longitudinal acoustic wave), use the low-temperature specific heat of eq. (11.19), and assume that $\Lambda_{eff} \to$ const. at low T (for example, phonons traveling along a wire with diffuse boundaries will all experience $\Lambda_{eff} \approx D$). The total phonon thermal conductivity is simply the sum of the thermal conductivities of the three subsystems. By equating your result to the analogous lumped expression for the total k assuming a single averaged velocity, show that eq. (11.16) is indeed the correct averaging rule. If

$v_{s,T} \approx 5200\,\text{m/s}$ and $v_{s,L} \approx 8840\,\text{m/s}$ (approximate average values for Si), what is v_s?

11.11   Velocity averaging for specific heat.   Derive an averaging rule analogous to eq. (11.16) to combine $v_{s,T}$ and $v_{s,L}$ into a single value of $v_{s,sp.ht.}$ that gives the correct low-temperature specific heat for the combined system of longitudinal plus transverse acoustic phonons.   Evaluate the numerical value of your $v_{s,sp.ht.}$ for $v_{s,T} \approx 5200\,\text{m/s}$ and $v_{s,L} \approx 8840\,\text{m/s}$ (approximate average values for Si), and compare to the value of v_s found from eq. (11.16) as well as to the simple weighted average $v_s = \frac{2}{3} v_{s,T} + \frac{1}{3} v_{s,L}$.

11.12   Equation (11.30a) gives the effective mean free path for the Fuchs-Sondheimer solution in the asymptotic limit of a thick film. Combine this result with Matthiessen's rule, eq. (11.26), to derive an expression for the boundary scattering mean free path Λ_{bdy} in this limit, as a function of L and/or Λ_{bulk}.

11.13   Equation (11.30b) gives the effective mean free path for the Fuchs-Sondheimer solution in the asymptotic limit of a thin film. Combine this result with Matthiessen's rule, eq. (11.26), to derive an expression for the boundary scattering mean free path Λ_{bdy} in this limit, as a function of L and/or Λ_{bulk}.

APPENDIX A: ORDINARY DIFFERENTIAL EQUATIONS

(1) Second Order Ordinary Differential Equations with Constant Coefficients

Second order ordinary differential equations with constant coefficients occur often in conduction problems. Common equations and their solutions are presented below. Detailed treatment can be found in the literature [1,2].

(i) Example 1:

$$\frac{d^2 y}{dx^2} - m^2 y = c . \qquad \text{(A-1)}$$

Solution:

$$y = C_1 \exp(mx) + C_2 \exp(-mx) - \frac{c}{m^2}, \qquad \text{(A-2a)}$$

or

$$y = C_1 \sinh mx + C_2 \cosh mx - \frac{c}{m^2} . \qquad \text{(A-2b)}$$

(ii) Example 2:

$$\frac{d^2 y}{dx^2} + m^2 y = c . \qquad \text{(A-3)}$$

Solution:

$$y = C_1 \sin mx + C_2 \cos mx + \frac{c}{m^2} . \qquad \text{(A-4)}$$

(iii) Example 3:

$$\frac{d^2 y}{dx^2} + 2b \frac{dy}{dx} + m^2 y = c . \qquad \text{(A-5)}$$

The solution to this equation depends on the magnitude of b^2 relative to m^2:

(a) $b^2 < m^2$

$$y = C_1 \exp(-bx) \sin(\sqrt{m^2 - b^2} \; x) + C_2 \exp(-bx) \cos(\sqrt{m^2 - b^2} \; x) + \frac{c}{m^2} .$$

$$\text{(A-6a)}$$

(b) $b^2 = m^2$

$$y = \exp(-bx)[C_1 x + C_2] + \frac{c}{m^2}. \qquad\qquad \text{(A-6b)}$$

(c) $b^2 > m^2$

$$y = C_1 \exp(-bx + \sqrt{b^2 - m^2}\ x) + C_2 \exp(-bx - \sqrt{b^2 - m^2}\ x) + \frac{c}{m^2}. \text{(A-6c)}$$

(iv) Example 4:

$$\frac{d^2 y}{dx^2} + 2b\frac{dy}{dx} + m^2 y = f(x). \qquad\qquad \text{(A-7)}$$

The solution to this equation depends on the magnitude of b^2 relative to m^2:

(a) $b^2 < m^2$

$$y = C_1 \exp(-bx)\sin(\sqrt{m^2 - b^2}\ x) + C_2 \exp(-bx)\cos(\sqrt{m^2 - b^2}\ x) +$$

$$\frac{\exp(-bx)}{\sqrt{m^2 - b^2}}\left[\sin(\sqrt{m^2 - b^2}\ x) \int f(x)\exp(bx)\cos(\sqrt{m^2 - b^2}\ x)\,dx - \right. \text{(A-8a)}$$

$$\left. \cos(\sqrt{m^2 - b^2}\ x) \int f(x)\exp(bx)\sin(\sqrt{m^2 - b^2}\ x)\,dx \right].$$

(b) $b^2 = m^2$

$$y = \exp(-bx)\left[C_1 x + C_2 + x \int f(x)\exp(bx)\,dx - \int x\,f(x)\exp(bx)\,dx \right]. \text{(A-8b)}$$

(c) $b^2 > m^2$

$$y = C_1 \exp(-bx + \sqrt{b^2 - m^2}\ x) + C_2 \exp(-bx - \sqrt{b^2 - m^2}\ x) +$$

$$\frac{\exp(-bx + \sqrt{b^2 - m^2}\ x)}{2\sqrt{b^2 - m^2}}\left[\int f(x)\ \exp(bx - \sqrt{b^2 - m^2}\ x)dx \right] - \text{(A-8c)}$$

$$\frac{\exp(-bx - \sqrt{b^2 - m^2}\ x)}{2\sqrt{b^2 - m^2}}\left[\int f(x)\ \exp(bx + \sqrt{b^2 - m^2}\ x)dx \right].$$

(2) First Order Ordinary Differential Equations with Variable Coefficients

A linear differential equation of the form

$$\frac{dy}{dx} + P(x)y = Q(x), \tag{A-9}$$

has the following solution:

$$y = e^{-\int P(x)dx}\left[\int Q(x)e^{\int P(x)dx}dx + C\right], \tag{A-10}$$

where C is constant of integration,

REFERENCES

[1] Hildebrand, F.B., *Advanced Calculus for Applications*, 2nd edition, Prentice-Hall, Englewood Cliffs, New Jersey, 1976.

[2] Wylie, C.R. and L.C. Barret, *Advanced Engineering Mathematics*, 5th edition, McGraw-Hill, New York, 1982.

APPENDIX B

INTEGRALS OF BESSEL FUNCTIONS

In the following formulas $Z_n(x)$ represents $J_n(x)$ or $Y_n(x)$. Note that the integral $\int Z_0(x)dx$ can not be evaluated in closed form.

1. $\int Z_1(x)dx = -Z_0(x)$

2. $\int x Z_0(x)dx = x Z_1(x)$

3. $\int x Z_1(x)dx = -x Z_0(x) + \int Z_0(x)dx$

4. $\int x^2 Z_0(x)dx = x^2 Z_1(x) + x Z_0(x) - \int Z_0(x)dx$

5. $\int x^m Z_0(x)dx = x^m Z_1(x) + (m-1)x^{m-1} Z_0(x) - (m-1)^2 \int x^{m-2} Z_0(x)dx$

6. $\int x^m Z_1(x)dx = -x^m Z_0(x) + +m \int x^{m-1} Z_0(x)dx$

7. $\int \dfrac{Z_0(x)}{x^2} dx = Z_1(x) - \dfrac{Z_0(x)}{x} - \int Z_0(x)dx$

8. $\int \dfrac{Z_0(x)}{x^m} dx = \dfrac{Z_1(x)}{(m-1)^2 x^{m-2}} - \dfrac{Z_0(x)}{(m-1)x^{m-1}} - \dfrac{1}{(m-1)^2} \int \dfrac{Z_0(x)}{x^{m-2}} dx$

9. $\int \dfrac{Z_1(x)}{x} dx = -Z_1(x) + \int Z_0(x)dx$

10. $\int \dfrac{Z_1(x)}{x^m} dx = -\dfrac{Z_1(x)}{mx^{m-1}} + \dfrac{1}{m} \int \dfrac{Z_0(x)}{x^{m-1}} dx$

11. $\int x^n Z_{n-1}(x)dx = x^n Z_n(x)$

12. $\int x^{-n} Z_{n+1}(x)dx = -x^{-n} Z_n(x)$

13. $\int x^m Z_n(x)dx = -x^m Z_{n-1}(x) + (m+n-1) \int x^{m-1} Z_{n-1}(x)dx$

APPENDIX C: Values of Bessel Functions

x	$J_0(x)$	$J_1(x)$	x	$J_0(x)$	$J_1(x)$
0.0	1.0000	0.0000	4.0	- 0.3971	- 0.0660
0.1	0.9975	0.0499	4.1	- 0.3887	- 0.1033
0.2	0.9900	0.0995	4.2	- 0.3766	- 0.1386
0.3	0.9776	0.1483	4.3	- 0.3610	- 0.1719
0.4	0.9604	0.1960	4.4	- 0.3423	- 0.2028
0.5	0.9385	0.2423	4.5	- 0.3205	- 0.2311
0.6	0.9120	0.2867	4.6	- 0.2961	- 0.2566
0.7	0.8812	0.3290	4.7	- 0.2693	- 0.2791
0.8	0.8463	0.3688	4.8	- 0.2404	- 0.2985
0.9	0.8075	0.4059	4.9	- 0.2097	- 0.3147
1.0	0.7652	0.4401	5.0	- 0.1776	- 0.3276
1.1	0.7196	0.4709	5.1	- 0.1443	- 0.3371
1.2	0.6711	0.4983	5.2	- 0.1103	- 0.3432
1.3	0.6201	0.5220	5.3	- 0.0758	- 0.3460
1.4	0.5669	0.5419	5.4	- 0.0412	- 0.3453
1.5	0.5118	0.5579	5.5	- 0.0068	- 0.3414
1.6	0.4554	0.5699	5.6	0.0270	- 0.3343
1.7	0.3980	0.5778	5.7	0.0599	- 0.3241
1.8	0.3400	0.5815	5.8	0.0917	- 0.3110
1.9	0.2818	0.5812	5.9	0.1220	- 0.2951
2.0	0.2239	0.5767	6.0	0.1506	- 0.2767
2.1	0.1666	0.5683	6.1	0.1773	- 0.2559
2.2	0.1104	0.5560	6.2	0.2017	- 0.2329
2.3	0.0555	0.5399	6.3	0.2238	- 0.2081
2.4	0.0025	0.5202	6.4	0.2433	- 0.1816
2.5	- 0.0484	0.4971	6.5	0.2601	- 0.1538
2.6	- 0.0968	0.4708	6.6	0.2740	- 0.1250
2.7	- 0.1424	0.4416	6.7	0.2851	- 0.0953
2.8	- 0.1850	0.4097	6.8	0.2931	- 0.0652
2.9	- 0.2243	0.3754	6.9	0.2981	- 0.0349
3.0	- 0.2601	0.3391	7.0	0.3001	- 0.0047
3.1	- 0.2921	0.3009	7.1	0.2991	0.0252
3.2	- 0.3202	0.2613	7.2	0.2951	0.0543
3.3	- 0.3443	0.2207	7.3	0.2882	0.0826
3.4	- 0.3643	0.1792	7.4	0.2786	0.1096
3.5	- 0.3801	0.1374	7.5	0.2663	0.1352
3.6	- 0.3918	0.0955	7.6	0.2516	0.1592
3.7	- 0.3992	0.0538	7.7	0.2346	0.1813
3.8	- 0.4026	0.0128	7.8	0.2154	0.2014
3.9	- 0.4018	- 0.0272	7.9	0.1944	0.2192

APPENDIX C: Values of Bessel Functions (Continued)

x	$J_0(x)$	$J_1(x)$	x	$J_0(x)$	$J_1(x)$
8.0	0.1717	0.2346	11.5	- 0.0677	- 0.2284
8.1	0.1475	0.2476	11.6	- 0.0446	- 0.2320
8.2	0.1222	0.2580	11.7	- 0.0213	- 0.2333
8.3	0.0960	0.2657	11.8	0.0020	- 0.2323
8.4	0.0692	0.2708	11.9	0.0250	- 0.2290
8.5	0.0419	0.2731	12.0	0.0477	- 0.2234
8.6	0.0146	0.2728	12.1	0.0697	- 0.2157
8.7	- 0.0125	0.2697	12.2	0.0908	- 0.2060
8.8	- 0.0392	0.2641	12.3	0.1108	- 0.1943
8.9	- 0.0653	0.2559	12.4	0.1296	- 0.1807
9.0	- 0.0903	0.2453	12.5	0.1469	- 0.1655
9.1	- 0.1142	0.2324	12.6	0.1626	- 0.1487
9.2	- 0.1367	0.2174	12.7	0.1766	- 0.1307
9.3	- 0.1577	0.2004	12.8	0.1887	- 0.1114
9.4	- 0.1768	0.1816	12.9	0.1988	- 0.0912
9.5	- 0.1939	0.1613	13.0	0.2069	- 0.0703
9.6	- 0.2090	0.1395	13.1	0.2129	- 0.0489
9.7	- 0.2218	0.1166	13.2	0.2167	- 0.0271
9.8	- 0.2323	0.0928	13.3	0.2183	- 0.0052
9.9	- 0.2403	0.0684	13.4	0.2177	0.0166
10.0	- 0.2459	0.0435	13.5	0.2150	0.0380
10.1	- 0.2490	0.0184	13.6	0.2101	0.0590
10.2	- 0.2496	- 0.0066	13.7	0.2032	0.0791
10.3	- 0.2477	- 0.0313	13.8	0.1943	0.0984
10.4	- 0.2434	- 0.0555	13.9	0.1836	0.1165
10.5	- 0.2366	- 0.0789	14.0	0.1711	0.1334
10.6	- 0.2276	- 0.1012	14.1	0.1570	0.1488
10.7	- 0.2164	- 0.1224	14.2	0.1414	0.1626
10.8	- 0.2032	- 0.1422	14.3	0.1245	0.1747
10.9	- 0.1881	- 0.1603	14.4	0.1065	0.1850
11.0	- 0.1712	- 0.1768	14.5	0.0875	0.1934
11.1	- 0.1528	- 0.1913	14.6	0.0679	0.1999
11.2	- 0.1330	- 0.2039	14.7	0.0476	0.2043
11.3	- 0.1121	- 0.2143	14.8	0.0271	0.2066
11.4	- 0.0902	- 0.2225	14.9	0.0064	0.2069

APPENDIX C: Values of Bessel Functions (Continued)

x	$Y_0(x)$	$Y_1(x)$	x	$Y_0(x)$	$Y_1(x)$
0.0	- ∞	- ∞	4.0	- 0.0169	0.3979
0.1	- 1.5342	- 6.4590	4.1	- 0.0561	0.3846
0.2	- 1.0811	- 3.3238	4.2	- 0.0938	0.3680
0.3	- 0.8073	- 2.2931	4.3	- 0.1296	0.3484
0.4	- 0.6060	- 1.7809	4.4	- 0.1633	0.3260
0.5	- 0.4445	- 1.4715	4.5	- 0.1947	0.3010
0.6	- 0.3085	- 1.2604	4.6	- 0.2235	0.2737
0.7	- 0.1907	- 1.1032	4.7	- 0.2494	0.2445
0.8	- 0.0868	- 09781	4.8	- 0.2723	0.2136
0.9	0.0056	- 0.8731	4.9	- 0.2921	0.1812
1.0	0.0883	- 0.7812	5.0	- 0.3085	0.1479
1.1	0.1622	- 0.6981	5.1	- 0.3216	0.1137
1.2	0.2281	- 0.6211	5.2	- 0.3313	0.0792
1.3	0.2865	- 0.5485	5.3	- 0.3374	0.0445
1.4	0.3379	- 0.4791	5.4	- 0.3402	0.0101
1.5	0.3824	- 0.4123	5.5	- 0.3395	- 0.0238
1.6	0.4204	- 0.3476	5.6	- 0.3354	- 0.0568
1.7	0.4520	- 0.2847	5.7	- 0.3282	- 0.0887
1.8	0.4774	- 0.2237	5.8	- 0.3177	- 0.1192
1.9	0.4968	- 0.1644	5.9	- 0.3044	- 0.1481
2.0	0.5104	- 0.1071	6.0	- 0.2882	- 0.1750
2.1	0.5183	- 0.0517	6.1	- 0.2694	- 0.1998
2.2	0.5208	0.0015	6.2	- 0.2483	- 0.2223
2.3	0.5181	0.0523	6.3	- 0.2251	- 0.2422
2.4	0.5104	0.1005	6.4	- 0.1999	- 0.2596
2.5	0.4981	0.1495	6.5	- 0.1732	- 0.2741
2.6	0.4813	0.1884	6.6	- 0.1452	- 0.2857
2.7	0.4605	0.2276	6.7	- 0.1162	- 0.2945
2.8	0.4359	0.2635	6.8	- 0.0864	- 0.3002
2.9	0.4079	0.2959	6.9	- 0.0563	- 0.3029
3.0	0.3769	0.3247	7.0	- 0.0259	- 0.3027
3.1	0.3431	0.3496	7.1	0.0042	- 0.2995
3.2	0.3071	0.3707	7.2	0.0339	- 0.2934
3.3	0.2691	0.3879	7.3	0.0628	- 0.2846
3.4	0.2296	0.4010	7.4	0.0907	- 0.2731
3.5	0.1890	0.4102	7.5	0.1173	- 0.2591
3.6	0.1477	0.4154	7.6	0.1424	- 0.2428
3.7	0.1061	0.4167	7.7	0.1658	- 0.2243
3.8	0.0645	0.4141	7.8	0.1872	- 0.2039
3.9	0.0234	0.4078	7.9	0.2065	- 0.1817

APPENDIX C: Values of Bessel Functions (Continued)

X	$Y_0(x)$	$Y_1(x)$		x	$Y_0(x)$	$Y_1(x)$
8.0	0.2235	- 0.1581		11.5	- 0.2252	0.0579
8.1	0.2381	- 0.1331		11.6	- 0.2299	0.0348
8.2	0.2501	- 0.1702		11.7	- 0.2322	0.0114
8.3	0.2595	- 0.0806		11.8	- 0.2322	- 0.0118
8.4	0.2662	- 0.0535		11.9	- 0.2298	- 0.0347
8.5	0.2702	- 0.0262		12.0	- 0.2252	- 0.0571
8.6	0.2715	0.0011		12.1	- 0.2184	- 0.0787
8.7	0.2700	0.0280		12.2	- 0.2095	- 0.0994
8.8	0.2659	0.0544		12.3	- 0.1986	- 0.1189
8.9	0.2592	0.0799		12.4	- 0.1858	- 0.1371
9.0	0.2499	0.1043		12.5	- 0.1712	- 0.1538
9.1	0.2383	0.1275		12.6	- 0.1551	- 0.1689
9.2	0.2245	0.1491		12.7	- 0.1375	- 0.1821
9.3	0.2086	01691		12.8	- 0.1187	- 0.1935
9.4	0.1907	0.1871		12.9	- 0.0989	- 0.2028
9.5	0.1712	0.2032		13.0	- 0.0782	- 0.2101
9.6	0.1502	0.2171		13.1	- 0.0569	- 0.2152
9.7	0.1279	0.2287		13.2	- 0.0352	- 0.2181
9.8	0.1045	0.2379		13.3	- 0.0134	- 0.2190
9.9	0.0804	0.2447		13.4	0.0085	- 0.2176
10.0	0.0557	0.2490		13.5	0.0301	- 0.2140
10.1	0.0307	0.2508		13.6	0.0512	- 0.2084
10.2	0.0056	0.2502		13.7	0.0717	- 0.2007
10.3	- 0.0193	0.2471		13.8	0.0913	- 0.1912
10.4	- 0.0437	0.2416		13.9	0.1099	- 0.1798
10.5	- 0.0675	0.2337		14.0	0.1272	- 0.1666
10.6	- 0.0904	0.2236		14.1	0.1431	- 0.1520
10.7	- 0.1122	0.2114		14.2	0.1575	- 0.1359
10.8	- 0.1326	0.1973		14.3	0.1703	- 0.1186
10.9	- 0.1516	0.1813		14.4	0.1812	- 0.1003
11.0	- 0.1688	0.1637		14.5	0.1903	- 0.0810
11.1	- 0.1843	0.1446		14.6	0.1974	- 0.0612
11.2	- 0.1977	0.1243		14.7	0.2025	- 0.0408
11.3	- 0.2091	0.1029		14.8	0.2056	- 0.0202
11.4	- 0.2183	0.0807		14.9	0.2065	0.0005

APPENDIX C: Values of Bessel Functions (Continued)

x	$I_0(x)$	$I_1(x)$	x	$I_0(x)$	$I_1(x)$
0.0	1.0000	0.0	4.0	11.302	9.7595
0.1	1.0025	0.0501	4.1	12.324	10.688
0.2	1.0100	0.1005	4.2	13.442	11.706
0.3	1.0226	0.1517	4.3	14.668	12.822
0.4	1.0404	0.2040	4.4	16.010	14.046
0.5	1.0635	0.2579	4.5	17.481	15.389
0.6	1.0920	0.3137	4.6	19.093	16.863
0.7	1.1263	0.3719	4.7	20.858	18.479
0.8	1.1665	0.4329	4.8	22.794	20.253
0.9	1.2130	0.4971	4.9	24.915	22.199
1.0	1.2661	0.5652	5.0	27.240	24.336
1.1	1.3262	0.6375	5.1	29.789	26.680
1.2	1.3937	0.7147	5.2	32.584	29.254
1.3	1.4693	0.7973	5.3	35.648	32.080
1.4	1.5534	0.8861	5.4	39.009	35.182
1.5	1.6467	0.9817	5.5	42.695	38.588
1.6	1.7500	1.0848	5.6	46.738	42.328
1.7	1.8640	1.1963	5.7	51.173	46.436
1.8	1.9896	1.3172	5.8	56.038	50.946
1.9	2.1277	1.4482	5.9	61.377	55.900
2.0	2.2796	1.5906	6.0	67.234	61.342
2.1	2.4463	1.7455	6.1	73.633	67.319
2.2	2.6291	1.9141	6.2	80.718	73.886
2.3	2.8296	2.0978	6.3	88.462	81.100
2.4	3.0493	2.2981	6.4	96.962	89.026
2.5	3.2898	2.5167	6.5	106.29	97.735
2.6	3.5533	2.7554	6.6	116.54	107.30
2.7	3.8417	3.0161	6.7	127.79	117.82
2.8	4.1573	3.3011	6.8	140.14	129.38
2.9	4.5027	3.6126	6.9	153.70	142.08
3.0	4.8808	3. 9534	7.0	168.59	156.04
3.1	5.2945	4.3262	7.1	184.95	171.38
3.2	5.7472	4.7343	7.2	202.92	188.25
3.3	6.2426	5.1810	7.3	222.66	206.79
3.4	6.7848	5.6701	7.4	244.34	227.17
3.5	7.3782	6.2058	7.5	268.16	249.58
3.6	8.0277	6.7927	7.6	294.33	274.22
3.7	8.7386	7.4357	7.7	323.09	301.31
3.8	9.5169	8.1404	7.8	354.68	331.10
3.9	10.369	8.9128	7.9	389.41	363.85

APPENDIX C: Values of Bessel Functions (Continued)

x	$K_0(x)$	$K_1(x)$	x	$K_0(x)$	$K_1(x)$
0.0	∞	∞	4.0	0.01160	0.012484
0.1	2.4271	9.8538	4.1	0.009980	0.011136
0.2	1.7527	4.7760	4.2	0.008927	0.009938
0.3	1.3725	3.0560	4.3	0.007988	0.008872
0.4	1.1145	2.1844	4.4	0.007149	0.007923
0.5	0.9244	1.6564	4.5	0.006400	0.007078
0.6	0.7775	1.3028	4.6	0.005730	0.006325
0.7	0.6605	1.0503	4.7	0.005132	0.005654
0.8	0.5653	0.8618	4.8	0.004597	0.005055
0.9	0.4867	0.7165	4.9	0.004119	0.004521
1.0	0.4210	0.6019	5.0	0.003691	0.004045
1.1	0.3656	0.5098	5.1	0.003308	0.003619
1.2	0.3185	0.4346	5.2	0.002966	0.003239
1.3	0.2782	0.3725	5.3	0.002659	0.002900
1.4	0.2437	0.3208	5.4	0.002385	0.002597
1.5	0.2138	0.2774	5.5	0.002139	0.002326
1.6	0.1880	0.2406	5.6	0.001918	0.002083
1.7	0.1655	0.2094	5.7	0.001721	0.001866
1.8	0.1459	0.1826	5.8	0.001544	0.001673
1.9	0.1288	0.1597	5.9	0.001386	0.001499
2.0	0.11389	0.13987	6.0	0.0012440	0.0013439
2.1	0.10078	0.12275	6.1	0.0011167	0.0012050
2.2	0.08926	0.10790	6.2	0.0010025	0.0010805
2.3	0.07914	0.09498	6.3	0.0009001	0.0009691
2.4	0.07022	0.08372	6.4	0.0008083	0.0008693
2.5	0.06235	0.07389	6.5	0.0007259	0.0007799
2.6	0.05540	0.06528	6.6	0.0006520	0.0006998
2.7	0.04926	0.05774	6.7	0.0005857	0.0006280
2.8	0.04382	0.05111	6.8	0.0005262	0.0005636
2.9	0.03901	0.04529	6.9	0.0004728	0.0005059
3.0	0.03474	0.04016	7.0	0.0004248	0.0004542
3.1	0.03095	0.03563	7.1	0.0003817	0.0004078
3.2	0.02759	0.03164	7.2	0.0003431	0.0003662
3.3	0.02461	0.02812	7.3	0.0003084	0.0003288
3.4	0.02196	0.02500	7.4	0.0002772	0.0002953
3.5	0.01960	0.02224	7.5	0.0002492	0.0002653
3.6	0.01750	0.01979	7.6	0.0002240	0.0002383
3.7	0.01563	0.01763	7.7	0.0002014	0.0002141
3.8	0.01397	0.01571	7.8	0.0001811	0.0001924
3.9	0.01248	0.01400	7.9	0.0001629	0.0001729

APPENDIX D

FUNDAMENTAL PHYSICAL CONSTANTS AND MATERIAL PROPERTIES

D-1 Fundamental Physical Constants

Quantity	Symbol	Value
Ideal gas constant	R_U	8.314 J/mol - K
Avogadro's number	N_A	6.022×10^{23} mol^{-1}
Boltzmann's constant	k_B	1.381×10^{-23} J/K
Planck's constant (reduced)	$\hbar$	1.055×10^{-34} J - s
Speed of light in vacuum	c	2.998×10^8 m/s
Stefan-Boltzmann constant	σ	5.670×10^{-8} W/m^2 - K^4
Electron mass	m_e	9.110×10^{-31} kg

D-2 Unit conversions

Electron volts [eV] to Joules [J]: $1\,eV = 1.602 \times 10^{-19}$ J

D-3 Properties of Helium Gas

Atomic weight: $M = 4.003$ g/mol

Thermal conductivity at atmospheric pressure:

T [K]	100	200	273.15	300	400	500	700	1000
k [W/m - $^\circ$C]	0.0730	0.1151	0.142	0.152	0.187	0.220	0.278	0.354

D-4 Properties of Copper at 300 K

Density: $\rho = 8933$ kg/m^3

Thermal conductivity: $k = 401$ W/m - $^\circ$C

Specific heat: $c_p = 385$ J/kg - $^\circ$C

D-5 Properties of Fused Silica (Amorphous Silicon Dioxide, SiO_2) at 300 K

Molecular weight: $M = 60.1$ g/mol

Density: $\rho = 2220$ kg/m^3

Thermal conductivity: $k = 1.38$ W/m-$^\circ$C

Specific heat: $c_p = 745$ J/kg-$^\circ$C

Sound velocities $v_{s,T} = 3764$ m/s, $v_{s,L} = 5968$ m/s

D-6 Properties of Silicon

Density ($T = 300$ K): $\rho = 2330$ kg/m^3

Specific heat ($T = 300$ K): $c_p = 712$ J/kg-$^\circ$C

Thermal conductivity of bulk silicon [4] Note: Low-temperature values are dependent on sample size.	
T	k
[K]	[W/m-$^\circ$C]
1	4.48
2	31.7
3	99.8
5	424
10	2110
15	3930
20	4940
30	4810
40	3530
60	2110
80	1340
100	884
150	409
200	264
300	148
400	98.9
600	61.9
800	42.2
1000	31.2

D-7 Measured Thermal Conductivity of a 56 nm Diameter Silicon Nanowire at Selected Temperatures [5]

T	[K]	320	300	250	200	150	100	60	40	30	20
k	[W/m -°C]	25.0	25.7	27.5	29.0	28.5	25.5	18.9	6.75	2.77	0.72

D-8 Calculated Thermal Conductivity of Single-Walled Carbon Nanotubes, Selected Values [6]

$T = 316$ K		$T = 31.6$ K	
L	k	L	k
m	W/m -°C	m	W/m -°C
1.06E-08	3.53E+01	9.97E-09	1.94E+00
2.25E-08	7.16E+01	3.77E-08	7.41E+00
4.13E-08	1.17E+02	2.91E-07	5.07E+01
8.48E-08	2.16E+02	1.64E-06	2.82E+02
1.61E-07	3.43E+02	5.95E-06	1.07E+03
4.16E-07	6.15E+02	3.67E-05	5.77E+03
9.20E-07	1.01E+03	2.16E-04	2.61E+04
3.09E-06	1.92E+03	6.85E-04	5.37E+04
8.25E-06	2.86E+03	2.71E-03	9.62E+04
1.83E-05	3.54E+03	9.78E-03	1.22E+05
4.70E-05	4.25E+03	4.22E-02	1.35E+05
1.00E-04	4.52E+03	2.48E-01	1.40E+05
3.23E-04	4.80E+03	1.07E+00	1.35E+05
1.70E-03	4.94E+03		
3.51E-02	4.93E+03		
1.01E+00	4.91E+03		

REFERENCES:

[1] D. N. Lide ed., *CRC Handbook of Chemistry and Physics*, 87th ed,. CRC Press, Boca Raton, FL, 2006.

[2] F. P. Incropera, D. P. DeWitt, T. L. Bergman, and A. S. Lavine, *Introduction to Heat Transfer*, 5 ed. (Wiley, 2006).

[3] C. Kittel, *Introduction to Solid State Physics* 8ed. (Wiley, 2004).

[4] Touloukian, Y. S. ed., Purdue University Thermophysical Properties Research Center, *Thermophysical Properties of Matter*. IFI/Plenum, New York, 1970.

[5] Deyu Li, Y. Y. Wu, P. Kim, L. Shi, P. D. Yang, and A. Majumdar, "Thermal conductivity of individual silicon nanowires". *Applied Physics Letters* **83**, 2934-2936 (2003).

[6] N. Mingo and D. A. Broido, "Length dependence of carbon nanotube thermal conductivity and the "problem of long waves"". *Nano Letters* **5**, 1221-1225 (2005).

INDEX

CPSIA information can be obtained
at www.ICGtesting.com
Printed in the USA
LVHW021006200720
661065LV00004B/45